BUGS in Writing

The BUGS Classification System

BAD: The example sentence contains a genuine error or mistake (for example, bad grammar); you should avoid the bad portion completely.

Making such a mistake, embarrassment is unavoidable.

UGLY: The example sentence is technically correct, but is not acceptable for one reason or another; you should avoid the ugly portion.

When one writes in an ugly manner, in order that one may sound like one knows what one is doing, one may lose credibility with the audience.

GOOD: The example sentence is acceptable and correct.

When you write well, you get your point across.

SPLENDID: The example sentence demonstrates an improvement over a given good solution or example, or represents correct application of several points under discussion.

Once you have developed ear, you will be able to tell without hesitation what constitutes correct syntax, style, and semantics; you will have competence in written English, and will be equipped to communicate effectively with your fellow humans.

BUGS in Writing

A Guide to Debugging Your Prose

lyn dupré

ADDISON-WESLEY

An imprint of Addison Wesley Longman, Inc.

Reading, Massachusetts ■ Harlow, England ■ Berkeley, California
Menlo Park, California ■ Don Mills, Ontario ■ Amsterdam
Bonn ■ Mexico City ■ Sydney ■ Tokyo

Library of Congress Cataloging-in-Publication Data
Dupré, Lyn.
    BUGS in Writing : A Guide to Debugging Your Prose / Lyn Dupré. --
Rev. ed.
        p. cm.
    Includes indexes.
    ISBN 0-201-37921-X
    1. English language--Rhetoric.   2. English Language--Grammar.
3. Report writing.   I. Title.
PE1408.D85   1998
808′.042--dc21

                                                                97-46656
                                                                CIP

ISBN   0-201-37921-X

2 3 4 5 6 7   CRS   01 00 99 98

2nd Printing October, 1998

BUGS in Writing

A Guide to Debugging Your Prose

~~A Catalog of Prosaic Catastrophes~~

~~The Text-Engineering Guide~~

~~Listening to Prose~~

~~Cataclysmic Blunders, Muffed Lines, and Slips of the Mouse~~

~~Writing for Nerds~~

~~How to Banish Fallacious Sentences~~

~~Diverse Diversions and Contrary Contretemps~~

~~Error Trapping and Extermination~~

Dedication

To Max, my transporting and transported angel.

This revised edition is also dedicated to my father,
Garrett Oppenheim,
who passed away on February 6, 1995.

Disclaimer

*The author takes no responsibility for the actions
and speech of the characters in this book.*

Illustration based on a gift from Peter Gordon

Contents

Foreword

I HAVE BEEN PUBLISHING books for more than 15 years, yet I have never before been given—nor have I requested—the opportunity to write a foreword to one of them.

In this special case, the author and I agreed, I was a natural choice. I have worked with Lyn Dupré as closely as anyone has during much of my publishing tenure, and I am therefore as familiar as anyone is with her unique ability to help writers communicate more clearly. She has been my most valued editor, valued not simply for cleaning up other people's prose, but for helping them to develop their own effective style. Lyn holds that most people will write more lucid prose if only they can learn to listen carefully to what they are writing, and to be aware of the most common errors that a writer can make. I have seen up close how she has taught so many writers—from graduate students preparing their first research paper for a journal to world-renowned authors developing their latest book—to pay closer attention to their writing. I am delighted that, with *BUGS in Writing,* Lyn is able to share with a much broader audience her knowledge of the language, her experience with writers across disciplines (but especially in technical fields), and her unmatched talent for making a guide to a subject so prosaic as writing style as enjoyable as it is practical.

As an editor for me and my colleagues, Lyn has worked with a "Who's Who" of computer-science authors, including Fred Brooks, Rick Cattell, Steve Feiner, Jim Foley, John Hughes, Don Knuth,

Nancy Leveson, Kirk McKusick, Carver Mead, Peter Neumann, Christos Papadimitriou, Yoh-Han Pao, Jef Raskin, Robert Sedgewick, Ben Shneiderman, Ted Shortliffe, Avi Silberschatz, Andy van Dam, Gio Wiederhold, Patrick Henry Winston—the list is long! These successful authors, to be sure, are among those writers who need the least help; each had already developed a style that works. As Lyn teaches, however, every author can learn to write more effectively. And the most skilled authors are also the first people to recognize the importance of lucid writing for the aspiring—or the established!—scientist or other professional.

Lyn has also worked closely over the years with scores of university students and faculty members. With graduate students in Ted Shortliffe's program at Stanford University, in Carver Mead's laboratory at the California Institute of Technology, and in other academic institutions around the country, she shaped and refined her techniques for helping would-be (and must-be!) writers. Correcting her students, instructing them, getting them ultimately to produce their own highest-quality results, she had a direct influence—anyone who has spent a moment with Lyn would never say "impact"—on their professional writing. Papers, theses, and proposals were never done until, as their writers said, they had been "Lynified." In the process, the editor was herself edified: Seeing the same mistakes made over and over, yet knowing that her clientele possessed fine minds and keen wits, Lyn devised the principles of lucid writing that have now been collected into this one handy guide.

I have often been impressed by Lyn's ability to teach good writing through her editing—an ability reflected on heavily marked manuscript pages, in spirited e-mail exchanges, in lectures and conversations. With *BUGS in Writing,* I find that the teacher and editor can herself write! In 150 short segments, each of which can be read and absorbed in a brief sitting (or standing, for that matter, if you are an urban commuter), she has boiled English down to the essence of what every (scientific) writer needs to know to write clear and effective prose. Amazingly, with a collection of offbeat examples— sometimes clever, often hilarious—she has managed to write a book about language that is fun to read! But if the examples in *BUGS in Writing* are funny, the instruction that they provide is decidedly practical. I personally hope that legions of new writers will find, as I have, that the good habits spread by *BUGS in Writing* are, like the author's own style, infectious.

PETER S. GORDON
Publishing Partner
Addison Wesley Longman, Inc.
Reading, Massachusetts
September 1994

Read Me: Ear

Given that you are reading this page, I assume that you want to know how to write well. Given that you have that desire, I assume that you already know a considerable amount about writing. Any person who pays attention to, and thinks about, the techniques of good writing already has the necessary basis for developing ear.* *Ear* is the ability to hear whether a given word order, sentence, or term is correct.

The simplest way to improve your expository writing substantially is to learn to avoid a limited set of extremely common errors. Because these errors are endemic in scientific literature, they sneak into the prose of even extremely bright people who write well. You probably have become so used to seeing them that you do not immediately recognize that they are problematic.

My purpose in this book is to teach you to recognize common errors instantly by describing as clearly and simply as possible what those errors are. My intention is to convey to you the basic principles of good writing, and thus to help you to develop ear, without subjecting you to the pain and boredom of absorbing the voluminous body of formal and technical knowledge on the English language. I aim to present information in a form such that you will read because the

---

* Chances are that you have the normal complement of ears—those appurtenances attached to the sides of the human head. *Ear* is different from *an ear*; most people lack ear, but almost anyone who wants to can develop ear.

material is amusing and interesting, and in the process will learn the principles that you need to know to improve your competence in written English.

When you have a smooth, intimate relationship with written language, you have at hand a tool of enormous power: You can communicate with and convince your fellow humans.

You write a manuscript to convey information, to communicate concepts, and to persuade people. You may have numerous opportunities for writing in your work, such as the following:

- You complete a research study, and want to convey your results and conclusions to your colleagues.
- You design a software package, and want to write a manual that will tell users how it works, or to write advertising copy that will tell people why they should buy it.
- You have a hypothesis that you wish to test, and want to write a grant or contract proposal in the hope that you will obtain funding to carry out your research.
- You have been invited to contribute a chapter to a book on a topic related to your area of expertise, and want to restructure your thinking so that your theories will be accessible to people who are not familiar with your field.
- You have completed a review of work on a specific topic, and now want to write an article for a specialty journal.
- You have been asked to give a talk at a conference, and want to develop the slides for your presentation.
- You have been asked to report the progress that your department has made on a project, and want to write a memorandum or business report.

Whatever your specific reason for undertaking a writing project, when you write, your goals are to communicate and to convince.

The principles of syntax, style, and semantics structure your writing to reduce ambiguity and to add clarity. Lucid, clean writing conveys thoughts rapidly and effortlessly; murky, obfuscating prose makes readers work hard to glean any sense at all. Many of the principles boil down to common sense; implementing them reduces potential or actual confusion. Patrick Henry Winston[†] suggests these reasons for learning to avoid common mistakes:

- Errors such as split infinitives and nonparallel fragments make reading difficult, because they place an unnecessary burden on that part of the brain that makes use of syntax. If you overload that part of the brain, there are fewer brain cells available to handle meaning.
- Errors such as word shifts — a shovel in one paragraph becomes a spade in another — place an unnecessary burden on that part of the brain that is trying to deal with semantics. They make your reader wonder whether the shift conveys meaning or merely reflects a misguided sense that word repetition is bad.
- Errors in general distract the careful reader, in particular a reader who writes and who cares about his writing.

I add only that errors in general detract from your credibility. Even if your reader does not herself know precisely what rules you are breaking, she will certainly notice whether your thoughts are easy to understand. If you write muddy, indecipherable prose, riddled with mistakes, your reader is bound to wonder how carefully you designed your study, collected your data, applied statistical techniques, validated your system, or otherwise behaved like a respectable scientist.

---

† Personal communication, 1994.

Many of the principles of syntax, style, and semantics are not strict, in the sense that experts may disagree about them. Anyone who studies language develops her own set of principles; the one presented in this book is my own, subject to my unique, opinionated biases. I have developed this set for computer people (a class that I shall identify in the next paragraph), but the principles are applicable to all technical writing; for that matter, they are applicable to any writing at all. Whether you are writing a doctoral dissertation or a love letter, your goal is to convey information to, or to convince, your reader; limpid clarity will get your point across.

I have written this book for all people who might plausibly be found wandering around in that section of a bookstore that has on the shelves computer-related books; such *Homo sapiens* I have dubbed *computer people.* You may be a computer scientist, or you may hark from any discipline that uses computers as an integral component of its work. I have explored challenges that are specific to such disciplines, such as how to set code or to style the names of the keys on a keyboard. I have also used numerous examples throughout the book to show you ways to use wisely the terminology of computers. If you are any breed of scientist, even one who does not differentiate RAM from a male sheep, you will still find this book highly applicable to your work. If you are a businessperson, you also will find in this book many techniques to help you communicate effectively and thus to work productively.

This book comprises numerous short, unordered,‡ unlinked,** easily digestible segments. That is not the usual structure for a book, and you deserve an explanation. When you begin reading, you may find the lack of hierarchy irritating. Because this book is so noticeably different from other texts whose purpose is to teach you how to write well, let us first examine the more usual approach.

Language is evolving stuff that exists in the world. We humans generally try to learn about stuff by observing it and seeking patterns that will let us classify it, and make predictions about it. Many intelligent people have observed English for ages; they have developed an immense body of knowledge, a classification system that codifies the patterns according to one model of language. The result is a huge hierarchical rule base; such rule bases are presented in style manuals.

A *style manual* is thus a highly organized exposition of this rule-based model, presented in a specialized terminology. Style manuals contain terms such as *possessive case, appositive, restrictive function,* and *antecedent.* Style manuals are comprehensive: They classify everything you might need to know about the rules governing the English language. I strongly urge you to purchase a good one—my personal recommendation is *The Chicago Manual of Style* [Chicago: University of Chicago Press], because it is used by most publishers in the United States—and to use it to look up answers to specific questions that arise as you write.

There is nothing wrong with that approach to modeling language; on the contrary, it is an admirable, scientific, comprehensive way of going about the business. Most editors, for example, use that rule-

---

‡ The segments are numbered, but the numbers are merely identifiers to help you find a segment that you want to read. They do not denote a preferred reading sequence.

** That is, the segments do not contain cross-references; each stands alone.

based system in their work. There is, however, a different way, and that way is what I will teach you in this book.

The alternative is a more intuitive, right-brain observation of language, the result of which is the development of ear. As I said earlier, *ear* is the ability to hear whether a given word order, sentence, or term is correct. Ear is based on numerous principles that have to do with the logic and rhythm of language. To develop ear, you learn the principles; once you have ear, you can make judgments without resorting to any specific principle. You are thus equipped to handle new, unexpected situations.

In this book, I give you a set of principles; the set is not comprehensive, but it does cover most of the problems that mistuned ears miss. I would like you to think of the segments of this book as, say, daily columns that might turn up on the opinion page of your newspaper, waiting for you at the breakfast table. Each segment covers one conceptual chunk of information, and each stands alone. You can read the segments in any order, as many at a sitting as you wish. Certain information shows up in more than one segment, for two reasons. First, because the segments are independent, I want to ensure that you do not need to refer to more than one at a time. Second, showing you how a notion applies in different situations is a good way to familiarize you with it.

I have written the book in this way, refusing steadfastly to organize it in the traditional manner, because I believe strongly that attempts to impose organization on the principles lead you to the analytic system of style manuals, rather than to the development of ear. Had I organized this book well, I would eventually have ended up with a structure identical to that of a standard style manual, and you would have in your hands a mediocre, incomplete book. I have, instead, purposely introduced chaos so that I can actively discourage you from trying to read this book linearly, memorizing the principles as you go. Instead, I want you to open the book randomly, or perhaps to find in the table of contents a segment that interests you today. No one can absorb more than a few segments at one sitting, and I do not want you to try to do the impossible. I want you to keep this book on the breakfast table, next to your hammock on the porch, on your nightstand, or wherever you will be inclined to pick it up for a bit of grazing when you have a quiet moment or two.

I want you to read around in this book, now and then, over a period during which you also do your own writing, because, as you write, you will understand more and more about how the principles apply. Every segment in this book describes a problem that many writers have; any individual writer has her own set of frequent glitches. You might want to identify your own personal subset of the principles in this book. As you read, flag those segments that describe the errors that you make most often. Then, if you learn to avoid that small set of errors, you will write smoothly and well.

Eventually, you will absorb a new model of language; you will start to understand a way of looking at language that lets you tell instantly whether the wording of a sentence sits well with you. You will develop ear, and that is my intention.

If you are teaching other people to write, this book will be a valuable tool. If you are supervising graduate students who write research reports and dissertations, for example, or are training your staff to write clear progress reports, or are helping colleagues to prepare papers for presentations, first ensure that each person has her own copy of this book. Then, help her to identify her own set of, say, her 20 most common errors. As she writes, have her check her writing for those specific problems, before she gives you the text. You will both profit from the exercise.

At the end of the book, I have included an Index by Category. You can use that index to find segments that discuss, for example, punctuation marks, or terms to be avoided. I have also included an Index of Principles. There, you will find the titles of the segments listed alphabetically, and annotated with the relevant principles. This index will allow you to relocate information when you want to refresh your memory.

No single principle is essential; once you have developed ear, you may well reject several of my ideas. If you do not yet have finely tuned ear, however, I recommend that you first simply follow the principles in this book. What sounds right to you now may well sound gratingly clumsy in the future. What sounds stilted and formal and overly correct to you now may well sound clean and pleasant in the future. Your ear will change.

Eventually, you will be able to forget the specific principles—both the ones given here and any that you discover for yourself—because you will have an accurate sense of the natural rhythm of language.

It is critical that you understand that the principles are derived from the language, rather than vice versa. That is, the principles are *descriptive,* rather than *prescriptive.* If you know how to apply the principles, you have competence in written English. Because the principles are descriptive, there is not always an answer to the question, "Why is the language that way?" There often is an answer, and that answer usually depends on logic and has to do with disambiguation; where the answer exists, I tell you about it. Sometimes, however, there is no apparent reason why the language behaves as it does: It just is that way.

Furthermore, in certain sections (for example, styling of numbers), I describe a set of conventions. The rationale for such conventions is that they are standards: You and other writers follow the standards to maintain consistency within and across documents.

When I write in this book

> *You should <follow a principle>.*

I mean that, if you wish to bring your writing into line with competent written English, then you should follow the principle that I describe. I might say, for example,

> *You should not split infinitives.*

I have used this construction because omitting you from the discussion would be unfriendly, and would leave me various unappealing options, such as simple commands (which are even more directive, in essence telling you that *you must,* rather than that *you should)*

> *Do not split infinitives.*

or passive voice

> *Infinitives should not be split.*

or verbosity

> *If you wish to write well, then you might consider adopting the principle of never splitting infinitives.*

or bad grammar

> *To write well, infinitives should remain unsplit.*

Thus, I want you to understand that I am not so much telling you what to do, as I am telling you what you *can* do if you want to improve your writing. I needed the construction *you can,* however, to tell you about those situations in which there are various — equally good — alternatives.

Furthermore, as I mentioned, language is *evolving:* Language is a form of human behavior, and it is never stagnant. Be aware that the rules of spoken language are substantially different from those of written language. Be aware also that, at any given moment, a person who has competence in the language knows where to place the boundary between incorrect usage and usage that has become correct over time. The precise location of that boundary is a matter of judgment; most of the boundaries[††] expressed in this book are reflections of my own judgment. They are here to guide you, rather than to bind you.

Because I am offering you an alternative model of language, I have avoided, to the extent practical, the terminology used to describe the rules of style and syntax. I do assume that you know, or can infer from context, the meaning of *verb* and *noun;* in general, however, I have used few such terms; when I do use them, I make sure that you and the term are properly introduced before you play together. For the most part, however, I encourage you to think of language as describing activities undertaken in possible worlds by various

---

[††] In particular, the location of the boundary between ugly and good, as I shall describe in a moment, is a matter of judgment and ear.

agents; objects or creatures can undertake activities or can be the recipients of activities.

A large portion of the pedagogy in this book is by way of *example*. My own experience in teaching writers has convinced me that people acquire ear by contemplating myriad examples. Two examples may seem excessive to the person who already understands the point being made, but six examples may be insufficient for the person who is trying to grasp a new pattern. If, after you read the first example, you comprehend the distinction, or structure, or other point, then by all means skip the others; they are there for other readers.

In most cases, I have given the examples in pairs or sets,‡‡ showing you first the problem, and then the solution. I have not used identical sentences in the sets, for two reasons. First, reading the same sentence repetitively would bore you. Second, I want to teach you to read for structure, as well as for content. Even though the words in the example sentences are different, the structures of the sentences in a given set are identical (or are sufficiently similar that they are identical for the point under discussion).

A good writer reads a sentence in two ways: for structure and for content. If you read sentences for content only, you will be unable to discern many of the reasons for the principles, and you will be unable to apply the principles to your own writing. If you read for structure only, you will be unable to understand what a sentence means. You must read for both. Thus, I have chosen to set up the examples such that you are forced to read for both structure and content, so that you will be able to apply what you learn. Your own sentences will certainly have content different from those in the examples, but they will have the same structure. Because you will

---

‡‡ The example sets are separated from each other by pathways frequently used by bookish felines, as evidenced by the spoor.

know how to read for structure, you will be able to see that your sentence is like the one on page *xx* of this book, and you will know how to write your sentence correctly.

Because, as I said, many principles of syntax, style, and semantics are not hard and fast rules, but rather are, to a degree, subject to opinion, I have developed the BUGS system to classify all the examples in this book. BUGS is a four-point scale, denoting

BAD: The example sentence contains a genuine error or mistake (for example, bad grammar); you should avoid the bad portion completely.

Making such a mistake, embarrassment is unavoidable.

UGLY: The example sentence is technically correct, but is not acceptable for one reason or another; you should avoid the ugly portion.

When one writes in an ugly manner, in order that one may sound like one knows what one is doing, one may lose credibility with the audience.

GOOD: The example sentence is acceptable and correct.

When you write well, you get your point across.

SPLENDID: The example sentence demonstrates an improvement over a given good solution or example, or represents correct application of several points under discussion.

Once you have developed ear, you will be able to tell without hesitation what constitutes correct syntax, style, and semantics; you will have competence in written English, and will be equipped to communicate effectively with your fellow humans.

Keep firmly in your mind, as you read, that the distinction between bad and ugly is one of kind, rather than one of degree. There are many bad (technically incorrect) sentences that are considerably less objectionable than are many ugly (technically correct but klutzy, horrific, stultifyingly dry, or otherwise unacceptable) sentences.

The distinction between ugly and good is, like that between bad and ugly, one of kind. It is also, necessarily, one of judgment, or of ear. That is, members of both classes are technically correct, but those in the ugly class are objectionable on a different, valuative measure. Thus, the border between ugly and good will vary across experts; the one presented in this book is mine, offered for your consideration.

In contrast, the distinction between good and splendid is one of degree, as well as one of ear. In most cases, a good sentence (one that solves the problem under discussion) is as good as it gets, so to speak; there is no splendid case. Splendid sentences are rare; they occur only when (1) there is a more graceful, more communicative, or otherwise more desirable solution, or (2) there is a solution that applies correctly more than one of the lessons in a given segment of this book.

I have classified every example so that there is no room for confusion about whether I am showing you a problem or a solution. In certain segments, I am pointing out that different word order, for example, results in different meaning; there is no bad or ugly example because all ways are correct—you just need to be sure that what you intend to write matches what you do write. In other cases, the topic of the segment is a beast that you should generally expel from your prose, such as a redundant brutish monster beast; in those segments, the only examples are ugly ones.

I have had great fun writing this book, and I hope that you will similarly enjoy reading it. I encourage you to swim into it playfully, and to let it lap over you without worrying about what you are learning. Soon enough, you too will be entertaining yourself by writing your own manuscripts.

LYN DUPRÉ
Woodside, California
September 1994

Do not read any further
until you have read the preceding
Read Me material!

Otherwise,
much of what follows may
make little sense to you.

1   *Passive or Missing Agents*

If you want to learn only one technique to improve your writing substantially, you should learn to *avoid using passive voice*. Passive voice is the form in which you can say only that an event or action took place in the world, without necessarily admitting what or who the causal agent was, such as

> ugly: Several of the data sets were lost.
>
> ugly: The Waterford crystal was dropped.
>
> ugly: The car was dented.

I recommend that you take responsibility in your writing, and always name the creature or object that is acting.

Naming the agent in *active voice* will give your prose a more vigorous rhythm.

> good: Lyn lost the data sets.
>
> *Lyn is the agent.*
>
> good: Dona dropped the Waterford crystal.
>
> *Dona is the agent.*
>
> good: Max dented the car.
>
> *Max is the agent.*
>
> good: Dona dropped the Waterford crystal on Max's foot, causing Max to drop his briefcase on the hood and thus to dent the car, but Lyn lost the accident report.
>
> *Dona, Max, and Lyn are all guilty.*

In contrast, passive voice is boring, even if you reveal the agent.

> UGLY: The Waterford crystal was dropped by Dona.
>
> *Dona is the agent, but the sentence is still in passive voice.*

Furthermore, using passive voice generally will expose you to the risk of failing to deliver on the promise of another part of your sentence—the promise to identify an agent in a specific location. We shall return to this point after we have examined the simpler case of correct (but ugly) use of passive voice.

Failing to name the agent in a sentence is uninformative and boring, and tends to lead you into all manner of grammatical errors. Yet, incomprehensibly, generations of scientists and technical experts have had drilled into them that keeping secret who performed actions is the correct way to undertake formal writing. You will read numerous scientific papers written predominantly in passive voice, and your colleagues may try to rewrite your prose to introduce passivity.[1] Fight back! Stick to *active voice*, except for the occasional sentence here and there, and *say who did it*, rather than *it got done*. Your writing will be exponentially clearer and more interesting.

You should *avoid passive voice* because it allows you to fail to inform your reader of who or what took action in, or otherwise influenced, the world. By switching to active voice, you give your reader possibly vital information.

> UGLY: The data were collected.
> GOOD: Brendan collected the data.
>
> *Perhaps Brendan's reliability is widely recognized.*

---

1. I am sad to say that even certain publishers' copy editors may try to *passify* your prose.

UGLY: When memory is so short that it cannot be freed suffi-ciently fast to satisfy demand, swapping can be used.

GOOD: When the operating system becomes so short of memory that the paging process cannot free memory sufficiently fast to satisfy demand, it can use swap-ping.

UGLY: In a stack, both insertions and deletions are performed at one end only; both popping and pushing are allowed.

GOOD: When you use a FIFO list queue, you make insertions at one end of the list, and deletions at the other end; you therefore must maintain rear and front pointers.

As I mentioned, you can name an agent in passive voice if you wish. Although that construction is not incorrect, active voice has consid-erably more verve.

UGLY: The book was written by Lyn.

GOOD: Max designed the software.

UGLY: The lamp was broken by BB.[2]

GOOD: Red pounced on the banana slug.

---

2. Contributed by Red.

UGLY: The critical-region construct can be used by people to solve the critical-section problem.

GOOD: Any programmer can use the conditional critical region to solve synchronization problems.

There is one justifiable use of passive voice. Passive voice emphasizes the receiver of activity (the object), rather than the actor (the subject), and there may be occasions when that emphasis is important. In the following example pair, we might have ample grounds for thinking that the critical information is that the book file (recipient of activity) was erased (activity), and that the identity of the eraser (the power outage, which is the actor) is immaterial. In such a case, the first example might well be more apt than the second.

UGLY?: The file for Lyn's book was erased unexpectedly by a power outage.

GOOD: A power outage erased the file for Lyn's book.

Nonetheless, because most people overuse passive voice to the detriment of their writing style, I urge you to avoid the construction as much as possible, until you have got into the habit of writing in active voice.

 You should not mix in one sentence parts that contain words ending in *ing* with parts cast in passive voice.

There is nothing wrong, grammatically speaking, with the previous examples of passive voice. You create an error, however, when you mix within one sentence a part that has no identified agent, and a part that contains a verb that requires an agent to keep that verb under control. *Such errors are perhaps the most common in formal writing*—they are so common that it probably will take time for your ear to adjust to the notion that the form is incorrect. One way that you can create the need for a matching agent is to use an *ing* word—that is, a word that ends with *ing*.

Consider this example:

> BAD: By pouring water on the flame, the fire can be extin-
> guished quickly.

What is wrong here?

The first chunk of the sentence—*by pouring water on the flame*—creates in your reader the expectation that you are about to reveal who or what is doing the pouring. The noun *fire* shows up, and grammatically you have made the claim that the fire is pouring water. Not only is the conjured picture illogical (in this example), but also there is another string of words—*can be extinguished*—of which your reader now tries to make sense. She backpedals, realizing that the fire is not doing anything at all; it is merely sitting there, about to become the victim of a homicidal act. Although your reader will probably figure out what you intended to say, what you have actually said is grammatical nonsense. Your sentence parts, or clauses, do not match.

Note that rearranging the order of the parts will not help you at all; the error lies in the failure of the parts to match.[3]

> BAD: The fire can be extinguished quickly by pouring water
> on the flame.

What you need to do to make your sentence work is to cast both parts in the same manner—with or without an agent. Thus, it is correct—but inadvisable—to leave out the agent entirely. Such technically correct sentences are remarkably awkward, as the following example demonstrates.

> UGLY: If water is poured on the flame, the fire can be extin-
> guished quickly.

---

3. I am grateful to Jan Clayton for suggesting that I mention this point.

Also correct, and decidedly preferable, is a sentence recast such that both parts speak about an agent.

> GOOD: By pouring water on the flame, you can extinguish the fire quickly.

There are other forms that you can use to avoid the problem.

> GOOD: To extinguish the fire quickly, you can pour water on the flame.

> GOOD: If you want to put out a fire quickly, [then] drown the flame in water.

> GOOD: Pouring water on the flame usually will extinguish the fire quickly.

Note that the final example in the preceding set contains no agent, and speaks of only an activity, but is still perfectly reasonable.

Thus, if one of your clauses has a form such as

- *By doing* an activity, …
- *Using* an object (or creature), …
- *In holding* an object (or creature), …
- *On realizing* an idea, …
- *Using* the principles in this book, …

then the rest of your sentence must inform your reader of who or what is doing, using, holding, realizing, or using.

> BAD: By using the same primary key, the implication is that the same real-world entity is being represented.

> GOOD: When combining user views, you should merge entities that have the same primary key.

BAD: Using a reallocation register, dynamic reallocation can be implemented.

GOOD: By loading the user process into high memory down toward the base-register value, you can ensure that all unused space lies in the middle.

BAD: Using the hysteretic differentiator, the output is primarily responsive to major changes in the sign of the derivative of the input voltage.

GOOD: Applying a waveform of a fixed shape to the input of the circuit, we can change the magnitude of the time derivative by changing the frequency of the wave, while maintaining the input amplitude constant.

BAD: Having already eaten my dinner, I think that you should at least refrain from gobbling up my dessert as well, as otherwise I shall starve.

GOOD: Having already used my washcloth, you should, I think, at least find me a dry towel.

BAD: Using great expertise, insight, and brilliance, the model was constructed by Max.

GOOD: Using her noodle, the hammer that Peter gave her, lumber, and nails, Lyn constructed the tree house.

Note that naming the agent does not get you out of the problem if the clause is in passive voice: The first example in the preceding pair asserts that the model used great expertise, insight, and brilliance in being constructed by Max.

You should not mix in one sentence parts that contain the verb form *to X* (to laugh, to cry, to write, and so on), called *infinitives,* and parts in passive voice.

> BAD: To get advice, an expert must be consulted.

The problem is the same as that in mixing passive voice with *ing* words: The infinitive creates the expectation of an agent. The first chunk of the sentence—*To get advice*—leads the reader to expect to know, from the next noun, who is getting advice. Instead, the writer creates informational havoc by plunking down an expert after the comma, whereas it is not the expert who requires the advice. What the writer intended to say was

> GOOD: To get advice, you must consult an expert.

If you attempt to squeeze out of this trap in passive mode—that is, if you cast both parts to exclude the expectation of and naming of an agent—you will create a convolution such as this muddled (albeit technically correct) statement:

> UGLY: For advice to be gotten, an expert must be consulted.

If your purpose is to provide an alternative to counting sheep, this style will serve you well.

Thus, if one of your clauses has an infinitive form—

- *To dive* in the deep end, ...
- *To run* a mile, ...
- *To fall* in love, ...
- *To scream* like a hyena whose foot is caught in a tube of toothpaste, ...

then you need to follow through by naming the diver, runner, faller, or screamer after that first comma.

BAD: To communicate a probability distribution over quantity x_1 conditional on the value of x_2, a two-dimensional graph can be used.

GOOD: To show two-dimensional uncertainties, you can use probability density functions, cumulative probability functions, or Tukey box plots.

BAD: To get to our house, Highway 280 can be used.

GOOD: To find our home, you must drive up a winding road through the redwoods, avoiding the deer and opossums on your way.

BAD: To proof a manuscript, sufficient sleep during the previous night and an appropriate blood-sugar level bin a goos idea.

GOOD: To check page layout, you need to have had sufficient food that your glucose level is passable.

SPLENDID: You should be well rested and well fed before you do any work on the final manuscript.

Note that, when the first part of your sentence contains an infinitive, you do not necessarily have to name the agent directly after the comma. Sometimes, another thought will intervene:

GOOD: To relax, assuming that doing so is safe, you should kick off your shoes.

Structurally, the chunk *assuming that doing so is safe* is merely a parenthetical remark inserted into the sentence *To relax, you should kick off your shoes.* The following two examples have the same structure as that of the preceding one:

GOOD: To reload the operating system, such as after an extremely nasty crash, you can boot from floppy disk.

GOOD: To retrieve information, in various applications, you can use an associative memory.

THE PRINCIPLE FOR LUCID WRITING here is that, to the extent possible, you should avoid using passive voice in your writing; you should instead use active voice and should name your agents. In addition, by using clauses such as this one, you promise your reader to name an agent: Keep your promise.

2   *You and Your Reader*

ONE GOAL in writing is to engage your reader, and a sure way to foil your attempts in that regard is to address your audience as *the reader* or to refer to yourself as *the author.*

 You should speak directly to your reader, declaring that you are the writer and addressing her as *you.* If you are the sole author of an article or book, do not be afraid to refer to yourself as *I.* If you have coauthors, then say *we.* In almost all cases, you should call your reader *you.*

> UGLY: The author wishes to remind the reader that it is often helpful to know the probabilities of class memberships, rather than knowing only the class memberships themselves.

> GOOD: I used bottom-up processing to obtain the following weighted sum, which you will recognize from the previous example.

> GOOD: We suggest here a novel neural-network algorithm for cluster formation.

> UGLY: The careful reader will wonder precisely what sort of couple comprises[4] Max and Lyn.

---

4. The whole comprises the parts; the parts constitute the whole.

GOOD: As you read this book, you will probably develop your own portraits of Max and Lyn.

 You should avoid vigorously the poisonously dry *one*; do not use *one* to denote yourself or anyone else.

UGLY: One should realize that there are several data structures that describe the state of a process.

GOOD: You can detect thrashing by observing the amount of free memory and the rate of memory requests.

UGLY: One has written previously about alternative scheduling priorities.

GOOD: I reported previously a scheduling algorithm that handles the assignment of priorities to tasks.

 In expository writing, when you are acting as a guide for your reader, you can say *we*—meaning you and your reader—rather than *I*. (When you are expressing your own opinion, however, you should stick to *I*.) Thus, I generally prefer to say, for example, *we shall examine this matter in detail in Section 7.4*, where I denote by *we* both myself and my readers, undertaking a joint endeavor.

UGLY: In this discussion, it is assumed that it is possible to get a closed form for at least one kind of sum.

GOOD: In this exposition, we shall assume first that $n = a^2$ is a perfect square.

UGLY: One can reduce entities to first normal form by removing repeating or multivalued data elements attributed to another, child entity.

> GOOD: We can use first normal form to organize data as flat structures with no repeating groups.

You should remember that, often, you do not need to use any pronouns when reminding or instructing your reader. However, you should avoid using passive-voice constructions.

> UGLY: The cat should be grasped firmly, with care taken to ensure that she feels secure when lifted.
>
> *No pronoun is used to denote the reader, but the construction is passive voice.[5]*

> GOOD: You should grasp the cat firmly, being careful to support her body and to tuck in her legs, so that she feels secure.
>
> *The pronoun* you *denotes the reader.*

> GOOD: Grasp the cat firmly. Carefully snuggle her close, to ensure that she feels secure.
>
> *The imperative form simply directs the reader to action, without recourse to a pronoun.*

THE PRINCIPLE FOR LUCID WRITING here is that I encourage you to speak directly to your reader: Call yourself *I* if you are the only author, or *we* if you have coauthors; call your reader *you;* and call the team comprising you and your reader *we.*

---

5. The pronoun denoting the feline is a red herring (Red cat?).

3   *So, So That, Such That*

S, so that, and such that have distinct meanings; failure to distinguish among them can lead you to convey a thought that you did not intend to convey.

- *So* means therefore.
- *So that* means in order that.
- *Such that* means in such a way that.

Thus, the following three sentences have notably different meanings, although all represent correct usage:

GOOD: Lyn arose at 5:00 A.M., so she was tired.

Lyn arose at 5:00 A.M.; therefore, she was tired. Because Lyn arose at 5:00 A.M., she was tired.

GOOD: Lyn arose at 5:00 A.M., so that she could drive Max to the airport.

Lyn arose at 5:00 A.M. to make it possible for her to drive Max to the airport. Lyn's reason for arising at 5:00 A.M. was that she planned to drive Max to the airport.

GOOD: Lyn arose at 5:00 A.M. such that she was grumpy all day.

Lyn arose at 5:00 A.M. in such a way that she was grumpy all day. An aspect of the way that Lyn arose at 5:00 A.M. —perhaps out of the wrong side of the bed —caused her to be grumpy all day.

Let us consider other examples.

GOOD: I love you, so I am happy.

I love you; therefore, I am happy.

GOOD: I love you, so that I am happy.

The reason I love you is that doing so makes me happy.

GOOD: I love you such that I am happy.

The way I love you makes me, or allows me to be, happy.

GOOD: Peter drove from San Francisco to Woodside, so he took Lyn out to dinner.

Because Peter drove from San Francisco to Woodside, he took Lyn out to dinner; perhaps he decided that, having braved the traffic, he might as well brave Lyn.

GOOD: Peter drove from Woodside to the beach, so that he could get a chestful of fresh air.

The reason that Peter drove from Woodside to the beach was that he wanted to do his deep-breathing exercises.

GOOD: Peter flew from San Francisco to Boston, such that he was tired and sore the next day.

An aspect of the journey from San Francisco to Boston — we can surmise that Peter was routed via Taiwan — caused Peter to be tired and sore the next day.

GOOD: Lyn wrote a book that introduced fluffy logic, so she nearly developed an ulcer.

Lyn's incipient ulcer was a result of her decision to write a book.

GOOD: Lyn wrote a book that immortalized the various creatures in her life, so that she would have an excuse to take a vacation from editing.

The reason for Lyn's decision was that Lyn wanted to avoid correcting other people's words on an individual basis (rather than that she wanted to do this special favor for the creatures mentioned).

GOOD: Lyn wrote a book about prosaic catastrophes such that Peter nearly developed an ulcer.

The manner in which Lyn wrote the book (rather than the subject matter of the text) caused Peter serious gastric distress.

Now we can look at examples of how you might mislead your reader by choosing incorrectly among *so, so that,* and *such that.*

You should use *so* when you mean *therefore.*

GOOD: Misha never slept, so he was occasionally tired.

If you wrote so that, *you would imply that Misha is simply a masochist;* such that *would imply that Misha never slept in a way that made him tired.*

GOOD: Holly adopted two goats, so she was in seventh heaven.

If you wrote so that, *you would make the reasonable assertion that Holly adopted the goats to increase her happiness quotient;* such that *would assert that the way that Holly adopted the goats — perhaps the journey home with them — made her happy.*

GOOD: Hard disks provide speed, efficiency, and convenience, so most people prefer them over floppy disks as the primary storage medium.

If you wrote so that, *you would ascribe a teleological status to the hard drive; if you were writing about the hard-disk developer, however,* so that *might be most appropriate.* Such that *would mean that the way that hard disks supply those features (rather than the features themselves) causes people to prefer hard disks.*

You should use *so that* when you mean *in order that*. As you read each example, think about what substituting *so* or *such that* would do to the meaning.

GOOD: Misha stayed up all night (alone in his office) so that he could finish the proposal.

GOOD: Holly stayed up all night (working on her paper while she waited) so that she could talk to DeeDee.

GOOD: Max came home early so that he could spend time chatting with Lyn by candlelight (while sipping the rum that Maria had brought back for them from Venezuela) before they were both comatose.

You should use *such that* when you mean *in such a way that*. As you read each example, think about what substituting *so* or *so that* would do to the meaning.

GOOD: Misha set the placecards such that no one was sitting near an enemy.

GOOD: Holly wrote her program such that patients would not feel threatened when they were asked to use it.

GOOD: Lyn greeted Max on his return such that he was exceedingly surprised and pleased.

THE PRINCIPLE FOR LUCID WRITING here is that you should distinguish among the three terms *so, so that,* and *such that,* so that you write more accurately, such that you use words correctly, so you feel confident when you publish your document.

4   *Two or More*

CERTAIN WORDS apply to only two objects or creatures; other words apply to only more than two objects or creatures.

You should use *between* when precisely two agents or objects are involved; you should use *among* when more than two are involved.

> BAD: Misha's latest innovative epicurean creation was shared out between the eight expectantly salivating dinner guests.
>
> GOOD: Misha's copious and astounding gastronomic marvels were distributed among the three groaning and sagging buffet tables.

> BAD: Devon, Sheila, and Lyn have between them but one common attribute.
>
> GOOD: Brendan, Kurt, and Malcolm have among them but one modem.

> BAD: The printer is shared among the two machines.
>
> GOOD: The data are shared between the two systems.

BAD: The data are passed among the server and the client.

GOOD: The messages are passed between the spy and the agency.

 Similarly, you should use *each other* when precisely two agents or objects are involved; you should use *one another* when more than two are involved.

BAD: Red and BB hissed at one another.

GOOD: Max and Lyn kissed each other.

BAD: The two men eyed one another warily.

GOOD: The two women hugged each other.

BAD: Drs. Owens, Henrion, and Provan forgot to send papers to each other.

GOOD: Drs. Rutledge, Detmer, and Middleton refer patients to one another.

BAD: After the diplomatic high jinks had trailed to an end, the various nations regarded each other with desperately deepening suspicion.

GOOD: After the light repast had ended, the various wombats regarded one another with rapidly growing fondness.

 You should use *either* when there are precisely two options; you should use *any one of* when there are more than two, or, in most cases, you can simply leave out the phrase entirely and thus create a less awkward sentence.

BAD: Either go to sleep, read a book in bed, or get up and work.

GOOD: Either fish or cut bait.

BAD: The flight attendant offered us either coffee, tea, or booze; when we protested, he proffered either orange juice, spicy tomato juice, or peanuts.

GOOD: You can order any one of fish, fowl, or floribunda; if you prefer, you may choose any one of the classes flora, fauna, or Technical Writing 207.

SPLENDID: You can use any one of these programs with either system.[6]

SPLENDID: The can might contain apples, peaches, or pears.[7]

You should use *a couple* only when precisely two objects or agents are involved; you should use *several* only when more than two are involved.

BAD: An automatic navigation system for a car must have at least a couple of capabilities: (1) recognize an obstruction, (2) decide when to steer right, (3) decide when to steer left, and (4) decide when the most appropriate choice is to slam on the brakes.

---

6. We classify this example as splendid because it uses correctly both terms under discussion.

7. We classify this example as splendid because leaving out the phrase *any one of* is a more elegant solution than is the correct (good) solution of using that phrase. The construction does not rule out the possibility, however, that the can might contain two of the mentioned fruits.

GOOD: The key equations have a couple of important conse-
quences: (1) the partial derivative of performance
with respect to weight depends on the partial derivative of performance with respect to the following
output, and (2) the partial derivative of performance
with respect to one output depends on the partial
derivatives of performance with respect to the
outputs in the next layer.

BAD: I have a couple of reasons for disliking you; they are
too numerous to list here.

GOOD: I have several reasons for loving you; three of them
have nothing to do with you personally.

Several does imply fewer than many, but there is no rule for how
many; you must use your judgment.

UGLY: An enormous crowd of several people struggled to
touch the star.

GOOD: When you want peace and romance, several people
constitute a crowd.

THE PRINCIPLES FOR LUCID WRITING here
are that you should use the terms *between, each
other, either,* and *a couple* to refer to precisely
two entities; and you should use the terms
among, one another, any one of, and *several* to
refer to more than two entities.

5 *Only*

Y̲ou ̲should use care when you use *only*. The location of *only* in your sentence determines the meaning.

The word *only* causes writers multitudinous problems. It is always getting itself into the wrong position, and mangling the meanings of sentences. Learn to watch its location carefully, to avoid making an extremely common error.

Quite simply, *only* modifies the term that follows it. Thus, the meanings of the following four sentences are substantially different:

> GOOD: Only I love you.
>
> GOOD: I only love you.
>
> GOOD: I love only you.
>
> SPLENDID: Only I only love only you.

To understand the meaning of a phrase that includes the troublesome *only,* you should posit the counterfactual:

> GOOD: Only I love you; John, Dick, and Harry do not love you.
>
> GOOD: I only love you; I do not respect you.
>
> GOOD: I love only you; I do not love King Kong.
>
> SPLENDID: Only I only love only you (only I am not certain that only that is true); no one else has just one feeling about you and about no one else (except that I am uncertain whether this truth is unique).

23

Because misplacement of *only* is so common, you will have to train your ear rigorously. Consider the following examples:

BAD: I only have a few minutes.

GOOD: I have only a few minutes.

BAD: I was only trying to help.

BAD: I only was trying to help.

GOOD: My only intention was to support you in your moment of need.

GOOD: Only this garbage-collection algorithm is slow when memory is nearly full; the other techniques will work quickly to free up memory.

GOOD: This garbage-collection algorithm is slow only when memory is nearly full; if you use it often enough, you should have no problem.

GOOD: This garbage-collection algorithm is only slow when memory is nearly full; it does not freeze up completely.

GOOD: Max ate only breakfast; he had no lunch or dinner.

GOOD: Max only ate breakfast; he did not throw it on the floor.

GOOD: Max had only an apple for lunch.

GOOD: Red wanted only to catch a rat; he did not want to eat one, given the availability of more palatable human food at the dinner table.

GOOD: Red only wanted to catch a crow; he knew that his desire was unrealistic, so he did not expect to fulfill it.

GOOD: Red wanted to catch only a mouse; he did not care about grandiose hunting schemes for catching crows.

THE PRINCIPLE FOR LUCID WRITING here is that, whenever you use *only*, double-check that you intend it to modify only the term that follows it directly.

6   *Redundant Terms*

THERE ARE times when it is pedagogically correct to repeat an idea, to ensure that your reader understands that idea. In general, however, it is cleaner to avoid repeating yourself.[8]

On the word level, you should avoid phrases that express the same notion twice (or thrice), and you should more generally avoid using duplicate (or triplicate, or quadruplicate) words to express one thought. Consider the following examples:

UGLY: The clock hands move in a right-handed, dextrorotary direction—that is, they move clockwise.

UGLY: The reeking, stinking pong of boiled and steamed cabbage permeated the house pervasively.

UGLY: Moisés at times could be hotheadedly incautious.

UGLY: When Doug had on his Halloween gear, he looked like a severely corrupted pervert.

UGLY: When Lyn asked Peter whether he wanted to contribute to this segment, he told her to stop worrying about all the little details.

---

8. This book contains redundant material. Each segment stands alone, and that already entails a certain degree of redundancy. In addition, I want you to see how principles apply across various situations, so I repeat material in different contexts. The disorganization of the volume, however, is highly unusual, and I do not recommend it as a template for your formal scientific writing. In what way is this note self-referring?

UGLY: Please remain seated, with your seatbelt fastened, until the airplane has come to a full and complete stop; when you deplane from the airplane, be sure to take with you all your personal belongings.[9]

UGLY: Lyn figured that life could be worse: She could be married to a prodigal wastrel.

UGLY: Shellee asked Lyn whether it seemed that perhaps they should maybe pay a visit to the prognosticating prophet.

UGLY: Lyn said that she would rather see the famously popular, consummately finished and perfected live, real-time performance of the accomplished virtuoso.

UGLY: When Dona tried to cook red tomato sauce in the pressure-cooker pot, the resulting explosive blast left a concave depression in the ceiling.

UGLY: If you are a member of the Christian church, you should avoid committing a blasphemous sacrilege.

UGLY: Greg resolved decisively to spend a day indulging gratifyingly in introspective self-examination.

UGLY: Nicola decided to invest financially her money in a dependable security, without a protracted, drawn-out discussion that would leave her lying down prone, prostrate with fatigue and exhaustion.

UGLY: Lyn's mood and humor are always inevitably manifestly self-evident.

UGLY: "Why don't you believe me?" cried Max. "I'm telling you the honest truth!"

UGLY: Paul takes monotone black-and-white photographs.

---

9. Based on a suggestion by Marina Nims, who has been wondering what to do with her impersonal belongings when she derides at the amusement park after the rides come to only a full stop.

UGLY: The burning candles were all lit and aflame.

UGLY: Maria pointed out to Geoff an attractively handsome man who was reputedly known to be a philandering womanizer.

UGLY: Madeline clapped her hands together to amuse her sister Sophia, to whom she was related by birth.

UGLY: This light-reflecting, shiny, round sphere will implode into itself in a short period of time.

UGLY: Geoff told Maria that he had heard that the man was merely a lying knave; from whence[10] the man came he knew not.

UGLY: Lyn was in such a furious rage that she stooped down to vituperative contumely.

UGLY: Max's response was to remain inarticulately mute; he had no desire to be assaulted, assailed, or attacked.

UGLY: Lyn decided that the home in which they lived was fast turning rapidly into a crazy madhouse.

UGLY: "Well," Max comforted her consolingly, "at least life isn't tediously boring."

THE PRINCIPLE FOR LUCID WRITING here is that you should express a thought only once, and should not express your ideas twice.

10. The phrase *from whence* is redundant because *whence* means *from what place, source, or cause.*

7   *Pronouns*

A PRONOUN is a little word that stands in for a name, or for a noun; a pronoun denotes an object or creature without naming that object or creature. You should be extremely careful that your pronouns denote (point to) what you intend them to denote. You cannot simply spread them about in a sentence and expect them to take on the meaning that you intend.

If the subject and the pronoun in a sentence match each other (in case, number, and gender), then the pronoun refers to the subject. If they do not match, then you are courting trouble.

> GOOD: Max admired the singer's dress; he had never seen so many sequins.

So far, everything is fine; we are speaking of Max's experiences in both portions of the sentence. However, consider the following:

> BAD: The singer swished past Max to the stage; he was blinded by the sparkles from the sequin-covered dress.

Here, *he* is the singer, rather than Max—a meaning perhaps divergent from the one that the writer intended to convey.

The next two examples have the same grammatical problem; however, once the singer is clearly identified as female, the male pronoun is free to denote Max. Thus, although the sentences are, strictly speaking, incorrect, most people would accept them because they are unambiguous.[11] The third example solves the problem correctly, by avoiding the pronoun for the second actor. You should strive to use the third construction, simply because writing correctly keeps you out of trouble.

> BAD: The singer moved her lips close to Max's ear and hit a high E; he had never heard such a noise.

> BAD: The singer brushed against Max as she passed him; he was overwhelmed by her perfume.

> GOOD: The lion moved his whiskers close to Lyn's nose; Lyn had never experienced such a sensation.

Similarly, if the pronoun matches the subject of the previous sentence, then it denotes that subject.

> GOOD: Gooch frisked down the stairs. He was euphoric.

We may not know what manner of creature Gooch is, but we have information about his psychological state (and his gender).

Because the pronoun refers to the subject, you can face the same problem, without the clues afforded by gender-specific pronouns, when you are speaking of objects.

> BAD: The ball bounced off the pavement and through the window; it was cracked already.

The sentence states that the ball, which bounced through the window, was cracked already. However, we might entertain the suspicion that the writer was careless, and intended rather the meaning of one of the following recasts:

---

11. Thus, these examples highlight that a sentence classified as bad may be (albeit rarely) acceptable, whereas a sentence classified as ugly is not acceptable, even though it is correct; that is, the distinction between bad and ugly is one of kind, rather than one of degree.

> GOOD: The ball bounced off the pavement and through the window; the window was cracked already.

> GOOD: The ball bounced off the pavement and through the window; the latter was cracked already.[12]

> GOOD: The ball bounced off the pavement and through the window, which was cracked already.

> Which *denotes whatever is named or otherwise identified before the preceding comma (in almost all cases); thus,* which *denotes the window here.*

Pronouns and their denotations can fail to match on several other dimensions. Consider the following abomination:

> BAD: One can easily use their modem to cruise the web.

My objection to this sentence is two-fold. First, I strongly urge you to avoid the pronoun *one* as though it had a peculiarly nasty communicable disease. The disease is stilted writing, and use of *one* is the fastest way to contract it.[13]

My second objection to the preceding example sentence is the singular–plural mismatch of the pronoun *one* (which is singular) and the pronoun *they* (which is plural). If, for whatever suicidal reason, you insist on using *one,* at least be consistent (albeit ugly):

> BAD: One can easily catch a nasty cold if they forget to bundle up.

---

12. Many people object to *the latter,* either because they think it is overly formal, or because they think it makes the reader work too hard. I like it and use it often; it is a precise term that our language offers to get you out of having to repeat what may be a long, cumbersome phrase. The same reasoning (from all parties) applies to *the former.*

13. Note that the first *it* in the paragraph correctly refers to the offending pronoun, and that the second *it* correctly refers to the disease. Do you understand why?

UGLY: One can easily contract a cold if one shakes hands with everyone within reach.

Using the plural pronoun *they* to denote a singular creature is one of the most objectionable errors that you can make.

BAD: When the office manager ran into the conference room, they found the board of directors asleep on the job.

BAD: When the user sits at their terminal, the world is their mussel.

GOOD: When the programmer sits at her desk, she works!

Remember that *each* is a singular; do not try to marry it to a plural pronoun. Furthermore, always respect creatures of gender by giving them a matching pronoun—do not insult them with an *it*.

BAD: Each woman at the conference was carrying a cellular telephone, which they had slipped into their briefcases.

BAD: Each tomcat on the block was boasting about its latest catch, which it had carried to the party in its mouth.

GOOD: Each man at the retreat was accompanied by a miniature teddy bear, which he had slipped surreptitiously into his pocket.

There are many possibilities for singular–plural mismatches.

BAD: A person should be careful when they drive at night on mountain roads, especially when it is raining.

GOOD: People should be careful when they critique the work of their partners, especially when it is art.

GOOD: A person should watch her manners when she edits another person's writing, especially when it stinks.

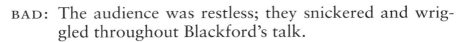

BAD: Lotfi created the set of fuzzy kittens, and handed them to the gardener.

GOOD: BB created the set of chewed-up snapshots, and laid out its members artistically on the living-room floor.

GOOD: Red created the set of chewed-up rats, and left it on the bedroom floor for Lyn to find in the morning.

BAD: The audience was restless; they snickered and wriggled throughout Blackford's talk.

GOOD: The audience was clearly fascinated; it sat quietly through 3 hours of presentations.

GOOD: The audience went wild; every member jumped to her feet and shouted "Bravo!"

In the latter two bad examples (Lotfi and the audience), the writer has switched without excuse from the singular set or audience to the plural members thereof. Note that collective nouns are singular.

 You should never use *it* in a sentence both as an anticipatory *it* (it is a far far better...) and as a pronoun *it*.

BAD: Adrienne thought that it was a good idea to write the book first, and then to worry about designing the various parts of it.

GOOD: Adrienne advised against spending hours designing a dingbat, only to reduce it so drastically that you could not see it anyway.

You can thus solve the problem by recasting the phrase that contains the first it; *here,* it was a (bad) idea *is replaced by* advised against.

GOOD: Adrienne thought that it was a bad idea to spend hours writing a figure caption, only to discover that the caption would not be needed anyway because the figure was irrelevant.

As another option, you can solve the problem by using an alternative form of identification in place of the second it; *here, rather than* it, *the second clause contains* the caption.

BAD: Brian believed that it would be alarming to release the software without testing it first.

GOOD: Brian was alarmed about the features list, so he asked Max to assign priorities to the entries in it.

Recast to avoid the first it.

GOOD: Brian thought that it was enlightening to examine his competitor's software, so he ordered the program by mail.

Recast to avoid the second it.

Whenever you use *gender-specific pronouns,* you face an obvious decision: How will you handle the denotation of the various generic actors (such as users, programmers, designers, or cat fanciers) in your text? How will you handle the *gender problem,* as it occurs in writing?

As we all know, for centuries, people simply used the male pronoun to denote all actors. I urge you not to maintain that tradition. Even if you do not find it offensive, you can be certain that at least one-half of your readers (of both genders) will so find it; why offend the people with whom you wish to communicate?

UGLY: The user reaches for his mouse.

There are several fixes for the gender problem that I find awkward. Many writers sprinkle their text with *she or he, her or him,* risking a form of stilted-writing disease, albeit in a good cause. Certain people[14] simply use female pronouns for all their characters, assuming that there cannot be any harm in redressing centuries of all-male pronouns holding sway, and that affirmative action is appropriate. A few people deteriorate into ungrammatical oblivion, using plural or neuter pronouns where such pronouns have no business being (as I just discussed).

> BAD: The user is likely to become grumpy when their software crashes the system.

> BAD: The administrator was late for work, as was its habit.

Certain writers go to extremes to write all their sentences with only plural actors, because the plural pronouns are not gender specific; alternatively, they avoid use of pronouns completely. Again, either approach is likely to leave you tangled in a number of silly sentences. In the following example, you can see that pronouns are useful in discriminating among actors; you can also see that correct use of pronouns is not sufficient to create a good paragraph:

> UGLY: When the decision analysts begin a job, they must interview the experts. The experts tell the analysts what variables they use, and the analysts try to understand the relationships among the variables. The analysts also ask the experts for values. The project directors meanwhile kibitz, and ask the analysts to hurry up. But the experts must also give the analysts probabilities, and the experts do not necessarily have the skills required, so the analysts may have to show the experts how to do such an assessment, even though the directors are impatient.

---

14. I am one such person.

GOOD: When a decision analyst begins a job, he must interview an expert. The expert tells him what variables
she uses, and he tries to understand the relationships
among them. The project director meanwhile
kibitzes, and asks him to hurry up. But the expert
must also give him probabilities, and she may not
have the skills required, so he may have to show her
how to do such an assessment, even though the
director is impatient.

As shown here, the simplest solution is simply to switch about even-
handedly between the two genders. Once you have assigned a gender
to an actor, let that actor keep that gender at least for the duration
of the paragraph. If you talk about *the user* frequently, you can let
that character be sometimes a male and sometimes a female.

Alternatively, depending on the subject matter of your writing, you
can assign genders to roles at the outset, explaining in a footnote that
you are doing so for convenience of notation, rather than out of a
prejudice regarding what genders tend to be correlated with what
roles. In such cases, it is good practice to use the *less-expected
gender*. Doing so cues your reader to your sensitivity to the gender
problem; failing to do so is likely to turn off a number of your
readers, even though you are trying to be sensitive by switching
between gender-specific pronouns. In a medical book, for example,
if you assign genders, let your nurses be male and your doctors be
female. In a book on business, speak of male secretaries and female
presidents. In an article on process reengineering in a software
company, make all your senior researchers female, and give them
male peons to carry out the implementation.

If you simply switch back and forth between genders, you can still
use the unexpected gender more often than not. People your world
with female coal miners and truckers, and male nannies.

THE PRINCIPLE FOR LUCID WRITING here is that you need to watch your pronouns with a hawk's eye, to ensure that <u>they</u> denote what you intend.[15]

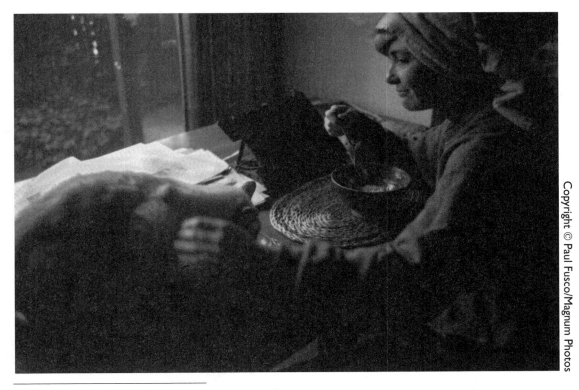

15. Patrick Henry Winston [personal communication] points out that you may need to pay extra attention to this matter, "especially if your mother (father?) tongue is a language with gender, which makes antecedents easier to figure out, thus leading you to overuse pronouns when you switch to English."

8   *Undefined This*

FAILURE TO use an accompanying noun leaves words such as *this, these, that,* and *some* undefined, masquerading as nouns that know what they are about. Just because *you* know what you mean does not guarantee that your reader will prove to be a psychic gymnast. Be kind, and specify your meaning.

You should always use a noun after words such as *this, these, that,* and *some.*

> BAD: Lyn asked Max to turn the garbage can upside down. This proved to be ill-advised.

What was ill-advised? Perhaps asking Max was a bad idea, because Max was in an exceedingly grumpy mood and responded by throwing the garbage out the window. Or perhaps turning the can upside down was a bad idea, because it resulted in the obvious debacle. Or perhaps Max slipped a disk[16] attempting to please Lyn. Clarification is in order:

> GOOD: This intrusion on Max's privacy proved to be ill-advised.
>
> GOOD: Making this request proved to be ill-advised.
>
> GOOD: This maneuver proved to be ill-advised.

---

16. There are two possible spellings of the word: *disk* is the general preferred spelling, whereas *disc* is the variant.

You should also provide a noun in situations where the ambiguity is nonspecific.

UGLY: Some become distraught when they cannot exercise.

Some whats become distraught? Hamsters? Mynah birds? If you mean athletes, then say so.

GOOD: Certain athletes get the heebie-jeebies when they are unable to exercise.

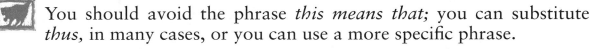

UGLY: This does not grow in the garden.

GOOD: Certain trees do not grow in the garden.

SPLENDID: Decision trees do not grow in the garden.[17]

You should avoid the phrase *this means that*; you can substitute *thus*, in many cases, or you can use a more specific phrase.

UGLY: All the stuff that you need is given in just 100 pages; this means that you can learn to program in 1 week.

GOOD: All the information that you need is given in just 100 pages; thus, with almost no pain, you can learn to program in 1 week.

UGLY: The simple relation between $F(z)$ and $A(z)$ is due primarily to[18] C. Jordon; this means that C. Jordon is the person who wrote about the relation first.

---

17. The specificity of *decision trees* is higher than is that of *certain trees*.

18. The term *is due to,* used to mean *was developed by, was invented by, was written about by,* and so on, is extraordinarily clumsy; limit your use of *is due to* to mean *is caused by,* if you must, and to the notion of *quid pro quo.*

GOOD: The complex relation between Jim and Lauralee is the result of two highly complex, interesting, and creative personalities; this assertion implies that Lyn likes both Jim and Lauralee.

THE PRINCIPLE FOR LUCID WRITING here is that you should not leave running around in your text words that are defined insufficiently. Rather than writing that this is a bad example, you should write that this example is a good one.

9 *Motivate*

O<small>NLY ANIMATE</small> objects possess motivation. *Motivation* is commonly misused to mean *rationale, basis, foundation,* or *justification*. For example, people often make claims such as

> BAD: Our research was motivated by an intense desire to breed guppies.

What they mean is more along these lines:

> GOOD: We were motivated to undertake this study by our intense desire to discover how quickly wombats reproduce.

> SPLENDID: My intense desire to learn how to teach people to write well motivated me to undertake the project of creating this book.[19]

Or consider these examples:

> BAD: The current study was motivated by the Brillig algorithm for scrubbing pots.

> GOOD: We report a study that was based on the Slithy algorithm for cleaning slimy toes.

---

19. Based on a suggestion by Richard Rubinstein; the switch from passive to active voice is the basis for the reclassification.

BAD: Our work was motivated by greed.

GOOD: Greed provided the basis for our work.

GOOD: We were motivated by greed to perform this work.

BAD: This hypothesis is motivated by previous work on possible worlds.

GOOD: This hypothesis is based on previous work on possible worlds.

SPLENDID: Our previous studies of possible worlds led us to formulate this hypothesis.

THE PRINCIPLE FOR LUCID WRITING here is that you should never attempt to motivate any entity that is not a living creature.

Copyright © Paul Fusco/Magnum Photos

10 Oxymorons

OXYMORONS are terms that contain inherent contradictions; a logically coherent oxymoron is an oxymoron.[20]

Of course, you may write intentional oxymorons, where the contradiction is a source of humor. Unintentional oxymorons, however, can make your writing look moronic.

The point is that many people fail to pay attention to the meanings of their words, and end up writing nonsense when they intended otherwise.

UGLY?: Richard told his boss that he would need a massive vacuum in which to run the experiment.

UGLY?: "You're an honest liar, Richard," his boss replied.

UGLY?: "It's a fact," insisted Richard—"those guys in military intelligence asked me to set it up."[21]

UGLY?: Our law courts often apply the standard of a reasonable man, but Lyn insists that the standard has an intrinsic problem.

UGLY?: Misha presented Holly with a necklace of genuine fake pearls.

UGLY?: "Devon!" called Max from the conference room, "Can you get this underfoot overhead out of here?"

---

20. Based on a suggestion by Peter Gordon.
21. Based on a suggestion by Max Henrion.

UGLY?: "Why not replace it with an artificial-reality teleconferencing system?" replied Devon.

UGLY?: Oxymorons are the bugbear of intelligent machines.[22]

UGLY?: Gwen was disappointed to find that she had wasted her money on a lying soothsayer.

UGLY?: Garrett told Gwen not to worry, as perhaps the experience had been a fair-value ripoff.

UGLY?: When Lyn shops for silk, she spends with parsimonious prodigality.

UGLY?: Peter was propelled by inertia to fall alertly asleep on the couch while listening (in one ear and out the other) to the soporific wake-up broadcast.

UGLY?: Red told BB to stay away from that clumsy cat down the block.

UGLY?: The party was a resoundingly successful failure.

UGLY?: Certain people believe that they have a right to defend themselves with assault weapons.[23]

UGLY?: The smart bomb exploded without reason.[24]

UGLY?: *BUGS in Writing: A Catalog of Prosaic Catastrophes*

A related class similarly sets up impossible relationships between two terms, because one term admits of no modification.

UGLY: "There are ten to the almost literally infinite power."[25]

---

22. Contributed by Peter Gordon; machines were originally, by definition, not intelligent.

23. Here the sentence, rather than a specific phrase, is oxymoronic.

24. Based on a flash of brilliance by Peter Gordon.

25. Spoken by Walter Pidgeon on Altair-4 in *The Forbidden Planet,* MGM 1956.

UGLY: When Elizabeth was extremely pregnant, she went on a tortuous 7-hour hike with Lyn and Jeff.

UGLY: The cottage in which Max and Lyn live is moderately ideal, given the odd combination of a perfect location and a near–falling-down construction.

UGLY: Lyn wondered whether hiking up Windy Hill in her bikini might be slightly illegal.

UGLY: Numerous people have told Lyn that she is very unique.

And while we are in the vicinity of misused and abused words, there are puns, which are worth the groan *when they are intentional.*

GOOD: In an attempt to attract customers, the airline reduced its fare; passengers, however, complained that they were served almost nothing to eat.[26]

GOOD: The program's sponsor, a maker of insecticides, interrupted frequently with routines to get rid of bugs.

GOOD: The model, caught napping on the painter's palette, arose hastily and red faced.

GOOD: Peter's play on words opened with a pun, characterized each part of speech, and ended after a period.

GOOD: Lyn saw Red every morning.

GOOD: The promotion manager for StickyStuff styling gel has come up with another hair-brained idea.[27]

GOOD: Lyn told Brendan that, at her house, a plum tree reached its branches across the upper deck, allowing her and Max to pluck ripe fruit as they relaxed; "Gosh," sighed Brendan, "How decadent!"[28]

---

26. This example and the three that follow were suggested by Peter Gordon.

27. Based tangentially on a suggestion by Patrick Henry Winston.

28. Based on a remark by Brendan Del Favero.

SPLENDID: Lyn explained to Carver that certain people find it easier to hear what she has to say if she writes a parable, instead of just blurting out her point. "You mean," suggested Carver, "that the story is sufficiently parabolic?"

THE PRINCIPLE FOR LUCID WRITING here is that you should not use phrases that are internally inconsistent unless you do so knowingly, with the intent to amuse your reader. If your words are intelligibly incomprehensible, your reader may become clearly confused.

11 Shall Versus Will

THE TERMS *shall* and *will* have usefully different meanings. You should distinguish between them, so that you can convey information precisely.

You should use *shall* when you are simply *predicting the future* and are speaking either as *I* or as *we*.[29] Certain people regard *shall* as archaic, but I think it is a respectable word with a job to do; that is, it conveys a meaning quite different from that of *will*. You should always use *shall,* for example, when you are speaking about what comes later in your manuscript, because presumably, by the time someone reads your words, whatever comes later is no longer subject to your will.

GOOD: We shall discuss creative uses for CD-ROM drives in Section 12.4.

GOOD: In Chapter 5, I shall analyze the differences between Jim and Lauralee's constitutions.

GOOD: After we work through the proof, we shall discuss how the theorem can be applied.

GOOD: We shall define *a* to be the number of ducks in the living room, and *b* to be the number of spots on the carpet.

---

29. I do not myself use *shall* with *he* or *she* or *they*, or generally in talking about other people; my concern in this section is to encourage you to use *shall* when speaking about yourself, alone or in company.

SPLENDID: We define s to be the speed (in mph) at which Max travels, t_c to be the current time, and $t + t_t$ to be the time of Max's first appointment, where t_t is the time that it takes Max to travel to the office.

In such cases, it is correct to use shall, *but it is cleaner to use nothing at all.*

GOOD: I shall find the article in *Excerpta Medica*.

GOOD: I shall be crushed if Avi does not like this book.

You should use *will* when you mean to imply *intentionality;* that is, a creature's will should be involved. *Will* is much stronger than *shall*. In certain situations, the choice is not clear cut, and you must use your judgment.

GOOD: After we dance the night away, we will make a champagne toast to the sunrise.

I am determined to drink champagne at an ungodly hour of the morning.

GOOD: After I stay up all night working, I shall be too exhausted to play.

I am merely predicting that I shall be burned out after a long all-nighter.

GOOD: I shall scream if I see a mouse.

I shall emit piercing shrieks, perhaps without volition, if a mouse shows its furry little face.

GOOD: I will scream if I see your face again.

I will choose to emit piercing shrieks if you show your face.

GOOD: I shall cry if I have to edit another book on DOS.

GOOD: I will refuse to edit another book on FORTRAN.

SPLENDID: I *will* drown! No one shall stop me![30]

SPLENDID: I shall never love anyone as I loved you; however, I will not sit home alone.[31]

SPLENDID: I will never love anyone as I loved you; therefore, I shall be lonely.

I recommend that you always use *will* when you are writing a research proposal. You are contracting to do the work, so you should sound as though you intend to carry it out. Consider the contrast among these examples:

GOOD: We shall complete the project within 2 years.

We predict that the project will take 2 years to complete; use this form in, for example, a report to your supervisor.

GOOD: We will complete the proposed work within 2 weeks.

We will do what it takes to complete the work within 2 weeks; use this form in a proposal.

SPLENDID: We commit to completing the proposed experiment within 2 months.

---

30. Contributed by Wynter Snow.

31. The prediction and threat were contributed by Peter Gordon; the next example is mine.

Note that *shall* (when used in the first person in certain circumstances) can be stronger than *will,* even though—or precisely because—it does not bring volition into the picture.

GOOD: "Whatever you decide to do, Moisés," Tania explained firmly, "I shall go dancing tomorrow night."

Tania is saying that there is no question about what she is going to do tomorrow night, no matter what Moisés does. Tania is not even asserting that she wills her dancing; she is simply stating that she is going to dance.

GOOD: If you drop my cat, I shall drop you out the window.

A simple warning, and a serious one.

THE PRINCIPLE FOR LUCID WRITING here is that you should use *shall* to predict the future (when you are speaking of yourself, alone or with other creatures), and *will* to imply intentionality. I shall teach, and you will learn.

12 Key Terms

I N A N Y document that you write, you should consider applying the concept of a key term. A *key term* is a word or phrase that is critical to your exposition or discussion.

In most cases, you should *define key terms* in your text. In almost all cases, you should define them the first time that you use them. The exception occurs when you use the term only parenthetically or tangentially in your discussion, and later will use it as central to your presentation; in such cases, you should define the key term where it is essential, provided that its meaning is sufficiently clear that your reader will get the idea on first mention.

Certain elaborate textbook designs define key terms in the margins, essentially maintaining a running glossary. The key term is highlighted in the text, and the definition is set in a marginal text column in the vicinity of the term.

Note that your definition may be implied rather than explicit.

> GOOD: **Expert systems** give advice about specialized subjects (such as finance, mineral exploration, and medicine).

> GOOD: We define **question-answering systems** to be systems designed to answer queries posed in a restricted but large subset of a natural language.

> GOOD: A **theorem-proving system** verifies that the computer programs and the digital hardware meet the stated specifications.

You should *choose a typographical distinction* for key terms, and should use it when you define the term. If you do not define the term, use the typographical distinction the first time that you use the term in the discussion of which the term is a critical part.

As I stated earlier, a **key term** is a word or phrase central to your exposition. For example, in a discussion of basic hardware, your key terms might include *central processing unit, hard-disk drive, monitor, keyboard, mouse,* and *CD-ROM drive.* In the sentence that begins this paragraph, the phrase *key term* is a key term.

Whenever you set text in a typographical style different from that of your base text, remember to set the accompanying punctuation marks in the same style. Set a boldface (or italic) comma, period, quotation mark, parenthesis, question mark, exclamation mark, and so on next to a boldface (or italic) letter.

In almost all cases, you should use **boldface type** for key terms. Boldface type is easy to spot. If your reader is looking for the discussion that describes what a term means, she will find it easily if you have set the term in boldface type. *Italic type,* in contrast, tends to disappear into the page, and is thus much less well suited for use on key terms.[32] Certain textbook designs that use color pick up key terms with colored screens,[33] which can work exceptionally well. Under-

---

32. Note that, to the extent that this book has key terms, those terms are set in italic type (except for the example key term on this page), I made this decision because too much boldface type becomes distracting, and the short segments and numerous critical terms would have created a text page that resembled a Rorschach test.

33. A *screen,* in book publishing, is a block in which colored or black dots are laid down uniformly; a screen is described by the percentage of the area that is covered by ink, as controlled by the dot size. Thus, a 20-percent screen has 20 percent of the area covered by black or colored dots; a 100-percent screen is solid color or black.

<u>lining</u> and other highlighting techniques are possible, but be certain not to use one that will detract from your reader's ability to comprehend your text. You should never, for example, use ***<u>boldface, italic, and underlined type</u>***.

UGLY: The publisher and the author sign a **<u>CONTRACT.</u>**

UGLY: The author submits a ***<u>final manuscript.</u>***

UGLY: The manuscript is torn apart and then reconstructed by a *copy editor.*

UGLY: A ***<u>castoff</u>*** is a count of the book pages that will be made up from a manuscript.

GOOD: Publishers use **blue lines** to check the film created from the electronic file.

GOOD: The individual sheets of film are made up into **signatures,** from which the plates are created.

GOOD: The author then takes a **vacation.**

Note that you should set the key term in boldface type (or however you are highlighting it, except when you use italic type) *only once.* Thereafter (or before then), you should use *italics* when you wish to emphasize the term.[34]

GOOD: A **guardian angel** is a being who deflects sorrow and pain that threaten you. Guardian angels are believed to result from cross-fertilization of fairy godmothers and genii. *Transporting angels* and *guardian angels* bear little resemblance to each other.

GOOD: **Plan-It Venus** is the weltanschauung held by Lyn. *Plan-It Venus* is a complex set of interrelated imperatives and valuative functions that permits navigation of life's tricky decisions.

---

34. You should, in general, use italics for emphasis of any terms, rather than for only subsequent use of key terms.

GOOD: Max, on the other hand, adheres to <u>Plan-It Pluto</u>, a
        set of views that is way out there. Max's custom-
        tailored version of *Plan-It Pluto* includes advice from
        Mars, in addition to secret midnight sessions with a
        consultant from Venus.

GOOD: **Dial-a-solution** allows you to choose the planning
        perspective of your advisor: press 1 for Plan-It Venus,
        press 2 for Plan-It Pluto, press 3 for Plan-It Mars, or
        press 0 for further options.[35] *Dial-a-solution* is similar
        to *dial-a-confessor,* a system designed and imple-
        mented by Christopher Lane.

GOOD: We suggest <u>mediation</u>[36] to solve insoluble difficulties;
        you can also use <u>meditation</u> to ponder imponodera-
        bles. Note that *mediation* and *meditation* bear super-
        ficial resemblances to each other.

THE PRINCIPLES FOR LUCID WRITING here
are that you should consider organizing your
presentation based on **key terms,** which are
terms critical to the meaning of your discussion; that you should usually define key terms
in text; and that you should use a form of typographical distinction (usually boldface type) to
highlight key terms.

35. Note that the text after the colon constitutes a list, and therefore the first
letter is not capitalized.

36. Be careful and sparing in your use of underlining. In many typefaces, under-
ling is exceptionally messy. Nonetheless, it is useful when you have assigned
other semantic meanings to the alternative forms of typographic distinction.

13   *Proven Versus Proved*

PROVED is a form of the verb *to prove,* whereas *proven* is an adjective.

BAD: The theory has proven to be correct.

BAD: The theory was proven to be correct.

GOOD: The theory has proved to be correct.

GOOD: The theory has been proved correct.

GOOD: This theory uses a proven technique.

BAD: A sheep flock is a proved approach to lawn mowing.

GOOD: A kangaroo in a crystal shop is a proven recipe for change.

GOOD: Red and BB proved that a brace of determined cats can hold armies of rats at bay.

BAD: For the problem of simulating an elevator system, Knuth offered a proved algorithm.

GOOD: For his treatise on truth propagation, Winston proved that a truth-propagation network is a type of value-propagation network.

GOOD: For their adventure in space, Richard and Betsy used a proven algorithm.

BAD: Various studies have proven that exposure to danger increases the probability that you will fall in love with the next person you meet.

GOOD: The scientists nonetheless proved that it is impossible to fall in love while diving from a tall building.

GOOD: A proven alternative is nude duo bungee jumping.

THE PRINCIPLE FOR LUCID WRITING here is that we have proved that you should use *proven* only as a modifier.

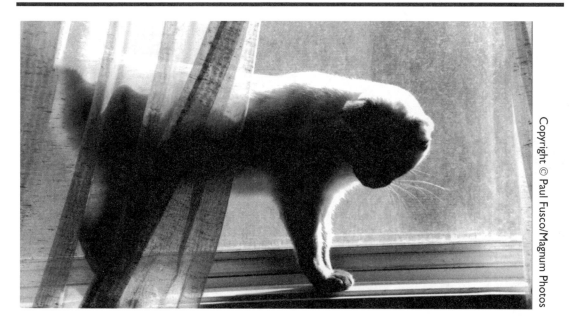

14 Everyone, Someone, No One, None

Bʏ ᴄʜᴀɴɢɪɴɢ the styling, you can change the meaning of various words that indicate how many *ones* you intend.

 You should set *someone* (meaning a person) and *everyone* (meaning all people under discussion) without a space, as one word.

> BAD: I wish I had some one with whom to play.

> GOOD: Someone just called to ask me out for dinner.

> SPLENDID: Max just called to ask me to go dancing.

> *Someone is a slightly fuzzy word; whenever appropriate, specify which someone you mean.*

> BAD: Not every one in the state of California appreciates good writing or good weather.

> GOOD: Everyone prefers to read well-written prose, given the alternative of reading garbage in a downpour.

You should set *every one* (meaning not just certain of the individuals, but rather every one of them) as two words.

> BAD: Everyone of the files was a mine of information.

> GOOD: Every one of the files was a text, source, or object file.

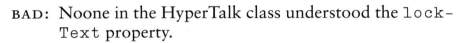

BAD: Max made everyone of his minutes count.

GOOD: Lyn made every one of her cats fat and sleek and well content.

 You should set *no one* (meaning not one person) as two words.

BAD: Noone turned up at the meeting.

GOOD: No one knew that the meeting had been called.

BAD: Noone in the HyperTalk class understood the `lock-Text` property.

GOOD: The teacher told no one that `wideMargins` applies to only fields, whereas only `style` is common to both fields and buttons.

You should understand that *none* can mean *not any* or *not one.* If your phrase would be recast as *none of them,* then you should use a plural form of the verb; if the phrase would be recast as *none of it,* then you should use a singular verb.

BAD: None of the rats is squeaking.

You would say none of them *about the rats, so the* is *is incorrect.*

GOOD: None of the kittens are lonely.

You would say none of them *about the kittens, so the* are *is correct.*

BAD: None of the raster-scan systems is up to the task.

You would say none of them *about the systems, so the is* is *incorrect.*

GOOD: None of the screen displays are comprehensible.

BAD: None of the meal were edible.

You would say none of it *about the meal, so the* were is *incorrect.*

GOOD: None of the pie is left; none of the ice cream has melted.

BAD: None of the audience are clapping.

You would say none of it *about the audience, so the* are is *incorrect.*

GOOD: None of the wire-frame model is visible.

SPLENDID: Everyone thought that every one of the fools had imagined that none of the water was wet—and none of them were correct.

THE PRINCIPLES FOR LUCID WRITING here are that everyone should remember that only *no one* and *every one* have a space in them, and that someone should remind you that *none* can be singular or plural.

15 Colon

A COLON signifies that what follows it expands on or explains what precedes it: This sentence is an example.

You should use a colon at the end of a sentence followed by a *list*. (Note that, even if the list constitutes a full sentence, you should not capitalize the first letter of the list. The list and the introductory tag together constitute a single sentence.)

BAD: Lyn was worried about Red for several reasons. Red was sneezing, Red's eyes were running, Red's nose was hot, and Red was drooping around the house.

GOOD: Although she had planned to take Red to the vet, Lyn decided to wait because she saw good signs: Red greeted her by rolling in the dirt; Red remembered how to purr; and Red left a shrew for her on the deck, so she figured he couldn't be all that sick.

BAD: This tutorial assumes that you know the basic techniques needed to use a Macintosh computer, how to point and click, how to double-click, and how to drag an object using the mouse.

GOOD: This book does not assume that you know the basic skills needed to write a manuscript: how to form a sentence, how to use words correctly and precisely, and how to laugh at your own errors.

You should use a colon at the end of a sentence followed by any text that *further explains* that first sentence. Note that, in such cases, either a period or an em dash usually will work also, and sometimes a semicolon will serve as well, but each punctuation mark has a slightly different meaning. The colon ties the two sentences together and designates their relationship to each other more tightly than do the other punctuation marks. Sometimes, the period, em dash, or semicolon will fail to convey the relationship, in which case the colon is clearly indicated.

GOOD: Lyn could tell that BB was going into heat. BB had been howling and yowling the entire night through.

The period works, in that it is correct; a semicolon would yoke the two sentences and would thus convey more information; a colon would be precisely right, because the second sentence explains *the first one.*

GOOD: Lyn could tell that Max had had a tough day—Max frowned dreadfully when she greeted him, after which he sat reading an upside-down newspaper for several hours.

The em dash works and is correct; a semicolon also would work, and so would a colon.

GOOD: Max could tell that Lyn had had a superb day: She was reading a newspaper upside-down, while practicing her headstands.

The colon is correct; an em dash would work also, but not as well.

SPLENDID: Lyn could tell that Red had been out hunting again: There were three mice neatly laid out on the upstairs silk rug (a display that Max failed to appreciate, until he confirmed the lack of stains).

Here, the colon is by far the most informative punctuation mark.

GOOD: Nicola was pleased that she had succeeded in her goal—making it through the first term.

The em dash is OK, but a colon would be preferable.

GOOD: Greg was happy with his talk because the audience clearly responded well. He received a standing ovation.

The period is OK, but a colon would be preferable.

SPLENDID: Greg and Nicola were delighted with the cottage because it had one exceptional feature: an apricot tree laden with fruit.

The colon is precisely right.

You should set a capital letter at the beginning of a sentence that follows a colon;[37] you should not use a capital letter if what follows is a fragment, or if what follows is a list, even if the list entries are full sentences or the list itself constitutes a sentence.

UGLY: Soren was fiddling with his Porsche yet again: this time, he was replacing the spark plugs and fixing the heater coil.

GOOD: Soren was fiddling in the kitchen once again: This time, he was making double-chocolate mousse and brewing lager.

UGLY: Shellee went wild on her shopping trip with Lyn: she brought home nine new blouses, but she took three of them back because she had no room in her closets.

---

37. If you prefer not to adopt this convention, do not do so; however, apply consistently whichever style you choose.

GOOD: Shellee promised to come back to California soon: She agreed that dancing to Dr. Loco is one of those necessities of life that cannot be denied for too long.

BAD: Dona had only one home: The house on Quadra Island (she had previously commuted between Florida and Canada).

GOOD: Dona and Lyn collected only one kind of wild mushroom: chanterelles *(Cantharellus cibarius)* (they had previously been content with store-bought snowcaps and tree ears).

You should use a colon only at the end of a clause that constitutes a sentence. One of the most common errors that writers make is to place a colon in the middle of a sentence merely because a list follows, which is an insufficient excuse, whether the list is intext[38] or displayed.

BAD: Jeff's few items of furniture were limited to: a couch, a chair, and a bed.

GOOD: Jeff's numerous cars included an enormous old Buick, a massive pick-up truck, and a Mustang.

BAD: To confine the display to columns A through C, you:

1. Move your cursor to the beginning of the range you will use for display—A1, say.
2. Type /Range Justify.
3. Highlight the cells in the range A1. .C1 with the right arrow key.

---

38. *Intext* is the editorial adjective meaning *set in the text line.*

GOOD: To create a documentation copy of a worksheet, you

    1. Use `/Worksheet Global Format Text` to set the global format to Text.

    2. Use `/Range Format Reset Pl . . X15` and `ENTER` to reset all the Range formats back to Global settings.

    3. Next, move your cursor to column R, and use `/Worksheet Column Set-Width`, and press the Right Arrow Key until column R is the desired number of characters wide.

    4. Remember to save your work, in case of unforeseen head crashes and other natural disasters.

UGLY: Max decided to change: his shirt, which had salad dressing on it; his tie, which had spaghetti sauce on it; and his beer, which had egg in it.

GOOD: Lyn decided to change her habitat, because the old one was too full of *Homo sapiens;* her sheets, because they were covered with white fur again; and her mind, because consistency is interesting only when dispensed in small doses.

THE PRINCIPLE FOR LUCID WRITING here is that you should use a colon at the end of a sentence to indicate that further explanation follows: An explanation might consist of a detailed list, for example, or of an example.

16 *Effort*

THE WORD *effort* is widely overused and abused in formal writing; it shows up so often in research papers in certain disciplines that it loses all meaning. Call on it to denote only its most narrowly defined meaning.

You should use *effort* to indicate strenuous physical activity of the sort usually accompanied by groans and perspiration, or to indicate strain of the mental-anguish variety.

> GOOD: The women huffed loudly at the effort it took to carry the piano upstairs.

> GOOD: Asking for a divorce took as much effort as did asking for a raise the previous year.

You should avoid using *effort* when you mean *project* or *program*.

> UGLY: The Teach Everyone Frisbee effort failed because of a worldwide shortage in plastics.

> GOOD: The Teach Everyone Singing project failed because of a worldwide shortage in decent voices.

> UGLY: The software-development effort was huge; it required enormous funding.

> GOOD: The implementation program was well thought out, and required only a small investment.

You should also avoid using *effort* when you mean *time* or *work* or *money;* that is, it is more informative to specify what sort of effort you mean.

UGLY: Eddie had invested a huge amount of effort in decorating the MRI room.

GOOD: Eddie had invested a huge amount of time in decorating the CAT scanner.

UGLY: Lyn had wasted only a little effort trying to get the car started, before she noticed that the distributor[39] was missing.

GOOD: Lyn had wasted only a little time trying to get the car started, before she noticed that the gas tank was empty.

SPLENDID: Lyn had wasted only a little time trying to get the car started, before she decided that she would have to make the effort to push it down the driveway.

 THE PRINCIPLE FOR LUCID WRITING here is that you should view *effort* as a tired, thin, overused word; omit it, or ~~make an effort to~~ substitute for it a more informative term that indicates more precisely what you mean.

---

39. When she went back to complain, she discovered that the dealer was missing as well.

17 *Which Versus That*

THE WORDS *which* and *that* have distinct meanings. If you are familiar with those meanings, you can convey information precisely; if you are not, you risk confusing your reader. There are several ways to understand the difference between these two words.

That identifies the objects about which you are speaking, whereas *which* merely provides further information about those objects.

> GOOD: The car that is speeding down the road is about to crash into a stuffed piglet.

> GOOD: The car, which is speeding down the road, is about to crash into a fricasseed mongoose.

In the first example, we can answer the question

> Which car is about to crash?
>
> (The one that is speeding down the road.)

The use of *that* picks out a single car, perhaps from among hundreds of staid cars that are toddling along at the speed limit. The second example, in contrast, does not answer the question

> Which car is about to crash?
>
> (We do not know.)

What we have instead is a parenthetical remark, an aside, that informs us that the car, in addition to being about to crash into a pseudocreature, is speeding.

Or consider this example:

GOOD: The bacteria that live on your skin can cause an infection if you are cut.

GOOD: The bacteria, which live on your skin, can cause an infection if you are cut.

In the first case, we know that the bacteria that cause infections are those that live on your skin. In the second case, we know that the bacteria—whatever they are, however they have been identified—live on your skin. We have knowledge; we do not have identification.

 A *that* clause picks out one among many, whereas a *which* clause often implies that there is only one.

GOOD: The ABC computer that has the most bells and whistles will sell fastest.

GOOD: The ABC computer, which has the most bells and whistles, will sell fastest.

In the first sentence, we are given sufficient information to draw the conclusions that ABC makes several computers, and that, among this set, the one that will sell fastest is the one with the most features.

In the second sentence, we might draw the conclusion that ABC makes only one computer, or at least that only one computer that ABC makes is relevant to our discussion. This ABC computer evidently is in competition with other brands of machine, yet we are reassured that it will sell fastest, because we have the added information that it has more features than do its competitors.

 If we remove a *that* clause from a sentence, we destroy the original meaning; if we remove a *which* clause, we leave the meaning intact.

GOOD: The machine that has the color monitor is your Christmas present.

GOOD: The machine, which has a color monitor, is your
Hanukkah present.

The identification of the machine as being the one that has a color
monitor is critical information in the first sentence; in the second
sentence, the information about the monitor's characteristics is a
remark that could be removed without substantial harm.

You should not set off *that* clauses by commas, whereas you should
set off *which* clauses by commas.

GOOD: The cave that is dark and dank is where we shall
spend the night.

GOOD: The cave, which is dark and dank, is where we shall
spend the night.

Note that commas may show up to confuse you even in the vicinity
of a *that* clause.

GOOD: The cave that is dark and dank, and the ledge outside
it, is where we shall spend the night.

GOOD: The cave, I fear, that is dark and dank is where we
shall spend the night.

Here, the extra commas are protecting their own clauses *(and the
ledge outside it; I fear)*; they have no truck with the *that* clause.

Note also that a *which* clause must have commas around it—
omitting the commas is an error.

BAD: Lyn's eyes which had been strained all day by reading
microfont were telling Lyn to knock off before they
gave up entirely.

GOOD: Max's arms, which had been supporting telephones
and lecture pointers all day, now were wrapped firmly
around Lyn, who happily gave up entirely.

A related point that you should understand is that *which* always refers to whatever happens to be sitting in front of the comma preceding the *which*. So now you have a new trap to avoid.

> BAD: Richard argued with the lamp, which was foolish.

The preceding sentence asserts that the lamp was foolish; we suspect that the intended meaning was rather along these lines:[40]

> GOOD: Richard argued with the lamp, which behavior was foolish.
>
> *Note that you can* change *the denotation of* which *with this device.*
>
> SPLENDID: Richard argued with the lamp; such is the habit of fools under stress.

> BAD?: Nick wrote letters addressed to Lyn's home, which was charming.
>
> *The sentence asserts that Lyn's home, rather than Nick's behavior, was charming.*
>
> SPLENDID: Render unto Caesar that that is Caesar's, which is everything secular.[41]

In summary, when you need to distinguish between *that* and *which,* you should ask yourself these questions:

---

40. A reviewer suggested as a solution the sentence *Richard argued with the lamp, which was a foolish idea.* This solution is no solution, as it asserts that the lamp was a foolish idea. In addition, there is no *idea* in the sentence, whereas there is a *behavior.* The suggested construction is extremely common; avoid it!

41. Contributed by Peter Gordon. Both *which* and *that* are used correctly; however, the repeated *that* is awkward, so you should avoid that construction except when you are making up silly examples.

1. Does the clause identify the object under discussion *(that)*, or does it merely add information *(which)*?

2. Does the clause pick out one among possibly many like objects *(that)*, or is there only one such object under discussion *(which)*?

3. Does the clause contribute information critical to the meaning of the sentence *(that)*, or is the contribution more in the nature of a parenthetical remark *(which)*?

4. Should the clause be left without commas *(that)*, or should it be set off by commas *(which)*?

Even if this distinction is new to you, you will eventually retune your ear, and every *which* that should be a *that*, every *that* that should be a *which,* will annoy you intensely. While you are in the process of retuning, consider using your word processor to help you. When you finish writing your document, check each instance of both words to make sure that the word is correct. The exercise takes only a few minutes, and ensures that you think about the distinction. You can use this technique for any phrases for which you are learning new or more precise applications.

THE PRINCIPLE FOR LUCID WRITING here is that you should use *that* to identify objects (which may be creatures), and should use *which* to make remarks that provide extra information about objects.

18 Spread-Out Phrases

YOU SHOULD keep intact, to the extent practical and graceful, phrases that belong together. Such phrases may be action descriptions comprising more than one word, or action-plus-object descriptions. Thus, you want to pick up tips, rather than pick tips up; look up tricks, rather than look tricks up; and make up examples, rather than make examples up.

As a general rule, you want to avoid forcing your reader to backtrack. If you make your reader wait to find out what you are doing, or to what you are doing it, until you have thrown in long strings of related information, you will force backtracking.[42]

> UGLY: Lyn hastily put Red, BB, and the mail, which looked like it contained several checks, down so that she could unlock the door comfortably.
>
> GOOD: Max eagerly put down his briefcase, laptop-computer case, suitcase, and LCD-screen case so that he could embrace Lyn affectionately when he got home from his trip.

---

42. To get a clear picture of this problem, as well as of numerous others, see Mark Twain's essay "The Awful German Language" for a description of the German habit of putting the verb at the end of the sentence.

UGLY: You can build the model using expert opinion, obtained through interviews or questionnaires, and the existing database as inputs.

GOOD: You can draw the network using as inputs the variables' values, obtained from published statistics, and the relations among variables.

UGLY: Anthony takes his work, which comprises acting the roles of numerous different characters, and for which he prepares carefully, seriously.

GOOD: Anthony played the character of a cantor, for which he took singing lessons, because he takes his work seriously.

UGLY: `StdPutPic` is the standard low-level routine for saving, as the definition of a picture, information.

GOOD: `StdGetPic` is the standard low-level routine for retrieving information from a picture definition.

UGLY: Tim stumbled as though he was drunk, even though he was not even slightly happy, down the stairs, taking as he went his jacket and tie and shoes off.

GOOD: Suresh pranced around the office as though he was a handsome, clever young man taking off a year to find out what he wanted to do with his life—and that's precisely what he was.

UGLY: Even though it was only 5 A.M., Lyn began to wake Max, who had a breakfast meeting scheduled and had not gone running in 3 days, and as a result was getting noticeably cranky, up.

GOOD: At 6:40 A.M., Red began to wake up Lyn, who loved to have her field of vision filled in the morning by 21 pounds of flame-point Siamese magnificence.

THE PRINCIPLE FOR LUCID WRITING here is that you should not force your reader to back-track by spreading relevant phrases, over the expanse of your sentence, out.

19 *While*

THE WORD *while* refers to *time;* it does not mean *whereas* or *although.*

 You should use *while* to mean *at the same time as.* The phrase *while at the same time* is redundant.

> BAD: Max could type on his laptop computer while talking on the telephone at the same time.

> GOOD: Lyn could brush her teeth while petting Red.

 You should not use *while* when you mean *whereas*—even though people often use *while* incorrectly in that way.

> BAD: Max could write impeccable reports, while he could not articulate his feelings about oatmeal.

> GOOD: Lyn could write impeccable reports, whereas she could not tolerate business meetings.

> BAD: A spreadsheet is used primarily for calculating and analyzing numbers, while a database is used for storing them.[43]

---

43. Peter Gordon points out that we could be talking about a multitasking system running the concurrent tasks of spreadsheet analysis and database input. If we meant to describe that situation, then *while* would be correct.

GOOD: Workgroup computing is designed to facilitate collaboration among team members, whereas electronic bulletin-board systems tend to encourage mental absenteeism.

You should not use *while* when you mean *although*, even though people often use *while* incorrectly in that way.

BAD: While Max could do almost any task himself, he was at the moment incapable of delegating the simplest grunt work without giving himself stomach cramps from anxiety.

GOOD: Although Lyn could give the most luxurious back rubs in the world, she was at the moment incapable of kneading bread without giving herself a finger cramp from wimpiness.

BAD: While we use the base-10 number system, we must translate it for computers, which have fewer fingers.

GOOD: Although computers use machine language, we must translate it for humans, who have less patience.

THE PRINCIPLE FOR LUCID WRITING here is that, ~~while~~ although people use *while* incorrectly, you should use it to mean only *at the same time as;* you should not use it to mean *whereas* or *although*.

20   Repeated Prepositions

Y OU SHOULD NOT be stingy with the little words that form relationships in your sentences—*of, by, with, to, from, for, on, in,* and other prepositions. You should use as many of them as you need to disambiguate the information that your sentence conveys. Prepositions make clear how far any given clause or phrase extends.

BAD: You can try to assuage your baseline guilt by overtipping, giving sandwiches to people on the streets, or calling your mother, or you can study the effects that guilt has on your blood pressure.

There should be two additional bys *in the sentence.*

GOOD: If you are upset, you can help yourself to feel calmer by taking a long walk, by lolling in a hot bubble bath, or by eating your favorite decadent food, or you can revel in your misery.

In this sentence, by *has been repeated twice, which is correct.*

SPLENDID: If you are lonely, you can cheer yourself up by talking to your cat, by calling a friend, or by taking yourself to a movie; alternatively, you can notice how peaceful and pleasant it is to be alone.

The preceding good sentence is fine, but all those commas and ors *have the potential to confuse the reader; the semicolon and* alternatively *make the meaning more readily apparent.*

BAD: Steve was getting flak from Judy for not doing the dishes, his boss for coming to work in neon purple cycling shorts, and himself for not eating breakfast.

There should be two additional froms *in the sentence.*

GOOD: Lyn got tips from men on how to cope with her mother, and from her mother on how to cope with men.

BAD: Now let us speak of networks that have learned how to hear as well as see, search tools that are able to locate and print an abstract, and administrators who know how to maintain security and to protect confidentiality.

There should be two additional ofs *in the sentence; note also the correct repetition of* to.

GOOD: Now let us speak of mice who have learned how to roar as well as to speak, of lions who are able to cuddle and purr, and of men who know how to cry and how to dance.

BAD: Have you thought of icons for the program and the company, dialog boxes for the interface, and a type style for the on-line help and the manual?

There should be two additional ofs *in the sentence.*

GOOD: Have you thought of flowers for the wedding and for the reception, of dresses for the bridesmaids, and of food for the guests and for the cats?

BAD: Thermal expansion occurs with strongly bonded solids and weakly bonded solids.

GOOD: With successive ambient-temperature increases, and thus with successive energy increases, the median interatomic spacing increases.

BAD: Judy can dance with Steve on Sunday, Mike Monday, and herself Tuesday, but then she will have to comfort Steve Thursday.

There should be two additional withs *and three additional* ons *in the sentence.*

GOOD: Max works with fools who know everything and with geniuses who know nothing, and on his business plan now and then.

GOOD: "Do you take this person..., to love and to honor, in sickness and in health, for better or for worse, until death do you part?" (As sanctioned by a government of the people, by the people, and for the people.)

THE PRINCIPLE FOR LUCID WRITING here is that you should repeat prepositions to indicate governance of the words that follow, or of entire terms.

21 Abbreviations and Acronyms

I F YOU USE numerous *abbreviations and acronyms*, you will make it difficult for your reader to understand your meaning. In computer science, an overabundance of abbreviations and acronyms will cause your writing to nose dive into utter incomprehensibility. Avoid the temptation to shorten every term in sight.

 You can, and usually should, use the *most standard* shorthand terms, defining them on first use. You define a term by spelling it out in full, followed by the shorthand term in parentheses:

> GOOD: You can define and use terms such as random-access memory (RAM), central processing unit (CPU), local-area network (LAN), and snafu.[44]

It is also perfectly reasonable to use standard abbreviations for units of measure, such as

---

44. In Great Britain, true acronyms *(BUGS, RAM, LAN, snafu*—words that are formed from the first letters of other words, and that are pronounced as they are spelled, rather than being pronounced as a string of individual letters (IBM)) are set with only the first letter capitalized. In the United States, however, acronyms are normally set in all-capital letters. A few terms, such as *snafu* and *radar,* have become accepted words, with their derivations as acronyms forgotten; they are set in all-lowercase letters. Note that you do not need to define *snafu,* as it is listed in the dictionary as an English word.

GOOD: You can give measurements in units such as MIPS,[45] MHz, Mbyte,[46,47] and sec;[48] you should define them on first use, unless you are writing in a forum where the abbreviations are common and are well known to your audience.

You can use shorthand terms when your subject is denoted by a cumbersome phrase that you use often in a given document.

GOOD: We report on the International Chest-Thumping Men's Back-to-Nature Summer-Camp Experiment (ICTMBNSCE).

In such cases, you may wish to avoid creating the abbreviation from first letters, and to use instead a more informative term.

SPLENDID: We participated yesterday in the International Chest-Thumping Men's Back-to-Nature Summer-Camp (Thumper) Experiment.

---

45. Note that in *MIPS* (an acronym for *million instructions per second)*, the S is part of the acronym, rather than an indication of a plural, so it is set as a capital letter.

46. Megabyte, MB, and Mbyte are all acceptable; what is not acceptable is mixing them within one manuscript. Also unacceptable is mixing, for example, KB and Mbyte.

47. I strongly recommend that you avoid using K and M alone as abbreviations, closed up to numbers, in manuscripts where you will also be writing about, for example, Kbyte and Mbyte. You will find it almost impossible to maintain a consistent style when you talk about 3 K and 3 Kbyte. Furthermore, you will tend to forget to specify the unit of measure, writing instead, for example, that your computer has 69K of storage available; you should instead always specify 69 Kbytes (or $69,000, or whatever you intend).

48. Note that you do not add a final s when you refer to more than one abbreviated unit of measure. That is, you can write either *3 seconds* or *3 sec,* but you should not write *3 secs;* you can write either *3 megabytes* or *3 Mbyte,* but you should not write *3 Mbytes.*

If you wish and if you are writing for an academic audience, you can use standard Latin abbreviations.

- *E.g.* [49] *(exempli gratia)* means *for example,* and should always[50] be followed by a comma.

- *I.e. (id est)* means *that is,* and should always[51] be followed by a comma.

- *Etc. (et cetera)* means *and so on,* and should always[52] be preceded by a comma.

- *Et al. (et alia)* means *and others;* you should use it only in reference citations—do not use it as a substitute in text for *and colleagues, and associates,* or *and coworkers.*

Use Latin abbreviations *only in parentheses,* however; do not use them in the regular text line. (Use instead *for example, that is,* and *and so on.)* Do not set these abbreviations in italic type; use roman.[53,54]

---

49. The Latin abbreviations in this list are set in italic type because the words themselves are under discussion. In all other cases, however, they should be set in roman type.

50. The exception occurs in sentences such as this one, where the term itself, rather than the denotation, is under discussion.

51. See previous note.

52. The exception is the unusual circumstance where the term begins a sentence.

53. It is a source of amusement to me that slanted type is italic (Italian), whereas plain type is roman (also Italian). Note that *roman type* is not set with a capital R, whereas *Roman toga* is; similarly, we write about *italic type,* but about *Italic branches of languages.*

54. The general rule is that you should set in roman type any foreign terms that have become so common in English that they are now considered to be English words. Look up the term in the most recent edition of *Webster's New Collegiate Dictionary* (Springfield, MA: Merriam-Webster); if the term is set in roman in the main body of the dictionary, set it in roman.

UGLY: Joe had had a tough day; *e.g.,* he had to drive all the way to Berkeley to pick up the fonts for the book.

GOOD: Joe also freaked Lyn out by going off line at the wrong moments (e.g., he forgot to mention that he was on his way to Berkeley).

GOOD: Joe did recognize that the day had a few silver linings; for example, he got to drive his brand new car.

UGLY: At three levels—*i.e.,* those of testing, confirmation, and denial of nodes of the causal network—CASNET's reasoning is probabilistic.

GOOD: Selection (i.e., of a diagnosis and associated plan) depends on the designer's understanding of the possible causal pathways through the network.

GOOD: Perfect accuracy in diagnosis thus is not an unrealistic goal; that is, there is no inherent reason why the system should not be able to perform flawlessly.

UGLY: Lyn gave John Gamache, the cover designer, various informational materials: the book manuscript, her mother's sketches, photographs of Red and BB, *etc.*

GOOD: John had to work under severe constraints (time, trim size, color, type of cat, etc.).

GOOD: Lyn was so delighted by John's interest in the project that she immediately started cogitating about other ways that they could use the design: in posters, on T-shirts, on coffee mugs, and so on.

UGLY: Dupré et al. [1994] reported a need to reengineer the publishing process to match technological advances.

GOOD: Lyn's team [Dupré et al., 1994] reported that conversion of an electronic file to film was an unexpected process, fraught with demons and saved by angels.

GOOD: Dupré and associates [1994] reported that producing a book at home is wildly satisfying, if a bit too exciting.

If you begin a list with *e.g.,* do not end it with *etc.*—both indicate an incomplete series, so using both is redundant.

BAD: There are several network-interrupt status bits (e.g., NETISR_RAW, NETISR_IP, NETISR_IMP, etc.)

GOOD: There are also several input queues (e.g., *rawintrq, ipintrq, and impintrq).*

GOOD: The values vary as well (PF_UNSPEC, PF_INET, PF_IMPLINK, etc.)

You should follow compulsively one simple rule with all shorthand terms: Define them on *first* use, and use them *always* thereafter. You define a shorthand term by placing it in parentheses directly following its parent term.

BAD: The ERROR (Editors and Readers Ridiculously Odd Reprobates) Association forgot to hold its annual meeting.

GOOD: The Smart Igloo-Living Leaping Yang (SILLY) Club held its annual meeting at Asilomar.

Note that you can define *on first use* and *thereafter* in various ways. First, in a journal article, anything that you write (or define) in the abstract does not count in the body of the paper—the abstract is considered to be a separate document. Thus, in most cases, you will not want to define any terms in your abstract; if you do define a term, however (say, one that you use many times), you will still need to redefine it on first use in the text. Second,[55] in a book, you may well want to define terms on first use *in each chapter,* rather than in the book, to keep your reader oriented. If you choose this approach, use it consistently across all shorthand terms.

You should never define an acronym or abbreviation that you will not use later. Furthermore, if you find that you use the shorthand term only a few times, chances are that you should not be using it at all; use the spelled-out version instead.

You should not capitalize the initial letters of the words constituting a phrase that you are about to abbreviate,[56] unless you have a separate reason for doing so.[57]

> UGLY: Lyn's Randomly Accessible Memory (RAM) is becoming entirely too random.

> GOOD: Lyn's random-access memory (RAM) gets dumped daily.

---

55. Whenever you write *first,* be sure to follow through by writing, at a minimum, *second.* Do not commit a common error by leaving your reader hanging, awaiting the next list item when no such item exists. Do not italicize numbers in lists, whether they are spelled out numbers or Arabic or Roman numerals.

56. You should generally avoid capitalizing any term unless circumstances (such as that the term is a trademark) force you to do so. Extraneous capital letters merely confuse your reader, and visually clutter the page. They can also lead to a Pooh-bear mode of writing, in which Everything takes on Great Importance, which may be all right for Children's Books, but is not Appropriate in Formal Writing. (Note that using capital letters to emphasize certain words is not the same as setting a line in capital and lowercase letters (cap/lc). In the former style, more words are left lowercase.)

57. Such a reason might be, for example, that the term is a proper name.

If the letters that you use in an acronym are not the leading letters in the words of the term, you can indicate the derivation of your acronym by using boldface, small capital letters, or underlining.[58] In most typefaces, you should avoid using italic type for this situation, as the combined roman and italic type may not set up well.

UGLY?:   Our new software, *qui*ck ou*t* (QUIT), is designed for people who want to switch careers fast.

GOOD: The system uses RAdio Detecting And Ranging (radar).

GOOD: Lumina sells **de**cision **mo**deling **s**oftware (Demos).

GOOD: The <u>So</u>phomoric <u>D</u>oggerel Comed<u>y T</u>our (SOD YOU) was a blistering success.

You should not use periods in abbreviations and acronyms.

BAD: Converting to R.I.S.C. architectures is fraught with unanticipated danger.

GOOD: It is foolish to ignore RISC architectures.

BAD: The A.A.A.I. Spring Symposium was held in the fall.

GOOD: The AAAI annual meeting was held in Seattle.

There are several exceptions, however:

BAD: We are US citizens.

GOOD: They are U.S. tourists.[59]

---

58. Underlining generally clutters up the page and is visually objectionable. However, there are cases (and typefaces) in which use of underlining may be your preferred solution.

59. Note that you should use the abbreviation only as an adjective; when you refer to the noun, spell out *United States*. Note also that *U.S.* is one of the few acronyms that you do not need to define.

BAD: The new popular drink in Geneva is the UN cola.

GOOD: The new popular song in China is "U.N. Forgiven."

BAD: People (eg, Lyn) do not like to find bugs in their hair.

GOOD: All programmers (e.g., Brian and Brian) do not like to find bugs in their code.

BAD: Max's office manager (ie, Devon) also answers the telephones.

GOOD: Our head of household (i.e., Red) never answers the telephone.

BAD: Pavel M, et al,[60] I cannot see the point. *Journal of Vision Research,* 14(2): 45–51, 1990.

BAD: Mulligan J, et. al., *Points of View.* Sunnyvale, CA: Technics Press, 1991.

GOOD: Ahumada A, et al., *The View from the Hot Tub.* Santa Cruz, CA: Sybarite, 1995.

         Because the term et al. *is short for* et alia, *only the* al *takes a period.*

BAD: Max was so tired that he could not distinguish 3 AM from 3 PM.

GOOD: Lyn was at her desk from 6 A.M. to 6 P.M.

Remember to use small capital letters.

BAD: The period of interest, for purposes of our study, ran from 40 BC to AD 1.

GOOD: Chinese medicine in 400 B.C. was more advanced than was Western medicine in A.D. 400.

Remember to use small capital letters, and to place the A.D. before the year.

BAD: BCE and CE offer an alternative notation.

GOOD: You can use B.C.E. instead of B.C., and C.E.[61] instead of A.D.

Remember to use small capital letters.

 You should be sure to remember what your shorthand term represents, so that you avoid redundancies.

BAD: We have implemented the systemic-infection system (SIS). The SIS system is a decision-support tool designed to save lives.

Error! SIS system *means the* systemic infection system system.

---

61. Because C.E. can be an abbreviation for *common era,* as well as for *Christian era,* certain people prefer it to A.D. (and similarly prefer B.C.E. [before common era] to B.C.), viewing it as a term without religious connotations. Even the latter meaning can be taken to be merely descriptive.

When you are quoting dialog, you should spell a term the way that it is said.[62]

BAD: Lyn asked, "What on earth is 20 MIPS?"

BAD: Mike, an experienced programmer, asked, "How many em eye pees does the system afford us?"

An experienced programmer would never mispronounce MIPS (which rhymes with slips), so we assume that the writer, rather than the programmer, is at fault.

GOOD: "The correct way to pronounce it," Max explained, "is twenty MIPS."

BAD: "Apparently," Soren told Lois, "the source of the problem was a discrepancy over 30 vs. 40."

GOOD: "Thirty versus forty whats?" asked Lois.

BAD: "The speed limit on this road," Betsy pointed out, "is 40 mph."

GOOD: "Well," replied Richard, "I am driving at sixty miles per hour."

BAD: "Let's meet at 2:00," Moisés suggested.

GOOD: "I'd rather make it three o'clock," replied Adnan.

GOOD: "Actually," murmured Max, "even I am usually asleep at twenty-three-hundred hours."

---

62. Similarly, you should always spell out all numbers in quoted dialog.

You should use *a* or *an,* depending on how the shorthand term is pronounced.

> GOOD: You need a CPU.
>
> CPU *is pronounced see pee you.*
>
> GOOD: I hooked up a LAN for an LIS.
>
> LAN *is pronounced lan;* LIS *is pronounced el eye ess.*

If, despite heroic struggles, you find that your text contains numerous shorthand terms (for example, many trade names and proper names are abbreviated letter strings, and you have to use them), consider setting the terms in SMALL CAPITAL LETTERS, rather than in REGULAR CAPITAL LETTERS. An abundance of terms set in regular capitals destroys the appearance of the page, because the capital letters break up the text line. Small capital letters, which are the same height as the lowercase letters,[63] do not create such ugliness and offer an acceptable solution.

> GOOD: If IBM and DEC join forces to bring out a new PC, the
> CPU may be confused, but the RAM should be OK.[64]

Somewhat tangentially, you should know that either *okay* or *OK* is self-denoting, whereas *Okay* is not (except at the beginning of a sentence that follows a period or a colon). You should not define *OK* as an abbreviation for *okay;* on the contrary, *OK* is an abbreviation for *oll korrect* (or *all correct),* and *okay* is merely a phonetic spelling of the abbreviation.

> UGLY: *Alright* is not Okay.
>
> GOOD: *All right* is OK, as is *okay.*

---

63. The classifications BAD, UGLY, GOOD, and SPLENDID in this book are set in small capital letters.

64. The implications for the power PC are unknown.

THE PRINCIPLE FOR LUCID WRITING here is that you should use a minimum number of abbreviations and acronym terms in your text (i.e., num abbrevs and ATs can be dfclt to read). Any that you do use, you should define on first use, and should use always thereafter.

22   *Verbize*

It is a nasty habit to *verbize* your words. Verbizing is taking an inoffensive noun that is just doing its job, and tacking onto it an *ize*, thus turning it into an offensive verb. The habit of creating verbs from nouns leads to highly jargonized speech.

BAD: We hypothesized that our algorithm would run faster than Laurel–Spitwater's algorithm does.

GOOD: We formulated the hypothesis that our machine would run faster than does Twithouse's invention.

SPLENDID: Our hypothesis was that our cats would run faster than Soren's would.

SPLENDID: We postulated that our water would run faster than Dona's cats.

BAD: We want to computerize our internal accounting system.

GOOD: We want to automate our dating system.

SPLENDID: We want to develop a computer-based love-life system.

BAD: Let's prioritize our tasks.

GOOD: Let's put our goals in order of priority.

GOOD: Let's order our wishes by priority.

GOOD: Let's assign priorities to our lovers.

GOOD: Let's determine a task ordering by priority.

BAD: Let's lunchize our leftovers.

BAD: Treat your staff well, so as not to mutinize them.

BAD: I think we should customerize our friends.

BAD: We need to productize the laboratory system.

BAD: It takes a long time to treeize an acorn.

BAD: Max decided to managerize one of his employees.

BAD: Lyn finally decided to let nature motherize BB.

Note also that you should avoid the term *customize*, albeit less violently, because there is a more elegant phrase:

BAD: Bob customized the software to meet his own needs.

GOOD: Max custom-tailored the model to suit his clients' needs.

GOOD: Lyn tailored her sleeping hours to meet Max's needs.

THE PRINCIPLE FOR LUCID WRITING here is that you should avoid verbizing, to avoid miserizing your reader.

23 Commas

COMMAS ARE useful punctuation marks. They give your reader considerable guidance in parsing your sentence. You generally should place them wherever a speaker would pause, or whenever there is a logical shift in your sentence.

You should use the *series comma*.[65] The series comma is the comma before the *and* or the *or* in a list of several items.

> UGLY: I don't want any ifs, ands or buts.
>
> GOOD: I don't want any coffee, tea, or milk.

> UGLY: The monitor, keyboard, mouse, printer and software are all included in the package deal.
>
> GOOD: Lyn, Red, and BB are all included in the package deal.

> UGLY: Please read pages 1, 2, 6, 7 and 39.

---

65. In Great Britain, writers use *open punctuation,* in which noncritical commas are left out (for example, the series comma is omitted). In the United States, it is correct to use *closed punctuation,* in which such commas are included. Certain magazines and various informal arenas choose to use open punctuation; in formal writing in the United States, however, it is correct to use closed punctuation. In addition, and to me more important, closed punctuation gives your reader more parsing information and makes your text easier to read.

GOOD: Please do Exercises 1, 2, 6, 7, and 39.

You should place a comma after an introductory word, phrase, or clause.

UGLY: Greg was worried; however he remained calm.

GOOD: Brendan was hungry; however, he remained calm.

SPLENDID: Lyn and Richard were still puzzled, however many times they reread the directions for assembling the stepper climber; however, they remained calm.[66]

UGLY: In summary we can deal with the problem of nonmonotonic reasoning by defining a logic that uses nonstandard, nonmonotonic inference rules.

GOOD: In summary, we can block unwelcome transitivity in one of two ways.

UGLY: By the time that Lyn reached Wall Street Gunnar had ridden back and forth across the Brooklyn Bridge 14 times.

GOOD: By the time that Gunnar found Lyn, she had already eaten blue-corn patties with jalapeño sauce, and she was settling down to the gumbo.

---

66. Eventually, after many days of sweaty hard work, not only was the stepper climber assembled, but also Lyn had written a letter to the manufacturer that resulted in her receiving a beta-test version of the reengineered design. After Lyn had solved the minor problem of the new machine's brakes catching fire when she exercised, with a contribution from Soren of a special heat-proof material developed for the Space Shuttle, she gave the second stepper climber to Marina, and Cosmo and Reno used it regularly.

You should use commas around (that is, on both sides of) clauses inserted in the middle of a *that* remark.

> UGLY: I want you to know that whatever happens, I will always think of you when I lounge in a papasan.

> GOOD: I want you to know that, whatever happens, I will never let another man put ice on my bruised fingers after I smash them with a pipe clamp.

> UGLY: Please remember that despite recent harrowing events, this software has always functioned without unexpectedly erasing the hard disk.

> GOOD: Please remember that, despite recent harrowing events, this woman has always functioned without pulling out her hair.

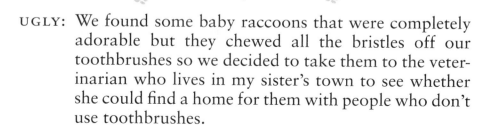

You should use a comma wherever there is a logical pause, or whenever you begin a new thought.

> UGLY: Misha wanted to buy black, squid-ink pasta so he went to Menlo Park.

> GOOD: Holly wanted to live with goats and rabbits and chickens and ducks and cats, so she moved to Oregon.

> UGLY: We found some baby raccoons that were completely adorable but they chewed all the bristles off our toothbrushes so we decided to take them to the veterinarian who lives in my sister's town to see whether she could find a home for them with people who don't use toothbrushes.

GOOD: We found four baby raccoons that were completely adorable, but they chewed all the bristles off our toothbrushes, so we decided to take them to the veterinarian who lives in my sister's town, to see whether she could find a home for them with people who don't use toothbrushes.

SPLENDID: We found four baby raccoons and took them home. They were completely adorable, but one day they chewed all the bristles off our toothbrushes. After a heated family discussion, we decided to take them to the veterinarian in the town where my sister lives. The town is called Gumsville, and we thought the vet might be able to place the babies in a household where no one needed toothbrushes. Such is what copy editors do for a living.

 Do not use *comma splices.*

BAD: A comma splice is a joining of two sentences (that should remain separate) by a comma and an *and,* and this sentence is an example.[67]

GOOD: Note that a semicolon can be used to join two sentences; this sentence is an example.

BAD: I had been typing all day in the airless office, and Red had given up on me and had gone to dream of bunnies, and I was ravenously hungry, and I got up and stretched wearily, and, suddenly, the office turned upside down, and I heard myself scream shrilly.

---

67. The sentence is incorrect because it contains a comma splice; it is properly two sentences, the first ending at the comma, and the second beginning with *This sentence.*

GOOD: I had been sleeping all night in the fragrant bedroom, with BB at my feet, and I was fully rested. I got up and stretched luxuriously; suddenly, the room turned inside out. I heard myself giggle idiotically.

 THE PRINCIPLE FOR LUCID WRITING here is that you should always include commas that are logical and helpful, such as those indicating pauses, or those after introductory remarks, <u>and</u> also remember that you should not splice together artificially two sentences simply by inserting a comma and an *and*.

24   Number Spelling

THERE IS no single rule governing when you should spell out numbers and when you should use Arabic symbols. It is generally accepted practice in scientific writing, however, to spell out zero through nine,[68] and to use numerals thereafter. There are, not surprisingly,[69] several exceptions to this principle.

You should spell out all numbers at the beginning of a sentence that follows a period or a colon.[70]

> BAD: 22 bar-code readers arrived at the supermarket.

> GOOD: Twenty-five people were shopping that day in the produce department.

> BAD: Max's briefcase was missing: 112 rats had spirited it away during the night.

> GOOD: Lyn's facial tissues were missing, too: Ten raccoons and two cats had joyously covered the house with a light dusting of confetti.

---

68. The *zero* and the *nine* in the text are examples of spelled-out numbers less than 10.

69. Every rule must admit of exceptions. In English, every rule pleads guilty to myriad exceptions.

70. Max Henrion suggests that the basis for this rule is that otherwise you cannot tell whether the word is set with a lead capital letter.

BAD: 2-dimensional drawings can be rendered in 3 dimensions.

BAD: 2D line drawings can form the basis for wire-frame models in 3D.[71]

GOOD: Two-dimensional graphing works for two variables, but you need a representation in three or more dimensions for this problem.[72]

You should use numerals for all units of measure.

BAD: The banana was six inches long.

GOOD: The banana slug was 7 inches long.

BAD: The distance from the primary host to the gateway is seven miles.

GOOD: The distance from our house to the top of Windy Hill is 7 kilometers.

---

71. Certain people argue that the acronyms 2D and 3D are acceptable for *two-dimensional* and *three-dimensional*. I find the terms ugly, but find them acceptable in documents that use them frequently. I do not find acceptable, however, use of 2D and 3D to mean *two dimensions* and *three dimensions*. This and the previous example would be classified as bad in any event, however, because they begin with a numeral: You should always spell out numbers that occur at the beginning of a sentence.

72. See previous note. I encourage you to spell out *two-dimensional, three dimensions,* and so on in all cases, however, rather than only at the beginnings of sentences.

BAD: The play started at eight-twenty, which was five or twenty minutes later than the announced time, depending on which newspaper you consulted.

GOOD: Max got home at 6:00 in the evening, which was either 22 hours late or 2 hours early, depending on Lyn's mood.

BAD: The intrinsic conductivity (ohm^{-1} · m^{-1}) of Silicon is five times ten to the minus four.

GOOD: The energy gap E_g of Silicon[73] is 0.176 10^{-18} J, 1.1 eV.

BAD: Max had been home for only two evenings in the preceding five months.

GOOD: Max had been home for only 2 minutes before he was snoring peacefully, because he had not had an 8-hour sleep in 1 year.

BAD: Marina noted a three-fold increase in the number of fleas in the house after Swix had been running gaily through the fields.

GOOD: Spud chewed on Swix sufficiently to cause a 2-fold decrease in the population of ticks on Swix's neck.

BAD: The company's income decreased by a factor of six during this quarter.

---

73. Note that, in certain disciplines such as materials science, the names of elements are set with lead capital letters.

GOOD: Lyn's mood lightens by a factor of 3 after she has taken a long, strenuous hike.

BAD: A seven-point scale that offers only six points is bound to confuse test subjects.

GOOD: The users were asked to rate the interface on various dimensions, using a 4-point scale.

Note that, in computer science, bits and bytes are used as units of measure and thus should take numerals.[74]

UGLY: Lyn freaked when she had only two megabytes of RAM left and could not save her file.

GOOD: The file requires only 5 kilobytes of disk space.

UGLY: The first seven bits are used to encode the data; the final bit is used in the error-checking protocol.

GOOD: There are 7 bits, plus 1 with glop on it.

Also note that, to many people, using a numeral below 10 with *decades* is jarring; I thus spell out numbers in this case, but I do not object when a writer uses the numeral.

GOOD: Cosmo and Reno are each less than one decade old.

GOOD: Cosmo and Reno together are still less than 1 decade old.

---

74. We sometimes speak of bits as objects (for example, toggle switches), or as containers for a value (the 8 bits encode information). However, if you attempt to spell out numbers in such cases, and to use numerals in the other cases (where bits are units of measure), your algorithm is likely to be too complex for your reader to understand it, and your writing will appear inconsistent. Thus, stick to the one style.

Similar reasoning applies to orders of magnitude.

GOOD: This relationship is three orders of magnitude more complex than were my previous friendships.

GOOD: Greg felt 2 orders of magnitude more relaxed after he had gone for a long bike ride.

You should use numerals when you are referring to the number itself.

BAD: Maria pulled the three and the zero off the birthday cake, and ate them happily.

GOOD: If Geoff adds 2 and 2, he may not get 4.

BAD: We use a_{12}, where the one and the two are indices to the matrix.

GOOD: In the string 1234, the 2 represents a dickybird.

You should use numerals when you are counting.

BAD: Lyn spilled her grapefruit juice on page four.

GOOD: The picture on page 3 shows Red and BB relaxing in a hammock.

BAD: Max tripped on Red, who was sprawled on step six, when he came home at two in the morning.

GOOD: The algorithm hung on step 5.

BAD: Use file two in folder seven.

GOOD: Insert disk 3 into drive A.

BAD: Steve and Judy left on September nine, 1993.

GOOD: September 23 is the day on which Max and Lyn celebrate their anniversary.

Note that, when you set a date in the form of the preceding example, you are *counting* the days of the month. You can also spell out the date, if you wish, as follows:

GOOD: For the twenty-third of September, 1994, Max and Lyn had outrageous plans.

You should use numerals when the number that is less than 10 is in a series that contains numbers greater than 10.

UGLY: The blue herons that built nests in the redwoods around Max and Lyn's home had three, seven, and 11 chicks.

GOOD: The rabbits that Max and Lyn saw on Windy Hill had varied litter sizes: 4, 11, and 18 bunnies.

GOOD: Lyn was wondering (and worrying) whether BB would have 1, 2, or 17 kittens.

You should use numerals when you are speaking of multiples of millions, billions, and so on.

BAD: Lyn found four zillion errors in the manuscript that she was editing.

GOOD: A cloud comprising 2 million butterflies settled around Max and Lyn as they lay in the field of spring flowers.

There are also exceptions to the rule that you should use numerals for numbers greater than nine.

You should spell out all numbers in quoted speech.

> BAD: Max moaned, "I have 43 reports to write tonight—
> and it is already 11 o'clock!"

> GOOD: Lyn bragged, "I wrote fifty pages before five o'clock
> today."

Note that, when you are quoting *printed* material, you should follow the style of the original, even if it does not match your style.

> GOOD: Calamity Jane's reports are suspect:
>
> > An earthquake of magnitude seven point two destroyed San Francisco at three ten in the morning yesterday. Many residents slept through the tremor, which caused little property damage (about twenty dollars' worth). Fortunately, only two people were killed, one fatally. [Calamity J, *Golden Gate Daily*, September 1, 1994, page A1]

You should spell out terms that are used to mean *approximate* amounts.

> UGLY: Doug was so nervous that he drank 45 cups of coffee.
>
> *This sentence would be fine if you wanted your reader to know precisely how much caffeine Doug imbibed.*

> GOOD: Lyn was so confident that she crushed dozens of competitors.[75]

> UGLY: Max and Lyn have 100s of skunks peeing on the plasterboard in the ceiling directly over their bed.

> GOOD: When two people live together, there are hundreds of opportunities for smooching.

---

75. Contributed by Patrick Henry Winston.

Note that you should use such terms[76] only when you are speaking generally. If you mean precisely 100, do not write *a hundred.* Do write *a day or two, about a week, almost a year,* and so on; that is, do use *a* rather than 1 when you mean *approximately but not necessarily precisely* 1.

 You should spell out *first, second,* and so on.

> BAD: The 2nd time that Max saw Lyn, she was suffering her 3rd migraine of the week and had just lost her breakfast for the 5th time; she was sporting an intriguing green tinge.

> GOOD: The first time that Max saw Lyn, she was sitting in the twelfth row in the second room; that event took place during the first hour of the gathering, which was the fourteenth annual laboratory retreat.

Note this trick while we are near the subject:

> GOOD: We are interested in the $[x + 5]$th number in the series.
>
> *The* $x + 5$th number, *if the term made sense at all, would be calculated in a different way.*

> GOOD: The $[2 - y]$th kangaroo jumped through the window.
>
> *The* $2 - y$th kangaroo *would be a different creature.*

> GOOD: Swix's $[x + 1]$st[77] puppy was a squeaky toy.[78]

---

76. That is, terms such as *hundreds* and *thousands* and *zillions*—rather than such as *smooching.*

77. Ronald Barry notes that "It is not uncommon for mathematicians to use the form [————]*st* for every such expression; admittedly, it sounds peculiar to the ear, but it looks natural to the eye." Following this alternative style, you would write, for example, $[x + 3]$st.

78. Malamutes are highly parental canines. Deprived of the opportunity to whelp, they tend to adopt, protecting the adoptee with ferocity.

In situations where you would use numerals for whole numbers, also use numerals for the fractions; in situations where you would spell out the whole number, spell out the fraction as well.

BAD: Brian ate $\frac{1}{4}$ of the cake; Adrienne had only 1 slice.

GOOD: Adrienne left one-half of the shrimp; Brian ate nine of them and saved the rest.

Note that spelled-out fractions take a hyphen.

GOOD: Adrienne had only $\frac{1}{4}$ hour left before she had to go to work.

THE PRINCIPLE FOR LUCID WRITING here is that, in general, you should spell out numbers from one to nine, and should use numerals otherwise; there are, however, many exceptions that you should learn. This area is one of the few in which ear will get you nowhere; rote memorization or frequent lookups are your choices.

25 Impact

THERE ARE only two pleasing uses of *impact*: to denote a forceful collision, and to mean packed or wedged in.

> GOOD: My car hit the curb with enormous impact.

> GOOD: My wisdom tooth was impacted.

Impact, however, has become one of the most overused, ugly words in the language. If you avoid using it, except in the two respectable cases just given, you will improve the quality of your prose.

You should avoid using *impact* to mean *influence* or *effect*.

> UGLY: Choosing a categorical-reasoning mechanism carefully permits you to have a big impact on decision making.

> GOOD: Choosing a limited domain where assumptions can be accepted with confidence will increase your chances of having a major effect on decision making.

> UGLY: Ursula's dancing had a major impact on Blackford.

> GOOD: Blackford's dancing had a major influence on Ursula.

> GOOD: Blackford, dancing, crashed into Ursula with major impact.[79]

---

79. Based on a suggestion by Joseph Norman.

UGLY: The impact of Max's grumpiness on Lyn's happiness is a bidirectional relation.

GOOD: The effect of Lyn's writing a book on the level of Lyn's housekeeping is appalling.

GOOD: The EPA requested an environmental-effects statement about the meteoric impact.[80]

Admittedly, use of *impact* to mean *effect* or *influence* has become so common as to be nearly unavoidable by all but the most vigilant writers. Using *impact* as a verb, however, will quickly loft you to an entirely different level of unacceptability.

BAD: The new billing system negatively impacted the accounting department.

GOOD: The new dispensing system affected positively the operations of the pharmacy.

SPLENDID: The new desktop-publishing system proved totally disastrous for the documentation department.[81]

BAD: Lyn is a superb editor; she impacted Doug's writing.[82]

GOOD: Doug is a superb friend; he influenced Lyn's mood.

BAD: Sarah's screaming impacted Max adversely.

---

80. Contributed by Peter Gordon.

81. Rather than writing about positive or negative effects, specify precisely what the effect is.

82. Contributed by Douglas K. Owens.

GOOD: Max's scolding affected Sarah adversely.

SPLENDID: Sarah's screaming made Max's ears ring and his nose run.[83]

THE PRINCIPLE FOR LUCID WRITING here is that you should not use *impact* when you mean *influence* or *effect,* and you certainly should not use *impact* when you mean *affect,* because impacting people is incredibly impolite.

83. Identifying the effect, rather than simply noting that it exists and is negative, supplies your reader with more—and perhaps more interesting—information.

26 Lists

THERE ARE numerous ways to set lists; you can set them in text or display them, and you can use delimiters such as numbers or bullets. Familiarity with the various types of lists will allow you to select and use those that are most appropriate to your content and presentation.

 You should use *intext*[84] *lists* for short items that do not require special emphasis. Intext lists all constitute one sentence, so each entry obviously cannot be itself a sentence.[85]

> BAD: There are many errors that you can make: (1) You can forget that entries in an intext numbered list must together constitute a sentence. (2) You can put periods at the end of the entries of intext numbered lists.
>
> GOOD: Instead, just remember two rules: (1) the entries are only clauses in a sentence, and (2) they should be separated by commas or semicolons.
>
> GOOD: The cat brought in several items: a mouse, an old sock, and what looked like old leftover roast chicken.

---

84. I am teaching you a small subset of the terminology of publishing; *intext* is a member of that subset. An intext list is one set in *running text,* as opposed to *displayed* lists, which are *broken out.*

85. Note that you should not capitalize the first entry of the list, whether or not the list could stand alone as a sentence. In all other cases, you should capitalize the first letter of a sentence that follows a colon.

Note that it is incorrect to set intext lists that have entries that are sentences. You can use displayed lists (discussed later in this segment) in such cases, or you can use a series of sentences starting with *First, Second,* and so on. I shall discuss such lists in a moment.

You should use *unnumbered intext lists* in most cases. You should use *numbered intext lists* when you specify the number of entries in the list, when you want to emphasize the number of entries, or when you want to refer later to the entries by number.[86]

> GOOD: There are three monkeys: (1) see no evil, (2) hear no evil, and (3) speak no evil.

You should always use *Arabic numerals,* rather than letters or Roman numerals, in an intext list.

> BAD: What would you prefer: (a) a pony, (b) a horse, (c) a goat, or (d) a car?
>
> BAD: I can get the references from (i) the library, (ii) my home office, or (iii) a review article.
>
> GOOD: Red enters the house in three ways: (1) through the cat door, (2) through the window, (3) through meows.

Note that, if you are writing numbered exercises that have parts, you should use letters for the parts. Such lists are not intext lists—they are parts. You should also use letters for parts of a figure caption.

> GOOD: Exercise 4.5. Give your opinion of the following animals: (a) cats, (b) dogs, and (c) hippopotami.
>
> GOOD: Figure 4. Artist's conception of different bugs. **a:** Syntax bug. **b:** Style bug. **c:** Semantics bug.

---

86. In the last case, however, a *displayed* numbered list is preferable, because your reader can find the entries more easily. A *display* is text that is not set in the text line; I describe displayed lists later in this segment.

You should always use *both a left and a right parenthesis* around a number in an intext numbered list. You should never use only a right (or only a left) parenthesis for any reason, and you should never use periods after the numbers in an intext numbered list.[87]

> BAD: I want to have these items on my desk by noon tomorrow: 1. a progress report, 2. your time sheets for the past year, and 3. a self-rating assessment report.

> BAD: If you fail to meet these requirements, I will 1) hang you up by your toenails, 2) gag you with a silk scarf, and 3) flog you with a daisy.

> GOOD: If you come home early tonight, I will (1) make dinner, (2) serve dinner by candlelight, and (3) keep you company while you wash the dishes.

You should use *commas* to separate entries in intext lists if no entry itself contains a comma. If any entry contains a comma, you should use *semicolons* to separate the entries.

> BAD: I would like you to do these chores for me: (1) wash all the dishes from dinner, and try not to break them, (2) make the beds in both your room and mine, and (3) scrub the bathtub.

> GOOD: In return, I will do the ironing, and try not to scorch your shirts and trousers; take out the enormous can of garbage; and feed the ravenous pet lion cub.

If you are setting a list where one entry must contain at least one sentence, and you do not want to emphasize the list items by displaying them (or do not want the visual effect of displays), then you should simply make the entries part of running text.

---

87. You can thus err in three ways by using a right parenthesis and a period in a displayed numbered list. Resist any temptation to sin so successfully.

GOOD: It was not a pleasant morning.[88] First, Cleo, Helen's cat, kept insisting that there was a mouse living under the fridge, even though Helen assured Cleo that she was not in the mood for a meal of mouse. Second, the milk curdled in Helen's coffee, reminding Helen that she had forgotten to shop on the previous day. Third, when she went to break eggs for breakfast, Helen managed to slop them all over the counter, which shorted out the toaster. Finally, Helen's boss called to say that Lyn had left so many voice-mail messages that the system was full and no one else could pick up mail, so would Helen please hurry into the office, given that she was the only person who could soothe Lyn at such times.

GOOD: Our marketing strategy will include several actions. We will organize a kickoff meeting. We will write to all the magazines in the field, and will try to get them to review our product. We will rent booths at several conferences. We will purchase two mailing lists, and will send out brochures.

Note also that you should never use a colon to introduce an intext list (whether or not the list is numbered) that has sentences as entries (that is, a list that has periods in it).

BAD: There are three activities that Red detests: (1) He does not like visiting the vet. (2) He does not appreciate having his ears cleaned. (3) He is not fond of baths.[89]

GOOD: Red has three particular sources of contentment in his life. First, he enjoys sharing Lyn's meals. Second, he enjoys eating the yeast and garlic treats that Lyn gives him every morning. Third, he enjoys eating catfood.

---

88. Note that there is no colon here.
89. Contributed by Red.

GOOD: BB also has favorite activities. She prefers to sleep on top of the telephone in the dining room (but she does not appreciate the telephone ringing). She likes to claw her way up the sleeves of Max's finest silk jackets, so that she can sleep on the top shelf of his closet. She enjoys sleeping on top of Lyn's computer monitor (but she hisses at Max if he uses the machine).[90]

You should use a *colon* before an intext or a displayed list *only* if the colon is logical and correct. Do not throw in a colon simply because a list follows.

BAD: If you so request, I can install: drop-caps, initial-caps, and rotating-text functions on the DTP software.

BAD: Be sure that you have both: a *kerning control,* which lets you make fine adjustments between the individual letters, and a *tracking control,* which allows you to specify a general setting for interletter spacing in a block of text.

GOOD: Both WP and DTP systems may allow you to import graphics, to set custom headers and footers, to link documents for printing, and to run text around graphics.

BAD: If you go shopping before lunch time, please pick up:

> A pastrami sandwich on sour corn rye
> A can of cherry-cola soda
> A box of extra hard, sour-dough pretzels
> A package of dark-chocolate–fudge cookies
> A bottle of indigestion pills

---

90. Contributed by BB.

GOOD: If you go upstairs, please bring down

> Ten or 20 candles
> A bowl of ice cream
> The open bottle of port
> Red's box of treats

GOOD: If you go shopping, please pick up the following seven items:

> 1. A bag of arugula
> 2. A quart of plain yogurt
> 3. A box of semolina pasta
> 4. A package of biscotti
> 5. A bottle of mineral water
> 6. A packet of vitamin supplements
> 7. A pocket calculator to count the number of items in this list

You should use *displayed lists* to emphasize list entries. There are several types of displayed lists.

An *unnumbered displayed list* has no markers before the entries. In certain designs, the first line of each entry is set farther to the left than are the subsequent lines—called a *hang indent*—so that the reader can see easily where each entry begins.

GOOD: The rule for unnumbered displayed lists follow:

> Neither numbers nor bullets are used to mark the entries.
> The entries are indented from the left margin.
> The entries may be hang indented, as these entries are.
> The entries can be of any length, and can include multiple sentences, as is true of any displayed list.

A *bullet displayed list* is the same as an unnumbered displayed list, except that each entry is marked with a bullet or other symbol. Because the bullets help to keep your reader oriented,[91] I recommend that you always use a bullet list (or a hang-indent unnumbered displayed list), rather than a plain unnumbered displayed list.

GOOD: We might wish to list the reasons for living in a city:
- Convenient shopping
- Great theaters
- Good restaurants
- Exciting nightlife

A *numbered displayed list* uses numbers, rather than bullets, to mark each entry. You should always use Arabic numerals, rather than Roman numerals or letters, in a numbered displayed list.

GOOD: Consider these three reasons for living in the country:
1. The sights, or lack thereof
2. The sounds, or lack thereof
3. The smells, or lack thereof

A *multicolumn list* is just what it sounds like.

GOOD: In the Chinese system, there are five key concepts, each of which may be designated in at least two ways, as follows:

water	winter
wood	spring
fire	summer
earth	late summer
metal	autumn

Remember to style the list entries consistently; I prefer all lowercase letters. (Try also to make your lists consistent with your tables.)

---

91. Icons serve the same purpose in certain books, such as in this one.

A *multicolumn list with heads* also is just what it sounds like.

GOOD: The system associates colors with emotions:

| | Emotion |
Color	Description
red	festive
green	jealous
blue	peaceful
white	mourning
yellow	energetic

Remember to style both the heads and the entries consistently across lists. I prefer to use a lead capital letter, or capital and lowercase letters, for the heads, and to use all lowercase letters for the entries, in a multicolumn list. A multicolumn list is a type of unnumbered table, so it is a good idea to style the entries in the same way as you style those in your tables. Note that you can center the items within each column, or can set them left justified.

An *outline displayed list* is set in the familiar outline format. You can use a numbered list with bullet subentries, or you can use the standard outline style, switching back and forth between letters and numbers. Generally, you should not use outline lists of the latter kind unless you absolutely have to do so—for example, because that format is required by a request for proposal (RFP).[92, 93]

You should indent each level of entry in an outline displayed list to the right of the level above it, so that your reader can see at a glance the structure of the list.

---

92. This example provides an exception to the principle that you should not define abbreviations or acronyms that you will not use frequently. When your readers are likely to recognize the abbreviation or acronym much more readily than they will the spelled-out version, then you may be justified in providing it.

93. If you work with RFPs, then chances are that you also know that it is critically important to follow meticulously the directions for preparing a proposal.

GOOD: Here are three tips for maintaining sanity when you are machine dependent:

1. Remember to back up your work often:
   - Do frequent saves while you are working.
   - Copy your files to a reliable storage medium on a regular basis.
2. Remember to use a surge protector.
3. Remember to protect your equipment:
   - Use a screen saver, and set it to turn on automatically after a few minutes of (your) inactivity.
   - Put your machine to sleep whenever you take a break.
   - Turn off your monitor when you are not using your machine overnight.

You can begin each entry in a displayed list with a *tag*. A tag is a sentence fragment that names a text object—in this case, it names the list entry.[94] Within a list, you should be consistent in setting tags on all or no entries. Across a document, you should be consistent in how you style tags and in what end punctuation you use with them.

You can style tags in *italic* or roman type; I recommend italic type, because it sets off the tag. You can also use **boldface** type, but only in cases where the entries are important enough to constitute subheads within your section, and only if you are not using boldface for key terms or have not given boldface other semantic meaning.

I strongly discourage you from setting tags with any capital letter other than for the first letter. Extra capital letters only clutter the page visually and disturb your reader.

---

94. A tag may also, for example, name a figure when used in a figure caption.

You can set at the end of the tag a colon, a period, or an em space. I recommend that you not use an em dash.

You should begin the list entry after the tag with a capital letter; the tag stands alone, and should not be considered to be the beginning of a sentence or fragment that carries on to the entry.

> UGLY: Lyn was proving to be a difficult author on many dimensions:
>
> - **Proofreading**—she found an error in the manuscript that the proofreader had missed.
> - **Interior Design**—she wanted to use fancy icons and graphics.
> - **Cover Design**—she wanted actual portraits of Red and BB on the cover, not to mention realistic bugs.
> - **Schedule**—she wanted to rush along at her usual breakneck speed, acting as though her publisher had no other books on its list.
> - **Size**—she had originally contracted to write a slim little book that would fit in someone's back pocket.

> GOOD: Lyn had to keep track of numerous tasks:
>
> - *Cover:* She had to tell the artist about the book and to send snapshots of Red and BB.
> - *Back cover:* She had to write or approve the copy, check the layout, and check the typeface.
> - *Interior design:* She had to develop and approve the final design pages.

- *Page makeup:* She had to arrange proofreading of laser printout, and to read and check the pages herself.
- *Film:* She had to arrange for a test run and to determine a schedule for filmout, in addition to ensuring that cover film was shot simultaneously.
- *Manufacturing:* She had to coordinate with her publisher for makeup of signatures, creation of plates, printing on the correct stock, and binding.

You should use a *period* after the number in a numbered displayed list. You should never use only a right or only a left parenthesis in any case, and you should never use pairs of parentheses in a numbered displayed list.

BAD: This example shows what not to do:

1). Do not use a single parenthesis, ever
2). Do not use a period and parentheses for list-entry numbers, ever[95]
3). Do use a period at the end of each entry when all the entries are sentences
4). Do not use numbers unless you have a reason to do so
4). Do not misnumber the entries when you do use numbers
5). Do not forget to indent your list entries from the left margin of your text column
6). Do not use boldface numbers (unless your designer mandates them, in which case use them consistently)

---

95. That is, do use parentheses around numbers in an intext list, but do not also use a period.

You should always begin each entry in a displayed list (unless the list is *multicolumn*) with a capital letter, whether the entry is a word, a sentence fragment, a full sentence, or numerous sentences.[96]

> BAD: The following items have been purloined from your bathroom:
>
> - toothpaste
> - toothbrush
> - shampoo
> - hairbrush
>
> GOOD: Peter requested this list:
>
> - Sushi
> - Ice cream
> - Chicken kabobs
> - Fermi[97]

You should use *no end punctuation* for those entries in a displayed list that constitute *less than one full sentence*.[98] Never use commas or semicolons in such cases, as you would in an intext list.

> BAD: The items in a displayed list
>
> 1. Are not the same as entries in an intext list,
> 2. Should never be followed by a comma or semicolon;
> 3. And should never include the word "and" in the final entry. Nor should they include a period on the final entry just because it is the final entry.

---

96. In an intext list, in contrast, you should not capitalize the first letter.

97. *Fermi* is Peter's name for a concoction made from milk, cornstarch, rose-water, pistachios, and almonds. It is unrelated to reactors.

98. That is, the entry constitutes less than a full sentence; a displayed list as a whole never constitutes a full sentence.

GOOD: The following is correct:
1. No end punctuation for fragments
2. No "and" at the beginning of the final entry

You should use a *period* as end punctuation on any entry that constitutes one or more sentences.

BAD: The entries in the following list are sentences, and therefore
- Each entry should end with a period[99]
- Each entry full sentence[100]

GOOD: In this case, the entries fulfill the criteria:
- Each one ends with a period.
- Both are full sentences.

The critical point for you to learn here is that, if one entry in a displayed list constitutes a full sentence, then *all* entries must be full sentences (and must end with periods). You should never mix fragments and sentences as entries in a single displayed list.

BAD: Many writers absolutely cannot handle displayed lists:

1). They make so many mistakes that copy editors cry,
2). Break all the rules—forget parallelism.
- Use bullets and numbers interchangeably!
3). Do they sometimes add questions as well?
4). Either make the point or stop writing

---

99. One reason why this example is classified as bad is that the entries do not end with periods.

100. Another reason why this example is classified as bad is that this entry is not a full sentence. If one entry ends with a period, then all should. Further, for an entry to end with a period, it must be a full sentence.

GOOD: You, however, are now equipped to handle displayed
       lists:

- Indent the list from the left margin.
- If you use numbers, use a period after them, rather than using one or two parentheses.
- However, do not use numbers unless you have a good reason to do so.
- If you do not use numbers, use bullets rather than nothing to mark each entry.
- If you use nothing to mark each entry, then use a hang indent if possible; otherwise, use extra space between entries.
- Align wraps (second and subsequent lines in an entry) under the first character of the first line of the entry, rather than under the marker or anywhere else.
- Be sure that all or none of the entries are complete sentences.
- If the entries are sentences, use periods; if they are not sentences, use no end punctuation.

THE THREE PRINCIPLES FOR LUCID WRITING here are that (1) at first blush, you may find the rules associated with the correct use of lists too numerous to swallow (but you should keep chewing); (2) you should learn to distinguish among the various types of lists; and (3) you should use the type and format that are most suited to the message that you want to communicate.

27   *Like Versus Such As*

THE TERMS *like* and *such as* have usefully different meanings that allow you to say precisely what you mean. *Like* and *as* also have different uses.

When you are referring to a set, the members of which have in common a given characteristic, and you wish to give an example that is a member of that set, you should use *such as*. When you are referring to a set that does not include your example, but that contains members that resemble your example, you should use *like*.

> GOOD: I do not understand baroque composers, such as Vivaldi.

> GOOD: I do not understand baroque composers like Art Garfunkel.

In the first sentence, the author is saying that she finds incomprehensible all baroque composers, of which Vivaldi is an example. In the second sentence, the author is saying that she finds incomprehensible those baroque composers who bear a resemblance to—are like—Art Garfunkel. She may find Art Garfunkel transparent. It may help you to think through your sentence with the implied words:

> GOOD: I do not understand baroque composers [who are] like Art Garfunkel.

GOOD: I cannot tolerate errors such as failure to distinguish *like* from *such as*.

I cannot tolerate a set of errors — say, common confusions about the meanings of terms — and a member of that set is the error named.

GOOD: I cannot tolerate people [who are] like my sponsoring editor.[101]

I cannot tolerate people who resemble my sponsoring editor; however, I may find my sponsoring editor delightful.

SPLENDID: Narcissus is like me, and doubtless loves his likeness as I love mine.[102]

GOOD: Mary wanted riches, such as a mansion or a yacht.

Mary's desired riches include a mansion and a yacht.

GOOD: Doug desires a wife [who is] like Moisés's.

Doug has no adulterous thoughts — he merely would like to wed a woman who resembles Moisés's spouse.

GOOD: If you use a compiling text formatter [that is] like Scribe, you will probably be reduced to tears before the document is in final form.

If you use Scribe, you may be happy; if you use another text formatter that is like Scribe, however, you will probably go berserk.

---

101. [Sponsoring editors, in contrast, tolerate a good deal more than their favorite authors think—ed.]

102. Contributed by my sponsoring editor and out of place, to confuse you.

GOOD: If you use typesetting languages such as T$_E$X, however, you can set equations without being institutionalized.[103]

If you use T$_E$X to set equations, then you will not change your sanity level; if you use, for the same purpose, a language that is like T$_E$X, then you will also remain baseline sane.

 You should use *as* for comparisons of activities (verbs), and *like* for comparisons of objects (nouns).

GOOD: Lyn's mother becomes excited and happy given the smallest excuse, much as Lyn does.

BAD: Lyn's mother runs up stairs and dashes about the house much like Lyn [does].

GOOD: Lyn's mother is like Lyn.

GOOD: You feel as I do?

BAD: You feel like I felt yesterday?

GOOD: You feel like Roger Rabbit?

GOOD: Tania walks across the room much as the most graceful ballerina pirouettes.

GOOD: Tania's walk is like the pirouette of a ballerina.

GOOD: Tania walks like a ballerina.

BAD: Tania walks like a ballerina does.

---

103. The implied assertion is that it may be worthwhile to subject yourself to the berserkogenic characteristics of a compiling program if that program allows you to set equations with equanimity.

In the second sentence in the final set, for example, you are saying that Tania is like a ballerina, at least insofar as her movement is concerned. The final sentence is wrong because the *does* means that we are now comparing the movements (Tania's and the ballerina's), rather than the people (Tania and the ballerina); in that case, we require *as*, rather than *like*.

THE PRINCIPLE FOR LUCID WRITING here is that you should always use *like* correctly to refer to likeness, or resemblance. You should use *such as* to indicate an example member of a group about which you are speaking. You should use *as* to indicate a likeness in activities. If you use your ear as I do, then your prose will be like mine, and you can compose sentences such as this one.

28   *Either and Both*

THE PLACEMENT of words accompanying *either* or *both* markedly alters the meaning of your sentence.

 You should always set exactly two arguments with *either* or *both*.

> BAD: Judy ate up both the chocolate truffles and the fudge and the chocolate-chip ice cream.

> GOOD: Steve declined both the milkshake and the cake.

> GOOD: Judy grabbed the plates, the glasses, and the silverware, and washed the dishes before Steve realized what had happened.

> BAD: You can either leave a message, telephone my assistant, or press 1 for further options.

> GOOD: You can either try to leave a message or scream in frustration.

> GOOD: You can press any one of several numbers, but the system will disconnect you regardless.

You should place before *either* any word that applies to the *two* alternatives. You should place after *either* any word that applies to *only one* of the alternatives.

> BAD: I want to travel either on a camel or a tiger.

BAD: I want to travel on either a condor's back or in a stork's beak.

Here, in the first example sentence, *on* applies to only the camel; because it should apply to both the camel and the tiger, it either should be repeated or should precede *either*. In the second example sentence, the *on* precedes the *either*, and thus the *in* is invalidated, making a mishmash of the meaning. The solutions are thus as follows:

GOOD: I want to travel either on a rickshaw or on a public bus.

The on *follows the* either *and is repeated.*

GOOD: I want to travel on either a bicycle or a soap box.

The on *precedes the* either *and is not repeated.*

We could also say

GOOD: I want to travel either in a coffin or on a white elephant.[104]

That is, when the preposition follows either, *you can designate more than one relationship* (in, on, atop, with, *and so on*).

We would run into trouble, however, if we tried to place in front of *either* a word that had no business spanning both terms.

BAD: I want to travel in either a car or a bicycle.

It would be difficult to travel in a bicycle.

GOOD: I want to travel either in a balloon or on a kangaroo.

Slightly tangentially, when you use *either*, you should use the verb that matches—singular or plural—the noun nearest to the verb.

---

104. All seats on the elephant are standby only, because of overbooking.

BAD: Either a widget or two wombats is required.

GOOD: Either I or they are being silly.

BAD: Either several cats or one lion are making that noise.

GOOD: Either several adults or one 3 year old[105] is making that noise.

As you should do with *either,* you should place before *both* any word that you want to apply to the two following terms, and should place after *both* any word that applies to only one of them.

BAD: Lumina both needs a message-handling service and a local-area network.

The relationship of Lumina to the local-area network remains woefully undefined.

GOOD: IDSR both needs a new computer and can afford one.

Here, needs *and* can afford *follow* both.

GOOD: Lumina invites both your suggestions for the product name and your attendance at the launch party.[106]

Invites *precedes* both *and thus modifies both the suggestions and the attendance.*

---

105. Note that there is no reason to hyphenate a compound noun such as 3 *year old*. If (and only if) you use the term as a compound adjective, then you should hyphenate it: 3-year-old sausage.

106. The sentence *Lumina invites both suggestions for the product name and you to attend the launch party* might seem to demonstrate how grammatical accuracy is not always sufficient to provide ease of parsing. There is a lurking error that explains the problem: *X for Y* and *X to do Y* are not parallel.

BAD: I want to ensure both confidentiality and to protect privacy.

Both *follows* to ensure *and thus invalidates* to protect.

GOOD: Our remote–log-in procedure both requests information and institutes a call-back routine.

Both *precedes* requests *and* institutes.

GOOD: The person logging in must supply both a unique user-identification code and a password.

Supply *precedes* both.

BAD: I want to eat both steak and to drink wine.

HUNGRY: I want to eat both steaks and to drink wine.[107]

BAD: The system can perform both linguistic strings and generate parse trees.

GOOD: A natural-language parser must contain both a lexicon and a grammar.

GOOD: To create our grammar, I both used a Backus–Naur metalanguage and built a restriction component.

BAD: I want both to cuddle Red and BB.

GOOD: Red wants to catch both bats and rats.

GOOD: Both Red and BB like both to snooze on the roof and to snuggle in bed.

---

107. Contributed by Joseph Norman.

THE PRINCIPLE FOR LUCID WRITING here is that a word placed before *either* or *both* applies to both alternatives, whereas a word placed after *either* or *both* applies to only (but not to either) one. You can either remember this principle or mark the page, but both remembering and marking would be redundant.

29 Hyphens

THE MOST IMPORTANT use of the hyphen is to tie together two words that modify a third, when the third word follows the first two. If you learn how to use hyphens in such *compound adjectives,* you will be able to tell your reader what you intend to modify what, and that is useful information. Without the hyphens, compound adjectives are ambiguous.

There are three kinds of dash; the hyphen is the little one:

- hyphen
– en dash
— em dash

Certain people prefer to follow the principle of hyphenating compound adjectives only when failure to do so would lead to confusion. My response is that failure to do so always leads to *grammatical* confusion, and grammatical confusion is bound to cause comprehension confusion; I have even classified unhyphenated compound adjectives as errors, as bad rather than as ugly. I strongly recommend that you take advantage of the richness of written language to disambiguate your sentences whenever possible. Hyphenating compound adjectives will reduce ambiguity enormously: Make a habit of it.

You should place a hyphen between two words that together constitute a single adjective.

BAD: The viscous forming process involves melting and shaping of a viscous silicate.

GOOD: Figure 8.2 shows sodium-fluoride sintering.

In the first example, *viscous* and *forming* have been transformed from two words, one modifying the other, into a single term that modifies a third, *process*. So, you should place a hyphen in *viscous-forming process*. Similarly, Figure 8.2 shows sintering of sodium fluoride; hence, there is a hyphen between *sodium* and *fluoride*.

You should not place a hyphen between two words that together constitute a noun or between two words that constitute an adjective–noun pair.

BAD: We need to develop a more intuitive front-end.

GOOD: This car has a dented back end.

Front end has evolved from an adjective–noun pair to become a compound noun; there is no good reason to hyphenate it. *Back end* is an adjective–noun pair. Note, however, that any compound noun can suddenly become a compound adjective, desperately requiring a hyphen transplant, simply because it has been dropped into a different sentence.

BAD: The front end code was written by a lunatic.

GOOD: The back-end dent was simple to fix.

You should not place a hyphen between two words that together describe an activity.

BAD: Decision-making under uncertainty can be stressful.

GOOD: Love making under a starry sky can be thrilling.[108]

---

108. The possible implication that decision analysts are subject to heart attacks during outdoor sporting events is neither intentional nor intended.

Both *decision making* and *love making* denote activities. Note, however, that the terms can also serve as adjectives, in which case they should be hyphenated.

> BAD: Decision making paradigms are scarce.

> GOOD: Love-making paradises are scary.

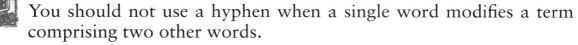 You should not use a hyphen when a single word modifies a term comprising two other words.

> BAD: The red-fire engine[109] took off with great commotion.

In this case, *red* modifies the compound noun *fire engine,* so the phrase should be set without any hyphen. We change *red fire engine* to *blue china dog,* and obtain

> GOOD: The green dragon lady waltzed down the street in the arms of the blue china dog.

You can think about hyphenation of compound adjectives as an English-language equivalent of the use of parentheses in mathematics to demonstrate the scope of the operation. Let us look at an example:

> GOOD: Internet-specific advisory information, such as ICMP error messages, also may be read from raw sockets.

Here, we have four words,

> *Internet specific advisory information*

We wish to show that the first two modify the third and fourth:

> *(Internet specific) (advisory information)*

---

109. You could stack up a huge compound adjective by writing *red–fire-engine smoke,* but that would take you into the realm of the en dash, which we shall get to in a moment.

That is, the first two constitute an adjective, whereas the third and fourth constitute a noun. So, we hyphenate the adjective, but not the noun:

Internet-specific advisory information

Now let us examine a second example:

GOOD: Written language is a rule-based system.

Here are the relevant words:

rule based system

Here is the parsing:

(rule based) system

Here is the result:

rule-based system

We shall work through several examples, to be sure that you get the hang of hyphenation.

GOOD: A 250-MB disk soon proves insufficient.

250 MB disk

(250 MB) disk

250-MB disk

GOOD: A moldy style manual can be soporific reading, although it may be a fabulous reference.

moldy style manual

moldy (style manual)

moldy style manual

GOOD: To the slightly disconcerted astonishment of Jeff and Elizabeth, Bruno will be a high-school student in less than two decades.

high   school   student

(high school) student

high-school student

GOOD: When Lyn first fell in love, she was dazed and drunk; she was a joyously high school student!

high   school   student

high (school student)

high school student

GOOD: The discussion starts at the system-call level.

system   call   level

(system call) level

system-call level

GOOD: Lyn offered Max a half baked potato for dinner.[110]

half   baked potato

half (baked potato)

half baked potato

You should not use a hyphen when the first word of a compound adjective ends in *ly*. In essence, the *ly* serves the same purpose as that served by the hyphen; therefore, inserting a hyphen would introduce redundancy.

BAD: You should avoid writing badly-formed sentences.

GOOD: You should never use grossly deformed pencils.

BAD: Max likes to work with a highly-motivated team.

---

110. Lyn's consideration was that Max had already eaten a huge lunch; had Lyn served a crunchy, partially cooked potato, a hyphen would be called for. Such behavior, however, would be evidence that Lyn was acting on a half-baked idea.

GOOD: Lyn likes to live in a sparsely populated (and sparsely road-signed)[111] area.

You should not place a hyphen in a compound adjective when the first word is *more, most, less,* or *least.*

BAD: Can you imagine a less-appealing creature?

GOOD: I cannot imagine a more attractive cat.

BAD: Only the least-used keys still had legible letters to identify them.

GOOD: Only the most sought-after fat and fluffy cats had hot dates on Saturday night to console them.

You should use an *en dash* as well as a hyphen when you stack adjectives. *Stacking adjectives* means stringing together a bunch of words to form a single compound adjective. Stacking adjectives is not good writing practice, and you should avoid it; however, when you cannot avoid it, an en dash will at least keep your meaning clear. (We shall not delve into all the intricacies of en dashes in this segment; we devote Segment 49 to that topic.)

BAD: The sales represenative emphasized that wide-area-network connections encourage open business communications across diverse sites.

        Both the mark between wide *and* area *and that between* area *and* network *are hyphens.*

GOOD: Local-area–network connections are easy to install.

        The mark between area *and* network *is an en dash.*

---

111. Contributed by Joseph Norman, who, like most novice visitors to the house, took an unanticipated tour of Woodside and Portola Valley.

Let us parse the terms as we did earlier:

> *local   area   network   connections*
> *[(local area) network] connections*
> *(local-area   network)   connections*
> *local-area–network connections*

So, you can think of the en dash as being the English-language equivalent of the square braces.

Be aware, however, that a more elegant solution is

SPLENDID: Connections for local-area networks are easy to install.

That is, whenever possible, unstack your adjectives.

Here are a few more examples:

GOOD: Be careful not to make an operating-system–command error.

> *operating   system   command   error*
> *[(operating system) command] error*
> *(operating-system   command)   error*
> *operating-system–command error*

SPLENDID: Be careful not to make an error in an operating-system command.

GOOD: Will you enter the grade-school–student spelling bee?

> *grade   school   student   spelling   bee*
> *[(grade school) student] (spelling bee)*
> *(grade-school   student)   (spelling bee)*
> *grade-school–student spelling bee*

SPLENDID: Will you enter the spelling bee that is being held for grade-school students?

GOOD: Soren and Lois were eagerly awaiting the car-seat–cover sale.

car   seat cover sale

[(car seat) cover] sale

(car-seat cover) sale

car-seat–cover sale

SPLENDID: Soren and Lois were eagerly awaiting the sale on car-seat covers.

You should not place a hyphen in adjectives that follow the noun that they modify.

BAD: This programming system is not user-friendly.

GOOD: This user-unfriendly interface is hypnotic.

GOOD: When Lyn has not had sufficient sleep, she is not user friendly.

BAD: Some people think that learning all the rules of hyphenation would be overly time-consuming.

GOOD: A time-consuming process is not necessarily a not-worthwhile process.

GOOD: Writing well is worth the effort, although it can be time consuming.

BAD: The changes in organizational structure were decid-
             edly profit-driven.

GOOD: A demand-driven operation may be the most cost-
             effective choice.

GOOD: The economy is currently modeled as demand driven.

SPLENDID: A cost-effective decision may be driven by necessity.

THE PRINCIPLE FOR LUCID WRITING here is
that you should hyphenate most compound-
adjectives terms, but should not hyphenate
terms that are compound adjectives that follow
the noun. A user-friendly book is not user
unfriendly.

143

30 Full Versus Incomplete Infinitives

IN SENTENCES that contain multiple infinitives—the verb form of *to be, to cough, to hiccough, to squeak,* and so on—you should know when to include the *to* in each infinitive.

When you are listing numerous activities, you can leave out the extra *to*s, although it is not incorrect (and can be preferable, because it is simpler) to leave them in.

GOOD: Lyn likes to play, to eat, and to sleep.

SPLENDID: Max likes to work, eat, and sleep.

SPLENDID: Cats like to sleep and eat.

GOOD: Nick likes to dream, to fantasize, and to imagine his youthful days of dancing in the shimmering translucent moonlight.

SPLENDID: Nick's old job required him to dream, fantasize, and imagine shoes dancing in the sparkling gold sunlight.

When you are using two or more infinitives that are modified in different ways, you should always use the *to*.

BAD: Max likes to eat sour cherries and rise early.

GOOD: Lyn likes to slurp chocolate sodas and to sleep late.

BAD: To install this bus network, you need to purchase a jabber controller and design a collision detector.

GOOD: The transceiver must be able to send and receive data to and from the cable, to detect collisions on the cable, and to provide electrical isolation between the cable and the interface electronics.

BAD: Max likes to use his laptop computer while traveling on airplanes, negotiate deals over lunch in Italian restaurants, and conduct meetings in his car en route to other engagements.

GOOD: Lyn prefers to take bubble baths, to dance in the rain, and to swim naked in the ocean by starlight.

SPLENDID: Max and Lyn like to spend quiet evenings lolling on the papasan together, talking of mouse devices and anthropoids, and burning zillions of candles but few calories, and to expend energy together, hiking on their favorite trails.

THE PRINCIPLE FOR LUCID WRITING here is that you should use full infinitives in a series in which the infinitives are modified in different ways. That is, to write well, or to avoid mistakes, you should learn when to repeat the *to*.

31 Titles

In GENERAL, there is no reason to capitalize titles except when they are accompanied (at least by implication) by the name of the holder of the office. When there is no reason to use a capital letter, do not use it; it simply clutters the page visually, and purports to have meaning when it has none.

 You should not capitalize a reference to the title or office itself.

> BAD: Max wished often that the President and the Chief Executive Officer in his company were roles held by different people.
>
> GOOD: Max learned quickly that being the president of a startup company can be hard on your personal life.

> BAD: The Pope lives in the Vatican.
>
> GOOD: The commander in chief should be well rested before she makes critical decisions.

> BAD: How are the Kaiser and the Führer different from the Queen?
>
> GOOD: When the subject of salaries was broached during the faculty meeting, the professors argued furiously with the chair of the department.

You should capitalize the title when you are referring to a specific person who holds it.

>   BAD: The sales representative tried to make an appointment with president Clinton and secretary of state Bilbo Baggins.
>
>   GOOD: "Come back tomorrow," Chief Quality-Assurance Specialist Suresh told the sales representative, "so that you can demo the system for President Henrion."

>   BAD: Chief financial officer Chaitin was wringing his hands over the $8 bill for the photographic film, and elucidator Dupré was rolling her eyes in disbelief.
>
>   GOOD: Senior Researcher Provan was running around the office in his gym shorts again, when he bumped into Office Manager Brown.

 You should capitalize the title if you are using it to refer to a specific person, even if you leave out that person's name.

>   BAD: "Hey president!" shouted Richard as he clapped his boss on the back, "How's life?"
>
>   GOOD: "You can pick out the Princess of Roses," encouraged the announcer, "as the little blue blur with the yellow spot above it in the upper-right corner of the display."

>   BAD: The ambassador, who had been dawdling at the buffet table, waddled across the room and tripped on her train.
>
>   GOOD: The Mayor, who had risen early, gave a rousing speech to his bathroom mirror while shaving.

You should not capitalize references to generic departments or divisions. If the name of the department or division is part of a title, you should follow the preceding rules to determine when to capitalize it.

> BAD: To install a laboratory information system, you need the cooperation not just of the Laboratory Department, but also of nearly all the other Divisions in the hospital, except perhaps for those such as Food Services, Maintenance, and Security.

> GOOD: To be granted tenure, you have to jump through hoops not merely in your own department, but also in what may appear to be every other division of the university, possibly including food services, maintenance, and security.

> BAD: Lyn had lunch now and then with her old friend, professor Dupré of the department of philosophy at Stanford University.[112]

> GOOD: Lyn spoke to Professor Bessel, Chair of the Department of Intriguing Sounds, about the habit of recording phonemes by forcing balloons down people's throats.

> BAD: The university decided[113] that it should create several new slots for Assistant Professors of Multimedia Networking and Communications.

---

112. Note that, unlike the department's, the university's name is capitalized.

113. This introduction would itself be sufficient for a classification of ugly, as universities do not have mental apparati that allow decision making, and must delegate such activities to the humans who hold administrative offices within them.

GOOD: The new assistant professors of computer science were given airless cubbyholes and antique terminals.

 You should capitalize names of companies, organizations, and government agencies, whether or not a title set with them is capitalized.

BAD: There have been several directors of the department of health, education, and welfare.

GOOD: The new Director of the National Institute for Science and Technology promised to trim lean meat.

GOOD: Three former directors of the Office of the Martyr have overdosed on port and stilton, and currently are snoring a cappella on the Asian carpet.

THE PRINCIPLE FOR LUCID WRITING here is that you should save capital letters for people, and should not waste them on titles of office not attached to people. Various presidents are quite different from President Lincoln.

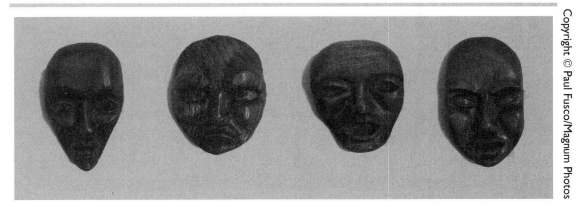

32 *Contractions*

ALTHOUGH contractions (terms formed of multiple words, in which missing letters are signified by an apostrophe) are respectable terms in casual writing (and in poetry), because they are part of our spoken language, they are out of place in formal writing. I advise you to strive for formal writing that is interesting and amusing, yet still is styled impeccably. Generally speaking, contractions are *unnecessarily casual*. Most publishers of textbooks and journal articles do not permit contractions.

You should generally spell out terms in formal writing, rather than using contractions.

> UGLY: Data are useful; to scientists, they're indispensable.
>
> GOOD: Cats are so delightful that they are indispensable.

> UGLY: He's so exhausted, he can barely get into bed.
>
> GOOD: She is so glad that he did, after all, make it to bed.

> UGLY: We've discussed an inference procedure based on the resolution principle.
>
> GOOD: We have described the simplified version of predicate calculus that is known as clausal form.

You should particularly avoid using contractions that include a negation. Even if you are writing in a casual voice, the negation is easy to miss if it is hidden in a contraction. Instead, be sure to spell out the *not*.

UGLY: You don't need a computer to write a book.

GOOD: You do not need a brain to write a book.

GOOD: Once you have managed to write a book, you do not need any accoutrement except for a sympathetic publisher.[114]

GOOD: After you have written a book, you must decide whether to have your publisher do page makeup, or to undertake that task yourself.

UGLY: Can't you understand that a user view is a model or representation of the data requirements for a single business function?

GOOD: Can you not grok[115] that an entity is an agent, location, object, or idea about which you want to record information?

SPLENDID: Can you use a different kind of programming language?[116]

UGLY: Red absolutely won't go near the water.

---

114. Based on a contribution by Peter Gordon.

115. The construction is faintly archaic (Canst thou not grok? Grok thou not?); hence, the following splendid example.

116. Often, recasting a negative into positive form will avoid the contraction while giving you a more powerful sentence.

GOOD: BB absolutely will not eat her rice pudding.

SPLENDID: Savannah absolutely refuses to eat beets.[117]

UGLY: Haven't you figured out that algorithm yet?

GOOD: Have you not yet prepared my dinner?

SPLENDID: I understand that you were too busy to cook dinner; we shall go out to our favorite restaurant instead.

UGLY: Isn't the diagram in the figure a typical establishment-wide LAN setup?

GOOD: Is not the rectangle in the middle of the figure a representation of a sitewide backbone?

SPLENDID: Is that odd blur on the lawn not a mongoose?[118]

SPLENDID: Was that white furry shape that just whizzed past the window Red doing his evening exercises?[119]

 Certain textbook publishers specify *let's* as the single allowable contraction. I prefer to use *let us,* but, if you wish, you can adopt the convention of contracting this term.

UGLY?: Let's set a equal to $(b - c) \pm \rho$.

GOOD: Let us walk through the algorithm.

UGLY?: Let's examine the results carefully.

---

117. *Absolutely refuses* is a stronger phrase than is *absolutely will not.*

118. Splitting the *is* and the *not* creates a more elegant rhythm.

119. Again, recasting to avoid the negative is a strong solution.

GOOD: Let us postulate that the moon is made of green cheese.

UGLY?: Let's forget about our date tomorrow, and get our work done instead.

GOOD: Let's forget all our troubles, kick off our shoes, and light the candles.

THE PRINCIPLE FOR LUCID WRITING here is that you should not use contractions in formal writing; it's OK to use them in casual writing.

33 *Per*

PER IS THE correct term to use when you mean *for* (or *on* or *during) each*. In formal writing,[120] you should use *per*, rather than *a*, in such cases.

BAD: I adore you 7 days a week.

GOOD: I respect you 7 days per week.

BAD: Shellee made the cookies last by eating only one a day.

GOOD: Shellee got rid of the cookies by giving them out at the party, one per person.

BAD: The PAD enters the data-transfer phase, once a call, after it has completed the virtual-call establishment procedure.

GOOD: The circuit-switched network provides one full-duplex, synchronous, data-transmission path per user call.

---

120. In casual writing, or even in a casual sentence, you might say "once a day," for example, without erring.

BAD: Lyn sold individually designed, home-dyed, hand-spun, and hand-knitted sweaters for $100 a piece.

GOOD: Lyn declined to sell her hand-carved masks for $500 per unit.

BAD: A least once a day, Lyn begins to wonder whether she has chosen the ideal career.

GOOD: At least once per day, Max enfolds Lyn in his arms and disperses all her cares.

THE PRINCIPLE FOR LUCID WRITING here is that you should use *per* to mean *for (or on) each*. If you read one segment per day, you will learn a useful principle on each day.

34 Number Styles

T HERE ARE various rules[121] about styling numbers and related writing. Most of these rules are simply conventions that permit you and other writers to style your numbers consistently. Such conventions may not have any raison d'être other than that they are standards that encourage writers to create documents that are consistent.

You should use a zero before a decimal point. The lead zero calls your reader's attention to the decimal point, and thus makes your writing clearer.

> UGLY: BB's perch in the tree was .1 kilometer higher than the maximum distance that BB could jump safely.

> GOOD: Red's nose is 0.6 inch high.

You should not use a comma in a four-digit number set in base text.[122]

---

121. Here, I use the word *rule,* in contrast to *principle,* to denote imperatives that have no particular rhyme or reason.

122. This convention provides a good example of one for which I know of no particular rationale. It is, however, a standard to which most publishers adhere. One theory I have is that the rule is a result of our having entered the computer age. A clever programmer was faced with a computer system unable to distinguish dates (for example, 1995) from other four-digit numbers. Wanting to avoid setting dates with a comma (1,995), and being unable to determine how to recognize a date easily, the programmer decided to institute a new convention: Drop the comma from *all* four-digit numbers.

UGLY: Malcolm wrote 3,876 words for his presentation.

GOOD: Alyssa giggled 3876 times in her first month.

You should use the comma, however, in tables, in displayed lists, and in any other columns where you want your four-digit numbers to align with other numbers.

BAD: The following are the debts incurred by the various countries (in $U.S.)

396,023
2389
418
32,592
204,592
491

GOOD: The following are the numbers of cats adopted per year, by cities C1 through C6:

2,387
248, 726
734
923,185
5,692
492

 You should use the comma in all numbers of greater than four digits.

UGLY: That day, Lyn had tried to telephone Max what felt to her to be 987654321 times.

GOOD: Lyn found 987,654,321 pennies in an old sock under the bed.[123]

---

123. Max Henrion, after performing loose calculations at 1:00 A.M. in the dark, roughly estimates that the pennies would take up 104 cubic feet. He adds that his estimates might be in error by a factor of a hundred or so.

You should spell out *million, billion,* and so on for numbers that are multiples (including fractions set as decimals) of these quantities.[124]

> BAD: Greg thought that 27,000,000,000 flowers might not fit into their garden.[125]

> GOOD: Nicola thought that she could see 2.7 billion stars.

> BAD: Can this machine perform 1,500,000 fast Fourier transforms without cracking up?

> GOOD: The highly parallel and pipelined system achieves a computation rate of 250 million 32-bit arithmetic operations per second.

You should use a dollar sign when you indicate dollar amounts.

> BAD: Jim gave Benjamin 1 dollar to walk in Wunderlich.

> BAD: Benjamin thought that one dollar was not nearly so interesting a reward as were the salamanders.

> GOOD: Lauralee bought Benjamin a new hat for only $1.

> BAD: The bill for the network implementation was 2 million dollars.

> GOOD: After several grueling months of ulcergenic negotiations, Max's company closed a $2-million contract.

---

124. In fields where the notation will make sense to your readers, you may prefer to use scientific notation. For example, you can write 10^{15} instead of *quadrillion,* or 10^{39} instead of *duodecillion;* you can also write, say, 3.46×10^4.

125. You are invited to submit your estimates of the acreage required, based on probabilities for each type of garden flower being present in a given proportion.

You should set monetary amounts of less than $1 in either of two ways: as whole-number cents or as decimal-number dollars. In general, you should use the latter style when you are comparing dollar amounts. Be careful, however, not to make the common error shown in the first example in the following set:

> BAD: Reno saved 0.40 cents to buy himself a candy bar.

> GOOD: Cosmo found 20 cents on the sidewalk.

> GOOD: The price for the earrings varied across jewelry stores, from $0.20 to $30.00,[126] with an average price of $5.20.

You should spell out *percent*, rather than using the percent sign (%), except in tables and figures, when space is constrained tightly, or when you are writing in a domain that uses myriad percentages (such as statistics).

> UGLY?:   Only 2% of journals use the percent sign in base text.

> GOOD: Fully 20 percent of the printer's cost is allotted to the toner package.

You should not repeat the *percent* when giving ranges; you should, however, repeat the percent sign (%) when giving ranges.

> UGLY: The sampled households varied widely, with 4 percent to 40 percent responding *yes* to query 1.

> GOOD: The interval from 5% to 95% covers the cases in which we are interested.

> GOOD: Lyn swears that 50 to 60 percent of graduate students at the top universities cannot distinguish a clause from a cause.

---

126. Note that *$30* would be incorrect in this case; when you give any number in a series to cents precision, give all numbers in that manner.

You should not use the en dash[127] for percent ranges in the text line (that is, you should spell out the *to* instead); however, you should use the en dash for ranges in tables (and in figure labels).

> UGLY: Doug reported that 4–40 percent of doctors cannot assess risk intelligently.

> GOOD: Doug told Lyn he was now feeling human during 80 to 90 percent of the day.

You should use *by,* rather than the times sign, to give dimensions.

> BAD: The mouse pad is 6-by-8 inches.

> UGLY: The wrist-support pad is 4 × 20 inches.

> GOOD: The screen is 12[128] by 8.5 inches.

> GOOD: To display the result, we use an 8- by 30-pixel array.[129]

You should spell out *degrees* in temperatures, *except* when you are specifying Celsius[130] or Fahrenheit *and* are using the abbreviation C or F. In the latter case, do use the degree sign and leave thin space[131] between it and the C or F.

---

127. An en dash is the middle-sized dash of the three: hyphen (-), en dash (–), and em dash (—). An en dash is used primarily to indicate ranges or equal-weighted pairs (see Segment 49).

128. In this case, it might be correct to write *12.0,* rather than simply *12.* If, for example, you give many measurements that include tenths of inches, then add the decimal point and the zero. If, however, your grain size is $\frac{1}{2}$ inch, then just *12* will suffice.

129. Note the space after the hyphen, which signifies a missing *pixel.*

130. The Celsius scale is used everywhere except in the United States and, I believe, in Burma.

131. *Thin space* is an editor's term for space that is less than the normal interword spacing; it is used without an article (that is, not *a thin space).*

BAD: Max and Lyn's living room is about 7° on most evenings in winter.

BAD: Lyn's toes are about −2° Celsius on most evenings in winter.

GOOD: Max and Lyn's living room is about 70 degrees Fahrenheit on most evenings in summer.

GOOD: Max told Lyn that, contrary to her suspicions, the temperature at the office was 98.6°F.

Note that, if you are giving temperatures in Kelvin, you should not set the degree sign, even with the abbreviation K.

BAD: Most people in the United States have no idea how hot 274°K is.

GOOD: Scientists who work with the metric system (all do) usually can translate without strain 274 K into degrees Fahrenheit or Celsius.

When you are giving a probability, you should use an italic lowercase p, and should use operators rather than spelling out the operators' names.

UGLY: The disparity between men's and women's pay scales is significant at p less than 0.1.

GOOD: The disparity that we identified between cohorts is significant at $p < 0.05$.

You should always set variables in italic type; you should usually set matrices and vectors in boldface type (there are other conventions that are correct as well, but we shall not detail them here). You should set constants (including letters that simply name an object) in roman type.

GOOD: We set $x + y = z$.

GOOD: $\mathbf{a} = [a_1, a_2, ..., a_n]$
        $\mathbf{b} = [b_1, b_2, ..., b_m]$
        $\mathbf{C} = \mathbf{a} \times \mathbf{b}$

GOOD: We see from Figure 8.19 that transistors Q_1 and Q_2 are diode connected.

GOOD: The A, B, and C companies competed in the software-publishing arena.

Note also that, when you set a subscript or superscript[132] that is a term or an abbreviation, you should set it in roman type.

GOOD: We denote by N_d the carrier density at the drain end of the channel; v_{drift} is the drift velocity.

GOOD: We denote by rp_{max}[133] the maximum retail price of the book, and by gm_{min} the minimum gross margin that Peter is willing to accept.

GOOD: $A = \text{Ao } e^{-RT\ln K}$.

 You should not throw in a spurious hyphen when you speak about an axis or test or other noun that accompanies a variable.

BAD: Use the x-axis to represent I_{in}.

GOOD: Use the y axis to represent I_{out}.

BAD: We use the y-coordinate to locate the point vertically.

GOOD: We can calculate the x coordinate for the probability-density graph as follows.

---

132. When you set a subscript or superscript, be sure to reduce the font (size of the type) if your word processor does not reduce fonts automatically.

133. The retail price and the gross margin are variables; in contrast, *max* is an abbreviation for *maximum*, and *min* is an abbreviation for *minimum*.

BAD: The data were analyzed using Student's t-test.[134]

GOOD: The results of a two-tailed t test on the Manx kittens were mildly encouraging.

SPLENDID: We applied the t test rigorously, and are delighted to report that Earl Grey won hands down.[135]

BAD: The p-value was highly improbable.

GOOD: The most common p value is 0.05.

SPLENDID: The p value should be related to the distance to the powder room and the length of the line.

Of course, you should use the hyphen when it is correct (whenever two words together modify a third, they should be hyphenated).[136]

BAD: The χ^2 test result was astonishing.

GOOD: The t-test value remains unknown.

BAD: The y axis value is off the graph.

GOOD: The x-axis label is missing.

---

134. This sentence contains two mistakes: It contains a disallowed hyphen, and it mixes a clause that requires an agent with a clause that is cast in passive voice. Even if the entire sentence were in passive voice, and thus grammatically correct, it would contain a common blemish: neglecting to mention just who analyzed the data.

135. Recasting in active voice earns the classification of splendid.

136. I discuss the important topic of hyphenation of compound adjectives in Segment 29.

THE PRINCIPLE FOR LUCID WRITING here is that you should learn the rules for styling numbers so that your text will be internally consistent and will be consistent with that of other scientific writers. Keep this book at your elbow, with this page marked, until you know all the rules in this segment.[137,138,139]

Copyright © Paul Fusco/Magnum Photos

137. Suggested by Patrick Henry Winston.

138. I added *in this segment* after contemplating the meaning of a sign in a local park that reads *All dogs must be on a leash*.

139. Once you know all the rules given here, you will find that there are many more. I have tried to cover those that I see plaguing writers most often. For the others, consult a style manual.

35 Quotations

THERE ARE several reasons why you should learn how to quote written material. You are setting someone else's words in your own text, so you need to know, for example, precisely how to identify the author.

When you quote written material, you should set the material exactly as set in the original.

> BAD: Shakespeare tells us that "All that glitters is not gold" [47, p. 185].
>
> GOOD: Shakespeare wrote that "All that glisters is not gold" [Shakespeare, *Merchant of Venice,* act II, sc. vii, l, 65].
>
> GOOD: *Non teneas aurum totum quod slendet ut aurum.*

The exception to this rule is that you can, if you wish, emphasize a portion of the quotation, provided that you indicate that you have done so. Note that any material set in square brackets is assumed not to be part of the original material.

> GOOD: President Wimp today agreed to allow industry to decimate or extinguish any species that got in its way, and to ignore all environmental effects such as pollution of water or air. This latest in a series of *compromise solutions* [emphasis added] was reached after 2 minutes of arduous negotiations [*Hibrow Times,* 19 June 1997, p. C4].

If you want to be sure that your reader knows that a particularly noticeably erroneous or distressing or funny portion of the quotation is set accurately, you can add *sic* in square brackets. Use *sic* sparingly, however, as it can have the flavor of childish humor.

> GOOD: The directions for using Red Panax Ginseng Extractum (Song Shu Pai, Pine Brand package insert) are as follows:
>
> 1. Remove the metal cap.
> 2. Lift the rubber plug.
> 3. Insert the plastic dropper.
> 4. Start dropping the tonic [sic].
> 5. Insert the inner kid [sic].
> 6. Screw on the plastic cap.

You should use square brackets to insert text into written quoted material when the insertion is required to clarify meaning.

> GOOD: Dr. Henrion reportedly said, "I am certain that she [Elucidator Dupré] has every aspect of the book-production process under control" *[Overconfidence,* 8 April 1993, p. 14].

 When you quote written material, you should avoid taking material out of context and thus changing the import of the words.

> BAD: "East is east and west is west, and never the twain shall meet" [42, p. 707].
>
> GOOD: Oh, East is East, and West is West,
> and never the twain shall meet,
>
> Till Earth and Sky stand presently
> at God's great Judgment Seat;
>
> But there is neither East nor West,
> border, nor breed, nor birth,

> When two strong men stand face to face,
>   though they come from the ends of the earth!
>
> *The Ballad of East and West*
> Rudyard Kipling [1889]

When you quote written material, you should always give a reference for the source, and you should include the number of the page from which the quoted material is taken. You should give the reference in whatever form you are using for other citations in your work.

GOOD: "It is an unsolvable crisis of unmanageable proportions" according to the insolubly enigmatic NASA spokesperson [*Newspeak*, 14 May 1992, p. 53].

GOOD: The new tool will "allow you to do everything you always wanted to do with and to your computer" [Juresco & Ernesti, p. 46].

GOOD: If you want to start a software company, you should "give up any thought of family life, leisure, relaxation, sports, sex, or sleep" [87, p. 1].

When you quote written material of five or fewer lines, you should set the quotation in the text line, and should use double quotation marks to delineate it.

BAD: The demands of a startup company can be intense, as illustrated by this statistic:

> "In Silicon Valley, 97 percent of startup presidents reported seeking psychiatric help within 1 year of the company being founded." [92, p. 73]

The quotation should not be set as an extract, because it is five or fewer lines. In addition, the extract itself is incorrect, because it includes quotation marks; quotation marks are redundant when the quotation is set as an extract.

GOOD: The demands of a startup can be intense, as illustrated by this statistic: "In Silicon Valley, 97 percent of spouses of presidents reported seeking psychiatric help within 1 year of the company being founded" [92, p. 75].

Note that you should not put a period before the citation with an inline quotation.

When you use double quotation marks in this manner, you should set any double quotation marks in the original material as single quotation marks.

GOOD: The dialog in the story is atrocious. Consider, "Moe said to Mae, 'Gee that's a nice smile.' Mae said to Moe, 'Gee, that's a nice thing to say.' Moe said to Mae 'Gee; well, gee.'" [Awk & Gawk, 1983, p. 27].[140]

When you quote written material of six or more lines, you should set the quotation as an *extract,* and should not use double quotation marks around it. You should not set the extract in italics.[141]

GOOD: Anna Quindlen writes with passion and insight, yet her prose is always deceptively simple and immaculately clean:

> Grief remains one of the few things that has the power to silence us. It is a whisper in the world and a clamor within. More than sex, more than faith, even more than its usher death, grief is unspoken, publicly ignored except for those mo-

---

140. Note that there is a period at the end of the quotation here, but that period is part of Mae's speech; the period ending the sentence that starts with *Consider* occurs after the final bracket.

141. If you have no control over the design and the design mandates italics, then you should use italics.

> ments at the funeral that are over too quickly, or the conversations among cognoscenti, those of us who recognize in one another a kindred chasm deep in the center of who we are. *[The New York Times, 4 May 1994, A17]*[142]

Note that you should put the period *before* the citation within an extract.

When you set a quotation as an extract, you should set any double quotation marks in the original material as double quotation marks, because you do not enclose the extract itself in quotation marks.

Also note that, if the material that you are setting in an extract is a displayed list (whether numbered, bullet, or unnumbered), then you should set the citation before the extract, rather than at the end of it. You should not use quotation marks on the list, of course, just as you should not use quotation marks on any extract.

GOOD: To explore a model in Demos, you can follow these steps *[Demos Tutorial and Reference, 1994, p. 7]*:

1. Double-click on the Demos folder, to open it.
2. Double-click on the models folder, to open it.
3. Double-click on the icon of the model that you wish to explore. This action will start Demos, and will open the model.
4. Explore the model at your leisure.

You should use *points of ellipsis* to indicate any material that is *missing* from the quoted material. Points of ellipsis are the three dots that indicate prosaic lacunae.[143]

---

142. Copyright © 1994 by The New York Times Company. Reprinted by permission.

143. In this segment, we shall not delve into the rules governing correct use of points of ellipsis; these rules are explained in Segment 86.

GOOD: The after-dinner speeches were predictable. Ron Garble stood swaying before us, and pontificated, "My fellow politicians, I am here tonight to say I will not mention that.... Nor will I mention.... I also will not keep you by telling the well-known story—the one that begins...." I slept.

When you quote written material of 50 or more lines (in total, even if you quote only a few lines at a time), you must obtain *permission* from the holder of the copyright. You also must set a permission or credit line as specified by that entity. If no permission or credit line is specified, you can use a format such as the following:

GOOD: <Long quotation—50 lines or more—set as extract.> [*Source:* Really TJ and True QR, *More Tall Stories,* Fish Press, Palo Alto, CA; used with permission.]

Note that the laws governing the circumstances under which you must, or are not required to, obtain permission are complex. For example, certain documents are considered to be public domain; in contrast, you would need permission to quote a 40-line poem. Consult your publisher (or an appropriate lawyer) to determine for what material you should apply for permission.

THE PRINCIPLE FOR LUCID WRITING here is that you should learn the rules associated with quoting written material, because "otherwise at best you will appear unknowledgeable, and at worst you could be sued" [Dupré L. It's safer to cite explicitly. *Cover It Up,* 9(3): 136].

36 Fuzzy Words

Fuzzy words have imprecise meanings and add little to the communication value of your writing. Examples of fuzzy words are *some, thing, very, truly, really, in fact,* and *actually*.

 You should avoid writing *some;* write instead a word that indicates how many you mean.

> UGLY: You need some utilities to make this system useful.
>
> GOOD: The addition of three utilities would make this system marketable.
>
> SPLENDID: The addition of a graphical editor, a text editor, and a formatter would make this system worth buying.[144]

> UGLY: Let's give this project some thought.
>
> GOOD: Let's consider carefully whether we should vacation in Tahiti.

 You should avoid writing *thing;* instead, name the entities under discussion.

> UGLY: There are a few things that might clutter the main logic of the algorithm.

---

144. Specifying not only how many, but also precisely which, items provides the most information.

GOOD: We can eliminate clutter in this algorithm by assuming that the string representing the infix expression contains only arithmetic operators, parentheses, the delimiter #, and operands that each consist of a single character.

UGLY: I discussed several things in the last section.

GOOD: In Section 5.4, I shall discuss how my research relates to work on dynamic interpersonal capitulation, and whether such relationships should be pursued.

 You should avoid writing *very;* either use a magnifier that packs punch, or leave out the magnifier entirely (omitting it often is more emphatic than is leaving it in).

UGLY: You look very impressive.

GOOD: You look exceptionally silly.

SPLENDID: You look remarkably ridiculous.[145]

UGLY: Max was very angry.

GOOD: Max was noticeably furious.

GOOD: Max was howling with rage.

SPLENDID: Max's wrath was volcanic.[146]

The phrase *a lot of* has a golly-gosh-gee-whiz flavor to it, and also is fuzzy. Instead of using it in formal writing, you should substitute more expressive and interesting terms.

---

145. Instead of magnifying a weak term, choose a stronger descriptor.
146. See previous note.

UGLY: You have to use a lot of scoring rules to evaluate a set of probability assessments.

GOOD: This book comprises myriad principles for lucid writing.

UGLY: Max was late because there were a lot of stationary[147] cars on 101 at rush hour.

GOOD: Lyn was delighted because her new stationery portrayed multifarious creatures in the margins.

UGLY: This circuit has a lot of responses, depending on the input signal.

GOOD: Lyn has multitudinous responses, depending on the body-language signal.

You should generally avoid writing *actually, in fact, truly,* and *really* when they fail to add meaning to your sentence. In most cases, you should simply leave them out. If you need a magnifier, use a magnificent one; if you need to emphasize actual versus virtual, then you can use one of these terms.

UGLY: Actually, differential files in fact have advantages, such as that recovery after a program error is fast.

GOOD: Trie hashing has benefits, such as that it preserves order, so sequential accessing is fast.

---

147. Be careful not to confuse *stationary,* which means immobile or unchanging, with *stationery,* which is the medium on which people used to write their correspondence, in the dark ages before electronic mail and facsimile transmission. Thus, history as well as language has demonstrated that stationery is not stationary.

UGLY: In fact, what I really want to know truly is whether you actually told your client that he is a dunderhead.

GOOD: I want to know whether you did tell your boss that she is a wimp.

THE PRINCIPLE FOR LUCID WRITING here is that you should always use the most specific, informative term available, so that you maximize the communication per word, all other considerations being equal. In fact, there really is something actually very annoying about truly fuzzy terms.

37 Parentheses

You will find it useful to know how to use parentheses correctly (to add remarks to your sentences, for example).

You should enclose in parentheses a remark that you want to downplay slightly. To emphasize a remark, you should enclose it in em dashes. (Both parentheses and em dashes are correct; they just have different effects. Note, however, that you can place in a given sentence only one remark set within em dashes; you can set any number of parenthetical remarks, although setting too many of them gives the cadence of a scatterbrained, self-interrupting writer.)

> GOOD: You need a custom-tailored delete rule for the `CURRENT-BOYFRIEND` to trigger creation of an `HISTORIC-BOYFRIEND` occurrence (which will inherit all relationships to old `DATES`).[148]

> GOOD: Holly—who was looking breathtaking in her red-and-black jumpsuit—was in a hurry to get home so that she could call DeeDee. Holly had not spoken to DeeDee—DeeDee was her daughter—in a long time. By the time that Holly had finished her dinner, she and DeeDee had not spoken in at least 2 hours.

---

148. Note this exception to the rule that you should set end punctuation in the same type style as that of the preceding letter. That is, when you are using an alternate typeface to differentiate code from base text, you should set in that typeface *only* code. Here, the closing parenthesis and the following period are set in the base-text face.

GOOD: Misha (who was looking especially handsome in his new all-black suit and black silk tie) was in a hurry to get home so that he could call Holly (his wife), to whom he had not spoken in ages (at least 2 hours).

 You should set in parentheses explanations and examples that are not critical to your sentence.

GOOD: Lyn and Richard used to hang out together (going on hikes, for example, or watching old movies on television); now, they catch each other on the fly (that is, they get together for lunch whenever their schedules do not forbid the meeting too strenuously).

GOOD: How many slices of Spam® can you obtain from the still-solid contents of a single can by making only five straight cuts? (The Spam® must not move while you are slicing it, and each slice must correspond to a plane.)[149]

GOOD: Gelareh is no longer nulliparous (she has had a baby).

 You should not set punctuation marks other than a period before an opening (left) parenthesis. Any punctuation marks that belong in your sentence at the place where you plunk down your parenthetical remark should follow the *right* parenthesis.

BAD: Jim and Lauralee, (who are clearly fun-loving people) rarely have time to go dancing these days.

GOOD: Jim does the shopping (at the market), and Lauralee creates a meal from whatever she finds in the kitchen.

---

149. Portions of Spam® that have been folded, bent, or stapled do not count.

BAD: When Moisés and Tania gave a barbecue—(to cele-brate the World Cup finals) on a sunny Saturday afternoon—Max brought his laptop, because he was worried about his work.

GOOD: Tania admitted (because she is a gracious hostess)—but not where the other guests could hear her—that she too had planned to work while pretending to watch the game.

The exception to this rule occurs when the parentheses are considered to be part of a term.

GOOD: Consider, for example, that (x, y, z) could easily be a term in mathematics.

GOOD: In 1993, (Goldstein et al.) reported that all authors experience episodes of acute, blatant psychosis as their books enter production.[150]

You should set the period outside of the right parenthesis at the end of a sentence if the parentheses occur within a sentence. You should set the period inside if the contents do not occur within a sentence. Note that, if the parentheses occur within a sentence, you must not enclose a period within them; if the parentheses do not occur within a sentence, you must enclose within them one or more complete sentences and a final period.

BAD: When Marina moved away from her house in Point Reyes, she had to find homes for Buttercup, Hershey, and Ruby (two goats and one horse.) (She also had to find a nursing home for Curly, a geriatric sheep who had turned up on her lawn one day).

---

150. I detest this particular citation style; however, if you are writing for a publisher that mandates the style, you should know how to use punctuation marks with the citations.

The first pair of parentheses contains a fragment and an incorrect period; the second pair contains a full sentence and no period.

GOOD: The new home was somewhat cramped for Swix and Spud (both of them are energetic, healthy dogs). (Swix is a talkative alpha-female malamute; she's not exactly a lapdog.)

BAD: One early system that handled uncertain knowledge was MYCIN (MYCIN aided physicians in diagnosis and treatment tasks within a limited medical domain. Another expert system that could handle uncertain knowledge was Prospector, which was designed to support geologists in mining exploration.), a system developed by Shortliffe [1976].

The pair of parentheses contains two sentences, and two periods; because it occurs in the middle of a sentence, it should contain no periods.

BAD: There are several nonprobabilistic approaches to dealing with uncertainty. (For example, modal operators, fuzzy logic, evidential reasoning, and certainty factors.)

The pair of parentheses purports to contain a full sentence, but the words constitute only a fragment.

GOOD: Heckerman gave certainty factors a probabilistic interpretation (Heckerman was at that time a doctoral candidate in Shortliffe's laboratory; he has undertaken various other interesting work projects since then). (Horvitz and Heckerman developed a framework that permitted comparison of several probabilistic techniques with nonprobabilistic approaches.)

You should use paired parentheses to enclose the numbers (or letters)[151] of an intext list. (You should use periods after the numbers [or letters] of a displayed list, and you should never use a single parenthesis.)

> BAD: Judy had 1) a bicycle accident, and 2) a skiing accident (the first one crushed her knee; fortunately, the second one was not so serious, although it did cause her considerable inconvenience).

> GOOD: Steve brought home to Judy (1) chocolate-fudge ice cream, (2) chocolate seven-layer cake, and (3) a bar of dark chocolate. (As you might deduce, Judy is fond of chocolate, and Steve is fond of Judy.)

> BAD: Betsy had (sequentially, rather than simultaneously) two husbands: 1. Byron, and 2. Richard.

> GOOD: Betsy had (sequentially, with a gap of about 5 years) two daughters: (1) Madeline, and (2) Sophia (one per husband).

You should choose one of two styles to use for nested parenthetical remarks, and you should use that style consistently. You can either alternate parentheses and square brackets, always setting the parentheses on the outermost edges of your remark, or you can use all parentheses. If you must unavoidably nest remarks heavily in your text, you may want to use a different delimiter for each level (for example, ([{< >}]) is a standard sequence), because doing so will reduce slightly the difficulty of the horrendous parsing task that your reader faces.

---

151. You might set letters, for example, for the parts of a numbered figure, example, or exercise.

GOOD: Greg (who was born in Jamaica [where he went to high school]) loves to hike in the hills ([partly] because they remind him of his homeland).

GOOD: Nicola (who was born in Canada (which is north of the United States)) loves to hike also (unfortunately, Austin (where she now lives) is too hot for hiking much of the time).

GOOD: Anthony (who is Marina's brother [and who hates {detests, loathes <choose a word>} to be called *Tony]*) lives in New York.

You should avoid using too many parenthetical remarks (especially long ones), and you should avoid nesting parenthetical remarks heavily. Long parenthetical remarks[152] force your reader to backtrack to pick up the sense of the enclosing sentence or paragraph; nesting compounds the problem.

UGLY: Most of the examples in this segment (for a good reason) contain too many (and not particularly logical) parenthetical (that is, set in parentheses) remarks (or asides, or explanations), which can make reading (well, parsing first) difficult, and can make the author (Lyn in this case) appear confused (or weak brained (or at any rate not well versed in the principles of lucid writing)).

UGLY: Pat (who is Marina's (Marina is the woman who owned the goats (Buttercup and Hershey (who once (accidentally) gored Marina)) and sheep (whose name I cannot remember (my memory is not what it used to be)) and horse (Ruby (who was in a book (called *Marina and Ruby)* that Paul (Paul is Marina's father) photographed (he is a world-famous photogra-

---

152. If you wish to add long remarks, use a footnote.

pher)))) (Marina also owned at that time several ducks (Adolph, Bugs, Wiggles, and CalTrans), a chicken (Darth Vader), a couple of geese (You Too (abbreviated U2) and Sandy), and a rooster (Foster (so named because he fell off the back of a Foster Farms® truck))) husband) was relieved when they found a new (perfectly comfortable) home (or cage (but undoubtedly a big, airy, pleasant cage (as cages go))) for the parrots (who were noisy (parrots often are) and jealous (parrots mate for life, so they become attached to their owners (and peck anyone else (if they are feeling insecure)))).

THE PRINCIPLE FOR LUCID WRITING here is that you should use parentheses to enclose asides (or tangential remarks), explanations (for example, of a word), or numbers (or letters) in an intext list.

38 *Split Infinitives*

An INFINITIVE is the *to be* form of a verb:

> GOOD: Examples of infinitives are *to giggle, to moan, to rave, to clobber,* and *to melt.*

A *split infinitive* is one in which text has been inserted between the *to* and the other word:

> BAD: There are people who like to impulsively giggle, to continually moan, to insanely rave, to fiercely clobber, and to romantically melt.

Certain people argue that split infinitives have become so common that there is now nothing wrong with split infinitives. I disagree strongly. A split infinitive constitutes a grammatical error, and rasps horribly on a well-tuned ear. Even if hearing a split infinitive does not yet vex your ear, using one will make you sound illiterate to a portion of your readers. Instead, avoid the error.

> BAD: You should learn to never split infinitives.
> GOOD: You should learn never to split infinitives.

> BAD: To always split infinitives is a hallmark of sloppy writing.
> GOOD: To split infinitives is always a sloppy choice, and thus careful writers maintain their infinitives intact.

BAD: To boldly go where no man has gone before.

GOOD: To go boldly where no one has gone before—and to return in one piece—would be exhilarating.

GOOD: An infinitive phrase—for example, *To boldly go to Hades*—be it split or intact, is not a sentence and should not end with a period.

BAD: To easily find a file block, the system must first bring an inode into memory.

GOOD: To convert easily a logical block number to a physical block number, *Bmap()* interprets the direct and indirect block pointers in an inode.

BAD: To easily delete a range name, enter `/<Range Name> Delete`.

GOOD: To delete all your range names quickly, use the `/<Range Name> Reset` command.

BAD: It is not good policy to incessantly criticize your spouse.

GOOD: It is good policy to encourage frequently your spouse.

SPLENDID: We encourage you to increase your degree of ecstasy gradually.[153]

---

153. Often, not only taking a word out from inside an infinitive, but also banishing it to the end of a clause or sentence, is the most rhythmic solution.

SPLENDID: I think that it is a bad policy to criticize your partner incessantly.[154]

SPLENDID: You will find it self-defeating to criticize your partner incessantly.[155]

THE PRINCIPLE FOR LUCID WRITING here is that there is never a good excuse for splitting an infinitive: Remember that to even occasionally split infinitives is a sloppy habit.

Copyright © Lyn Dupré

154. When you are faced with a sentence that makes a negative statement about positive entities or qualities, the strongest solution usually is to recast it as a positive statement about negative entities or qualities.

155. See previous note.

39 *Is Due To*

THE PHRASE *is due to* is subject to considerable disagreement. There are three possible uses for it; you should determine which are acceptable to you.

 You should certainly use *is due to* to indicate *just rewards*.

> GOOD: My thanks are due to Max, for tolerating with good humor the pride of felines in his bed.
>
> GOOD: Three dollars in change is due to me.

I find it acceptable (but there are people who do not so find it) to use *is due to* to mean *is caused by*. Alternative phrases are preferable, however.

> GOOD: The grapefruit juice on the window pane is due to a small breakfast altercation.
>
> GOOD: The stain is due to a chemical reaction.
>
> GOOD: The fur on the counterpane is due to BB.
>
> GOOD: That nagging sensation in your stomach is due to guilt about eating too much.

> SPLENDID: Max's headache is caused by overwork.
>
> SPLENDID: Lyn's flush is a response to excitement.
>
> SPLENDID: Jealousy is one of the sequelae of lack of trust.

SPLENDID: The display that you see is a result of the electron-beam sweep.

I do not find acceptable (but there are people who do so find it) use of *is due to* to mean *was* written, developed, invented, published, coined, or otherwise *originated by*.

UGLY: The term *coroutine* is due to M.E. Conway.

UGLY: The first tracing routine is due to Stanley Gill.

UGLY: TEX is due to Knuth.

THE PRINCIPLE FOR LUCID WRITING here is that you should use *is due to* to speak of reparation and perhaps to speak of causes, but should not use it to speak of origination. That much care is due to your reader.

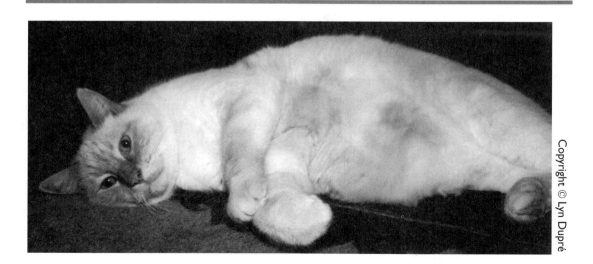

40 Center On

THE PHRASE *center on*[156] is overused and abused so often in formal writing that it is ragged and tatty. In addition, the meaning of *center on* is not particularly clear. Substitute a term that indicates more precisely what you wish to convey.

UGLY: The staff meeting centered on who should clean up the kitchen.

GOOD: The group therapy emphasized insight into custom-tailored personal self-empowerment and inner-idiot actualization.

UGLY: The tutorial centers on `QuickSort`.

GOOD: The book teaches you to avoid phrases such as *centers on*.

UGLY: We shall center on winning in this management class.

GOOD: We shall highlight breathing in this yoga class.

---

156. *Center on* is first cousin to *focus on,* and suffers from the same overuse syndrome.

UGLY: This chapter centers on the voltage output of the transconductance amplifier.

GOOD: In this chapter, we shall explore time-varying signals, and shall describe the resistor-capacitor circuit.

UGLY: During an uncomfortable period following a flight to the 'burbs, Nick's life threatened to center on the huge mortgages for two dwellings, eclipsing the purpose of the exercise.

GOOD: The equity problem finally solved, Nick's attention shifted happily to the joys of a gentle home and the challenge of complex multimedia production for a large pedal outfitter.

UGLY: Max's conversation tends to center on personnel problems, cash-flow problems, contract problems, and sleep deficiency.

GOOD: Lyn's conversation tends to stress wood carving, cat petting, and person nurturing.

THE PRINCIPLE FOR LUCID WRITING here is that you should use more interesting—and perhaps more meaningful—phrases in place of ~~centering on~~ tired, overused terms.

41 Quotation Marks

USING quotation marks (" ") for many purposes for which they were not intended is a common—and nasty—habit. Quotation marks have an important job to do, and you should not ask them to serve in other roles as well.

You should use quotation marks when you are actually *quoting a source*.

> BAD: Dupré [1994, page 77] wrote scathingly, This string of foolishness is a poor excuse for a sentence.

> GOOD: According to Spingholterusfoo, "Energy that is directed inward will eventually stultify the soul" [45, page 284].

> GOOD: *The New Yorker Times* reported "a calamity of severe proportions" [*NYer Times*, 8 May 1967, page F.1].

 You should use quotation marks when you are writing *dialog*.

> BAD: Lyn sobbed on the telephone, My computer screen has turned red and the system is flashing alarm messages at me!

> GOOD: "I don't have time," Max squeaked. "Can't you handle the disaster relief yourself?"

> GOOD: When Lyn walked in and saw what Red had done in her absence, she could only clutch her heart and gasp, "Whaaaa, whaaa, whaa?"

You should use *single* quotation marks only when they are nested inside of double quotation marks. (European style uses single quotations marks for numerous purposes, but in the United States you should reserve use of these punctuation marks for nesting.)

BAD: Max read from the paper, "The police spokesman[157] said, "this could be an inside job.""

GOOD: Lyn asked Max if he would mind her quoting him as follows: "Max read to Lyn, 'the sprain and strain is mainly on the wane.'"

You should use quotation marks to indicate *irony*.[158]

GOOD: Max and Lyn were careful not to step in the "gifts" that the neighbor's dog left in the driveway every morning.

GOOD: The president's "honesty" consisted of a series of contradictory press releases.

GOOD: Sondra needed to decide whether to tell Mary that Burt was busily being "faithful" in red-light districts all over the country.

You can use quotation marks to refer to a word or term *itself*, rather than to its denotation. It is also correct, and is usually more graceful and pleasing to the eye, to use italics in this situation.

GOOD: "Pollywog" starts with a "p."

GOOD: "Data" is a plural.

SPLENDID: *Pillowtalk* has 10 letters.

SPLENDID: *Data* is the name of a robot, or so my data tell me.

---

157. Unfortunately, many newspapers are not yet sensitive about using gender-specific terms.

158. When you use irony, you write the *opposite* of what you mean.

You can use quotation marks when you are referring to a *string*. You can also use them to denote the names of *keys* on a keyboard. Again, there are often more visually pleasing ways to handle the former case, such as by using an alternate typeface, and the latter case, such as by simply capitalizing the initial letter.

> GOOD: Enter the string "grievous_bodily_harm";[159] then, press "return."

> SPLENDID: Enter the string `dangerous_bodily_charm`; then, press Return.

You should *not* use quotation marks to indicate that you think that your use of a word is humorous (unless the humor is irony) or that you are using a word in an odd way, or that you are not sure that you have chosen the correct word, or just for no good reason at all.

> UGLY: As a "gold standard," we chose the measurements made by expert lice detectors.

> GOOD: Our gold standard was home-made, chunky rocky-road ice cream.

> UGLY: The "satisficing" conditions are as delineated in the next paragraph.

> GOOD: Lyn decided that minivacations—those lasting for several hours, rather than for several weeks—would have to be satisficing[160] until the company had grown.

---

159. Note that, if you use quotation marks to delimit a string, you should place the end punctuation after the closing quotation mark; otherwise, your reader cannot tell whether the punctuation mark is part of the string. Use of an alternate typeface avoids the problem entirely and is thus preferable.

160. Yes, *satisficing* is a word, albeit a technical one. Satisficing solutions are those that simply meet the constraints, in contrast to optimizing solutions.

UGLY: The computer "decides" which course to take.

GOOD: Lyn's machine finally started to cooperate, after arguing with her all morning.

You should *not* use quotation marks simply to call attention to a word. This bad habit is practiced by many sign writers in the United States, with hysterical results.

BAD: "Fresh" fish available here!

BAD: One-day "sale"—many "bargains"!

BAD: "Genuine" pearls at low prices!

BAD: "EAT" here!

You should determine what rules you will follow regarding the order of punctuation marks used with quotation marks. The standard U.S. formal-writing usage, adhered to by nearly every publisher, is as follows:

1. Place commas and periods inside quotation marks, no matter what the context.

   GOOD: "Lyn," said Geoff, "I'm absolutely certain that I'll be giving you another chapter of my dissertation by the end of next week."

2. Place colons and semicolons outside quotation marks, no matter what the context.

   GOOD: Maria said "how charming"; we weren't sure what she meant by that.

   GOOD: When Daphne fell ill most unexpectedly and inconveniently, Geoff prescribed "good stuff": antibiotics and an inhaler.

3. Place exclamation marks and question marks inside or outside of quotation marks, depending on the context:

GOOD: Maria asked, "Is that all there is to eat?"

GOOD: Did Maria say "I had enough to eat"?

GOOD: Geoff said, "I am so full I could burst a gut!"

GOOD: I could not believe that Geoff said "burst a gut"!

There are, however, good reasons to reject rule 1. British usage dictates that commas and periods should be placed according to context. Because that style provides more information and disambiguates your writing, I find it preferable; I do not use it in my work on books in the United States, however, unless I have the explicit approval of the publisher.[161] British style also avoids problems if you are using quotation marks to delineate strings, where inclusion of a period in a string, for example, might be fatal. (Philosophers also tend to use British style, regardless of their location.)

On the other side of the argument, placing the punctuation after the closing quotation mark does leave a nasty hole in the baseline of the type.

BETTER?: I will not shout "I am a walrus", unless you first clear the street.

BETTER?: Type the string "overwhelmed".

If you decide to adhere to British usage for rule 1, then you must be prepared to argue with your publisher and copy editor. I sincerely wish you luck.

---

161. Whether to place the period or comma inside or outside of the quotation mark is a matter about which people have marvelously vehement opinions. In 25 years as a professional editor, I have failed to detect any preference across the general population, or even across the population of computer people. Each person, however, is mightily assured that her view is the only correct one.

Somewhat tangentially, you should *obtain permission* to quote any text of which you quote, *in total, more than* 50 *lines* within your manuscript. Of course, what constitutes 50 lines will depend on the type size and column width. Furthermore, there are situations in which quoting fewer than 50 lines is not legal without permission, such as if you quote an entire document or a poem. You should consult your publisher or a lawyer if you are in doubt about whether you should obtain permission. If you quote a piece extensively, play it safe and be polite: Get permission. The permission usually will specify a credit line; if none is specified, be sure to provide one.

If you are doing your own typesetting for a document, you should use the curly or slanted quotation marks that allow you to distinguish left (opening) from right (closing) quotation marks; the two are slanted in opposite directions.

> UGLY: Max said, "I don't have time to sleep."
> GOOD: Lyn said, "I don't have time to worry."

> UGLY: Lyn sobbed, "We forgot to turn on the smart quotes!"
> GOOD: Jan comforted, "That's a simple matter to fix."

THE PRINCIPLE FOR LUCID WRITING here is that you should avoid using quotation marks when you have no good reason to use them, and should learn the rules governing correct usage. If you cannot see a problem with "quotation marks," reread this segment.

42 Remarks Inserted After That

IT IS CORRECT to place a comma after *that* when a separate clause follows. In essence, you are *inserting a parenthetical remark* that you could remove without ruining your sentence, grammatically speaking. So, you should delimit the remark by placing commas on either side of it. You should not place a comma after the *that* if what follows is not a separate remark.

BAD: Max feared that if he did not relax he would die.

GOOD: Max thought that, if he was alone, he would relax.

GOOD: Max thought that Lyn was not precisely relaxing.

BAD: Joe decided that because going to medical school was not sufficiently challenging, he would also pick up a doctorate in medical computer science during his sojourn at the university.

Here, because going to medical school was not sufficiently challenging *is a separate remark and should be preceded by a comma.*

GOOD: Joe decided that, because he was bored that day, he would read a manual and then replace his friend's car's muffler, even though he had never worked on a car before.

Here, because he was bored that day *is a separate remark and is correctly preceded by a comma.*

GOOD: Joe decided that going to medical school and graduate school, not to mention crawling around under cars, was still insufficiently exciting, so he shaved off all his hair as well.

Here, going to medical school and graduate school *is not a separate remark; the lack of a comma preceding it is therefore correct.*

BAD: C is a language that after its development in the early 1970s by Brian Kernighan and Dennis Ritchie at Bell Laboratories, became popular because of its integration with Unix.

GOOD: C was derived from B, a language that, in turn, was derived from BCPL.

GOOD: In 1971, Niklaus Wirth introduced a language that was named after the seventeenth-century French inventor Blaise Pascal.

BAD: Misha was convinced that given time, he could learn to love Portland's intellectual community and the constant rain.

GOOD: Holly hoped that, given time, Misha would learn to appreciate the excruciatingly cute pygmy goats that she was keeping in the backyard of their mansion, where they perched on the picnic table.

GOOD: DeeDee hoped that she would visit Holly soon, and that the goats would not eat her boots.

BAD: Lois decided that partly because she could get a great job on the Peninsula, she would move in with Soren.

GOOD: Soren was delighted that, contrary to his expectations, cohabitation suited him just fine.

GOOD: Soren and Lois decided that the only problem was where to park all their various Porsches.

BAD: Max hoped that despite the promises he had made, Lyn would forget that he was supposed to take a day[162] off that year.

GOOD: Lyn was ecstatic that, despite his many high-pressure commitments, Max kept his promise and spent the day moseying around on the coast with her.

SPLENDID: Max and Lyn agreed that, in every way, the day had been one of the happiest that they had spent together.

THE PRINCIPLE FOR LUCID WRITING here is that, when you insert a separate remark after a *that,* you should delimit the remark by placing commas on both sides of it.[163]

---

162. If the example occurred in the context of formal writing, the correct phrase would be *take off 1 day that year;* in this sentence, however, the casual phrasing is more appropriate.

163. Based on a suggestion by Ronald Barry.

43 *Figure Captions*

A FIGURE CAPTION is the text that accompanies a figure: It includes the figure number and title (if you use numbers and titles), and any further explanation of the figure contents. Complete and informative figure captions are critical to intelligent use of graphical information in your text. In addition, with the figure caption, you should set legends and credit lines.

You should *use tags* in *all* or *no* captions. A tag is a sentence fragment that *names* an item; in this case, a tag names a figure.[164] The tag can be (and usually should be) followed by full sentences. Examples of captions that begin with tags follow:

> GOOD: Red. An unusual cat on a variety of measures, Red weighs about twice as much as does the average male Siamese cat.[165]
>
> *Here,* Red *is a tag (whereas Red is a cat).*

> GOOD: Richard making a cynical aside heard by no one. An unusual man on a variety of measures, Richard possesses about twice the wits of the average male scientist.
>
> *Here,* Richard making a cynical aside heard by no one *is a tag.*

---

164. You can also use tags, for example, to introduce the entries in a displayed list.

165. Red weighs 21 pounds.

GOOD: Graphs showing how arbitrary waveforms can be built up from pulses. The response of the system to a waveform also is built up, based on responses to individual pulses.

Here, graphs showing how arbitrary waveforms can be built up from pulses *is a tag.*

GOOD: A cat on a hot tin roof. Note the energetic leaping and splayed feet; the cat uses these techniques to protect itself from further irritation.

Here, a cat on a hot tin roof *is a tag.*

A caption that has no tag begins with a full sentence (which may be followed by other full sentences):

GOOD: BB, who is an unusual cat on a variety of measures, weighs about one-half as much as does the average female Siamese cat.[166]

GOOD: We input the arbitrary waveform in the figure to a linear, time-invariant system. We approximate the input waveform as a sum of narrow pulses, each weighted by the value of the function at time t.

GOOD: A cat faced with a hot thin woof uses energetic leaping and splayed feet as techniques to protect itself from being eaten.

 You should provide *sufficient explanation* that the figure plus the caption are[167] self-explanatory and stand alone. You should not merely name your figures with a tag line. Often, readers browse through the figures, reading the captions, before they read the asso-

---

166. BB weighs 7 pounds.

167. A reviewer queried the form of the verb here, pointing out that 2 plus 3 is 5. The difference is that 2 plus 3 is [equal to] 5, whereas apples plus pears are the same as apples and pears.

ciated text—sometimes, they decide whether to read the text based on whether the figures catch their interest. So, do not give long-winded explanations of a figure in the text, and give inadequate information in the caption.

UGLY: A map of Silicon Valley.

GOOD: A map of Silicon Valley showing the failure rate of companies started between 1982 and 1992. Successful companies are shown as black dots; companies that went under are shown as red dots. The map may demonstrate a causal factor in the increased sales of antacid tablets in the area during this period.

UGLY: Lyn's office.

GOOD: Lyn's office. The apparent chaos is deceptive. The brightly colored slips of paper stuck on every conceivable surface are all notes about ideas for segments in the book. The business cards clipped to the lampshade contain telephone numbers for service providers who can help with production. The pillow at the corner of the desk, and the one on top of the computer monitor, are for her editorial assistants.

You should provide captions for all *figure parts*, and should label the corresponding parts in the figure. A figure part is a subfigure that is labeled **a, b, c,** and so on. You should never write one caption for two separate figures (figures that have different numbers).

BAD: Figure 2.1   A cat. Figure 2.2   A dog. Note the structural differences between these two creatures.

BAD: Figure 7 Max awakening to soft shafts of sunlight. Figure 8 Max awake, worrying about the day ahead.

Rather, you should label the two figures as **a** and **b** parts within a single numbered figure. You can use a period or a colon after the letter that denotes the part.

GOOD: Figure 2.1   Contrast in the structure of cats and dogs. **a:** A cat. **b:** A dog.

GOOD: Figure 2.2   Contrast in the body language of cats and dogs. **a.** Angry cat lashing its tail. **b.** Happy dog wagging its tail.

GOOD: Figure 7   Max's moods. **a.** At sunrise. **b.** At sunset.

For these subcaptions as well, you should consistently use or not use tags.

GOOD: Timeline and tasks for the proposed study. **a:** Interviews. We will interview four gorillas from each of six zoos to obtain preferences for banana types. The duration of this task is 6 months. **b:** Statistical analysis. We will run statistical comparisons for the different groups to detect significant trends in the data. The duration of this task is 6 months. **c:** Published report. We will write a report of our work, and will submit it to the appropriate peer-review journals. The duration of this task is 4 years.

Tags are used in the subcaptions.

GOOD: Timeline and tasks for the proposed study. **a:** We will interview four giraffes from each of six zoos to obtain preferences for necklace types. The duration of this task is 6 months. **b:** We will run statistical comparisons for the different groups to detect significant trends in the data. The duration of this task is 6 months. **c:** We will write a report of our work. The duration of this task is 4 days.

No tags are used in the subcaptions.

You should include at the end of your caption a *legend* for all symbols and abbreviations. (You can set legends for colors, dotted versus solid lines, and so on in the figure itself *or* in the caption.) Do not put explanations (such as of abbreviations) of labels (text within the figure) in the figure itself. You can put explanations of symbols in the figure in a legend box, or you can put them in the legend that goes with the caption. You can also put legend information in parentheses within the body of your caption.

> GOOD: Family tree showing current research interests within our laboratory. AI: artificial intelligence. DA: decision analysis. DS: decision support. ML: machine learning. NN: neural networks.

> GOOD: Graph of creatures brought home by Red for tea with cream and hot buttered crumpets (with lots of jam), 1993–1994.[168] Legend: ◊ rats; Δ bats; ♥ cats.

> GOOD: Bar-chart comparison of domestic violence in urban (outlined white bars), suburban (cross-hatched bars), and rural (black bars) settings, 1950 through 1990.

You should include a *credit line* for any material that you have copied or adapted from another source, or for any data on which you have based your figure. You must obtain permission to reproduce artwork, whether or not you adapt that artwork, from the holders of the copyright. In almost all cases, the publisher holds the copyright, and perhaps shares it with the author—a situation that surprises many authors. When you obtain written permission, the publisher (or other copyright holder) usually will specify how to set the credit line, and you should copy that line exactly. Here are examples of common credit-line formats:

---

168. Note that it is correct in figure captions, legends, and labels to indicate ranges by an en dash; in text, you should spell out the range in words (for example, *from 1993 to 1994,* or *between 1992 and 1995).*

GOOD: *Source:* From I. Sly and U. Wiley. Why you need our method. *Journal of Irresponsible Professionals,* 2(7): 43–45, 1994.

GOOD: *Source:* Based on visualization exercises suggested in Iversen, N. *Record Makers and Back Breakers.* Lost Youth Publications, New York, 1972, Chapter 69.

GOOD: *Source:* Adapted from Derek D. *An Analysis of all the Humors of the Body.* Meridian Press, Newton, CA, 1993, Figure 3.7, page 98. Used with permission.

GOOD: *Source:* Photograph courtesy of Paul Fusco/Magnum Photos, copyright © 1967.

If no credit line is specified, you can use any of the preceding formats. Or, you can style your credit line to suit the situation.

GOOD: *Source:* Based on data collected by Winnie Wilson and Wally Wow for a class project for CS 103, spring quarter, Eager Institute of Technology, Down South, CA; used with permission.

GOOD: *Source:* Madeline Adamo-Coons created these works (colored wax on construction paper) while she was waiting for Betsy to finish diapering Sophia, so that they could go bicycle riding in the park with Richard.

THE PRINCIPLE FOR LUCID WRITING here is that you should write figure captions that provide explanations, legends, and credit lines, and should set them consistently. Most important, make sure that each figure caption provides sufficient information that the figure plus the caption stand alone.

44 Data

PROBABLY THE most commonly misused word in science is *data*. *Data* is a plural; the singular is *datum*. Certain people think that, because the mistake of using *data* as a singular is so common, propagating it is now correct. I do not agree. The use of *data* as a singular grates on my ears as illiterate foolishness, and will undoubtedly strike at least a portion of your readers that way. Correct usage will offend no one, and will allow you to distinguish a single datum.

You should remember that *data* is the plural of *datum*.

> BAD: The scientist collected data, and put it in his hat.

> GOOD: The engineer collected data, and put them in her secret database.

> GOOD: One little datum was misread, and the satellite crashed.

> BAD: The data is suspect, because Lyn had had no sleep when she ran the experiment.

> GOOD: The data are accurate, because Lyn checked them umpteen times.[169]

> GOOD: "Where did you put that datum?" Max demanded of Red, who was looking contentedly thoughtful.

---

169. The assertion is manifestly idiotic.

You should write *volumes* or *numbers of data*, rather than *amounts of data*. You cannot have a large amount of objects; you can have only a large amount of (a form of) stuff. You can also use other phrases to indicate no dearth of data.

BAD: This useless spreadsheet program cannot deal with a large amount of data.

GOOD: You can run the program on any number of data.

GOOD: If you have large volumes of data, you may need to purchase extra memory.

SPLENDID: We have many data[170] to back up our wild assertions.

SPLENDID: Max's data were voluminous.[171]

You should write *fewer data,* rather than *less data.* You cannot have less objects; you can have only less stuff.

BAD: Could you please collect less data next time?

GOOD: The fewer the data, the less reliable the results.

THE PRINCIPLE FOR LUCID WRITING here is that you should disregard the rampant misuse of *data,* and should use *data* only to denote numerous discrete measurements (or other collections of values). When you wish to refer to a single item, use *datum. Data* is a plural, so data never *is* anything.

---

170. *Many data* is less cumbersome than is *large volumes (or numbers) of data.*

171. *Voluminous data* is both less cumbersome and less prosaic than are the previous alternatives.

45 Ensure, Assure, Insure

THE TERMS *ensure, assure,* and *insure* have notably different denotations. To avoid embarrassment, you should distinguish among them.

 You should use *ensure* to mean to make sure of a state of affairs, or to guarantee that an event occurs.

> BAD: Jim was careful to insure that the project went smoothly.
>
> GOOD: Carol was careful to ensure that the party went smoothly.

> BAD: Using this software will assure that you pay your taxes on time.
>
> GOOD: Using this software will ensure that you pay your rent on time.

You should use *insure* to mean take out insurance on.

> BAD: I need to ensure my car before I drive it.
>
> GOOD: I need to insure my car before I drive it.

BAD: Max was disappointed to discover that assuring his hard-disk had not protected the company against an earthquake-induced crash.

GOOD: Max was delighted to find that he had insured his laptop computer before it was stolen.

You should use *assure* to mean give assurance, or reassure.

BAD: I need to assure my car before I drive it.[172]

GOOD: I need to assure my passengers before I drive my car.

BAD: Lyn ensured Max that it was important that she know precisely when he would be home, so that the dinner party would not turn into a debacle.

GOOD: Max assured Lyn that he would be home in time to greet the guests.

GOOD: Max insured Lyn, so that he would be rich (in one respect) if she died.

GOOD: Lyn ensured that Max would not die young.

BAD: To ensure yourself that all is well, turn on the lights and check under the bed.

GOOD: To assure yourself that the proof is correct, you should take the time to work through it.

GOOD: To insure yourself, call a reliable company such as Lloyd's of London.

---

172. We can imagine an anthropomorphic world in which such behavior would be required. However, the following example presents a scenario that appears much more probable.

GOOD: To ensure that you are yourself, you might want to visit a psychologist.

SPLENDID: Rest assured that I have insured your home to ensure your peace of mind.

SPLENDID: Before I can insure your business, you must assure me that you can ensure that your programmers do careful quality assurance.

THE PRINCIPLE FOR LUCID WRITING here is that *ensure, assure,* and *insure* have three distinct and substantially different meanings; you should use each word correctly to ensure that your readers are assured that you know what you are doing, so that they feel no need to insure themselves against damage from bad prose.

46 *Foreword Versus Forward*

FOREWORD AND FORWARD have vastly different meanings. A *foreword* is a chunk of writing (often written by someone other than the document's author) that precedes the main text; its opposite is an *afterword*. *Forward* is a direction or a personality trait; its opposite is *backward*.

BAD: Certain people think that Lyn is a bit foreword.

BAD: You can turn a good marketing trick by asking a well-known expert to write the forward to your book.

GOOD: I have humbly asked an Excessively Important Person whether he would deign to write the book's foreword.

GOOD: When I get stuck in a rut, I just push forward.

BAD: Lyn was worrying again about the forward to her book.

BAD: Lyn was also worried because she could not get the cursor to move foreword.

GOOD: Lyn called Peter a dozen times that day to pester him about the foreword.

GOOD: "Oy, Lyn!" exclaimed Peter, "Quit worrying about it, and move forward with your writing instead!"

SPLENDID: Max told Lyn that asking that person to write the foreword was much too forward.

SPLENDID: The four-word foreword—"this author is weird"— made Peter look forward to reading the text.

THE PRINCIPLE FOR LUCID WRITING here is that you should find it astounding that many books open with a section entitled *forward*.

47 Blocks: Theorems, Proofs, Lemmas

Wʜᴇɴ your document contains *theorems, proofs, lemmas, hypotheses, definitions,* or other blocks of text that you use repeatedly for similar situations and to which you may refer back, you will need to choose and follow a consistent style for setting the blocks.

In general, you should number all blocks of text to which you may want to refer later. Use single (1, 2, 3) or double (1.1, 2.4, 3.7) numbers, depending on whether your level 1 heads use single or double numbers. You do not have to call out (refer to in text) these blocks of text, but you will have the option of referring to them precisely if you wish.

> Gᴏᴏᴅ: See Definition 1, where we described a particular threat to computer security, a *worm*.

> Gᴏᴏᴅ: Using Theorem 4.7, we obtain the following can of worms.

You should use a format that makes clear where the block ends. The most common device is to set an *em box* (□) at the end. You do not need to set the em box if the entire block is enclosed in a box, is screened (shaded), is set in a different typeface, or is otherwise distinguished from the base text. It never hurts, however, to set the em box; keeping your reader oriented is always a good idea.

GOOD: Definition 5.3. *Quadra.*

(1) A type of Macintosh computer. (2) An island off the coast of British Columbia.

You should set as the first line of your block the term to be defined, or the hypothesis, lemma, theorem, corollary, or whatever is appropriate. It is also a good idea to use typographical distinction for this line; for example, you can set it in italic or boldface type, and can follow it with a line break. I recommend strongly that you use only a lead capital letter on the line that describes the block, rather than setting combined capital and lowercase letters: Extra capital letters merely distract your reader, without adding information.

GOOD: *Hypothesis 6.7: The distribution for cat sizes is log normal.*

GOOD: **Definition 7: Arithmetic logic unit (ALU).**

If you are setting a theorem and a proof, be sure to indicate clearly where the proof begins. The simplest way to do so is to set the word *proof.* You can set expansions with theorems; set corollaries in their own blocks.

GOOD: **Theorem 1.1: Lyn is in love.**

Proof: Lyn thinks that every sunrise is an event to celebrate, Lyn sings while she tries to vacuum the house with a machine that merely redistributes the dust in an even patina, and Lyn overseasons her cooking (which behavior, her mother always told her, was pathognomonic of the disorder).

You should then choose a design for your blocks. If you are writing a journal or magazine article, or a book or book chapter, your designer will take care of the details of how your blocks are set, so all you need to do is to make sure that the different parts—title, body, proof, and so on—are clearly distinguished. If you are writing a business report, proposal, or other document that will not go through an external publisher, or are otherwise responsible for your own page makeup, you will need to design the blocks yourself. The most important advice that I can give you is this: *Keep it simple and consistent.* Do not draw your reader's attention away from the content by distracting your reader with unnecessary flourishes.

Theorem 1: *In a finite partition, the probabilities must sum to 1.*

In particular, for two complementary events A and $\tilde{A}$ (a partition with $n = 2$), it turns out that $\mathbf{P}(A) + \mathbf{P}(\tilde{A}) = 1$; that is, $\mathbf{P}(\tilde{A}) = 1 - \mathbf{P}(A) = {\sim}\mathbf{P}(A)$. Put another way, if $\mathbf{P}(A) = p$, then $\mathbf{P}(\tilde{A}) = \tilde{p}$. $\square$

Theorem 2: Let f and g be two algebraic curves of degree m and n, respectively. If f and g intersect in more than mn points, then they have a common component. $\square$

Theorem 3: (Fermat's theorem, 1640). If p is a prime number, then a^p a (modulo p).

Proof: If a is a multiple of p, then a^p o a (modulo p), so we need to consider only $a \bmod p \neq 0$. Since p is a prime number, a is relatively prime to p. Consider....

...and this equivalence proves the theorem. $\square$

Definition 1: Hamster. Any of various Old World rodents (*Cricetus* or a related genus) having exceedingly large cheek pouches. $\square$

THE PRINCIPLES FOR LUCID WRITING here are that you should set blocks such as theorems and definitions consistently, should distinguish the block's various elements, and should set off blocks from the base text. □

48 Above and Below

USING *above* and *below* when you mean *preceding* and *following* is both misleading and stilted. Many textbook publishers forbid this use of the terms. In addition, you should always tell your reader, as precisely as possible, where to locate items.

 You should use words such as *previous*, *preceding*, and *earlier*, rather than *above*. Often, the chunk to which you refer will not be set *above*—it may be set, for example, on the previous page.

> UGLY: The above argument proves that Carver's distaste for croutons in his salad is a symptom of a deep-rooted childhood trauma.
>
> GOOD: The preceding discussion indicates that Carver's frequent shifts in research emphasis reflect his brilliance and enthusiasm.
>
> GOOD: We described earlier how Carver could find his way in the woods, a skill that he learned as a child.

You should use words such as *following*, *next*, and *here*, rather than *below*. The chunk may well be at the top of the next page, and may not be *below* anything at all.

> UGLY: The complicated, fancy-looking algorithm given below is useless nonsense.
>
> GOOD: The following code will crash your coworkers' machines without giving you away.

GOOD: We provide the secret password here:

> Red fur at sunset makes BB fly
>
> BB greets dawn in roof catapult

GOOD: The correlation between anger and heart attacks in this population is given next:

Frequency of Rage (episodes per week)	*Mean Age at Sudden Death (in years)*
1	67
2	53
3	49
4	42

Whenever possible, you should give *explicit cross-references,* rather than simply indicating *preceding* or *following.* When a number is available, use it. Otherwise, pinpoint the location as precisely as you can, so that you help your reader to find the discussion, equation, example, or mongoose to which you refer.

UGLY: We shall describe Greg's struggle with his grant proposal later.

GOOD: We shall discuss Nicola's trial by fire—otherwise known as her first year of university teaching—in Section 4.3.

GOOD: We shall examine the possible etiology of Greg's weak ankle in a moment. First, we look at common aches, pains, sprains, and other sports injuries. Then, we look at the weak-knees syndrome.

GOOD: We shall analyze Nicola's desire for a garden later in this section.

UGLY: This equation is easier to solve than was the earlier one.

GOOD: The preceding equation is simpler than was Eq. 4.5.

GOOD: The following equation is more elegant than was the one that we discussed in Section 4.5.8.

GOOD: Equation 5.3 is similar to the one used to generate the graph in Figure 7.2.

THE PRINCIPLES FOR LUCID WRITING here are that you should avoid using *above* and *below* when you mean *preceding* and *following*, and that you should use specific cross-references whenever possible. The preceding principle is respected in both the foregoing and the later segments of this book.

49 *En Dashes*

EN DASHES are the correct dashes to use to indicate ranges and to denote pairs of equal weight that form an adjective.

An en dash is longer than a hyphen and shorter than an em dash; in certain typefaces, it is the width of the letter n. Learn how to create an en dash in your word processor or typesetting system. On many systems, the en dash and minus sign are the same.[173]

- hyphen
– en dash
— em dash

You should use an en dash when you use a dash to indicate a *range*.

BAD: Read pages 4-12.

GOOD: Read Chapters 4–12.

SPLENDID: Skip Exercises 4 through 12.

BAD: The monkeys weighed 40-70 pounds each.

GOOD: The wombats weighed 40–70 pounds each.

SPLENDID: The children weighed 40 to 70 pounds each.

---

173. Be careful that you know how to set a minus sign. The minus sign on the numeric keypad, for example, may not be a minus sign; it may be a hyphen.

Note that you should not use the en dash to indicate ranges in text; there, you should spell out the range, as in the third of the preceding examples. You should use the en dash to indicate ranges in tables and their titles, in figures and their captions, and in page ranges within reference citations.

 You should use the en dash to indicate an *equal-weighted pair* that is currently serving as an adjective.

> BAD: Did you check his acid-base balance?
>
> *The hyphen should be an en dash.*

> BAD: The cat person balance in the DePasquale Oxhandler household was incorrect.
>
> *The spaces in* cat person *and* DePasquale Oxhandler *should be en dashes.*

> GOOD: Have you heard of the work–play balance?

> BAD: Interrupts are posted by the controller as each input-output operation completes.
>
> *The hyphen should be an en dash.*

> BAD: To initiate a DMA input output operation, a device driver firsts sets up all the device-specific parameters for a transfer.

> GOOD: Certain input–output devices require exclusive use of the bus when they are doing data transfers.

> BAD: If you travel the New York-San Francisco route, you can take a direct flight.
>
> *The hyphen should be an en dash.*

BAD: If you travel the New York Miami Beach route, you can take a direct dive.[174]

GOOD: If you travel the New York–Los Angeles route, you can take a direct hit.

BAD: The read write head performs read/write operations, or read-write functions.

The example demonstrates three incorrect ways to set read–write: *with a space, with a solidus, and with a hyphen.*

GOOD: Lyn has the suspicion that the object between her shoulders has evolved into a read–write head.

BAD: The Dupré-Henrion algorithm is very[175]complex.

BAD: The Dupré Henrion equation is abnormally complex.

GOOD: The Dupré–Henrion relationship is remarkably, if not alarmingly, complex.

GOOD: Before we launch this project, let's commission a few detailed cost–benefit analyses.

GOOD: The number of shares traded and the price per share are represented by the trade–price node in the diagram.

GOOD: Lyn painted her toenails a deep brown–red shade.

---

174. You can get off the airplane and into the water, that is. Diving while still in the airplane is not recommended.

175. Avoid fluffy words; use precise terms.

Note that the term *equal-weighted pair* does not itself contain an equal-weighted pair; that is, the pair is not both equal and weighted. Rather, *equal* and *weighted* together make a single compound adjective that modifies *pair*. Thus, you set *equal-weighted pair* with a hyphen. In contrast, a *brown–red shade* is a shade that is both brown and red, so that term is set with an en dash. Similarly, the node represents both trade and price, the analysis takes into account both cost and benefit, the algorithm was developed by both Dupré and Henrion, the route is between both New York and San Francisco, and the electrolyte balance is affected by both acid and base.

THE PRINCIPLE FOR LUCID WRITING here is that you should use en dashes when you use the dash form of indicating a range (see pages 2–4), and when you have an equal-weighted pair serving as an adjective, such as a love–hate relationship.

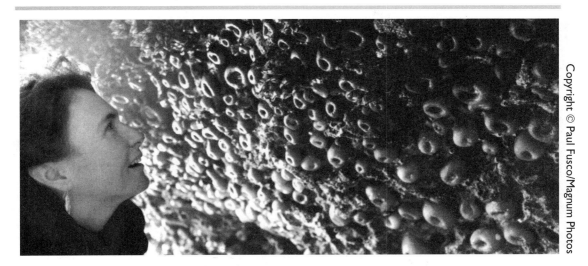

50 *As to Whether*

THE TERM *as to whether* is cumbersome and unnecessary; *whether* will suffice. *As to whether* often is coupled with peculiarly stilted phraseology—another reason to avoid its use.

UGLY: Provide an analysis as to whether the segment types being exchanged have typical sequence numbers.

GOOD: Discuss whether reliable stream services differ from basic message-transfer services.

UGLY: Granger enquired as to whether Max intended to finish the book.

GOOD: Granger enquired whether Max intended to teach any classes.

SPLENDID: Granger asked whether Max intended to go crazy, or had undertaken the journey accidentally.[176]

UGLY: Deborah could inform Lyn as to whether history repeats itself.

GOOD: Deborah could tell Lyn whether history (or Max) repeats itself (or himself).

---

176. The classification upgrade is based on the change from *enquired* to *asked;* the latter term is more vigorous and direct.

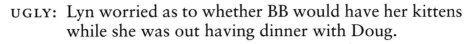

UGLY: Lyn worried as to whether BB would have her kittens while she was out having dinner with Doug.

GOOD: Doug worried whether Lyn had forgotten their date, when he arrived at the restaurant to find her absent.

UGLY: Could you write a report explaining as to whether you will undertake the project?

GOOD: Can you explore whether the alternatives are viable options?

UGLY?

GOOD: Peter asked Lyn whether the principle discussed in this segment applies to other cases; Lyn explained when, how, and why it might.[177]

THE PRINCIPLE FOR LUCID WRITING here is that, if you find yourself wondering *as to whether* you should use that term, you should simply wonder *whether* instead.

---

177. *As to when, as to how,* and *as to why* are all just as objectionable as is *as to whether.*

51 *Who Versus That*

THAT DENOTES nonhuman entities; *who* is the correct word to use to denote humanoids of various ilks.

You should refer to all humans as *who;* whether you refer to other creatures in your manuscript as *who* depends on whether you wish to afford them the status of honorary humans.

> BAD: People that never express emotions are boring; they are entirely too predictable.
>
> GOOD: People who laugh and cry and sulk and yell are, at least, interesting; they can also be exhausting.
>
> SPLENDID: Cats who purr loudly all night while draped over your ears are comforting; they are the most peaceful white-noise generators available.

> BAD: The scientist that designed this study method must have been a sadist.
>
> GOOD: The programmer who designed this software deserves to reach Nirvana.
>
> GOOD: The robot who delivered the hospital radiology films got lost and fell down the stairs.

BAD: You need to give your code to one of the hackers in this room; now, to what would you like to give it?[178]

GOOD: You should give your spare change to one of the people on this block; now, to whom would you like to donate it?

BAD: Max told Lyn about all the clients that had irritated him before noon.

GOOD: Lyn told Max about all the writers who had reduced her to tears before noon.

SPLENDID: BB told Red about the fluffy ginger tom who had just ensured that, in 9 weeks, she would add to the household census a litter of adorable kittens.

You should use *that* for all objects and creatures that you do not consider to be remotely human.

BAD: Suresh was trying to identify the computer on the network who was causing all the trouble.

GOOD: Boris was trying to identify the function that would create the most gorgeous graphs for the demo.

BAD: Max told Lyn about a book he had read who had changed his life.

GOOD: Lyn told Max about the book he had not read that was changing his life.

---

178. Various comments from the reviewers' peanut gallery have indicated that certain people are unwilling to afford to hackers the status of even *honorary* humans. Most of the hackers whom I know, however, are decidedly humanoid.

BAD: The Lumina telephone-answering system is the most annoying result of technical wizardry with whom Lyn have ever had the displeasure to converse.

GOOD: Max was again unable to retrieve his messages from the voice-mail system that he had had installed; he calculated that he had to enter 41 digits to retrieve his messages when the machine was in a cooperative mode, and to enter a multiple of 41 at other times.

THE PRINCIPLE FOR LUCID WRITING here is that you should use *who* to denote creatures <u>to whom</u> you grant human characteristics, and should use *that* for every other object <u>that</u> you discuss.

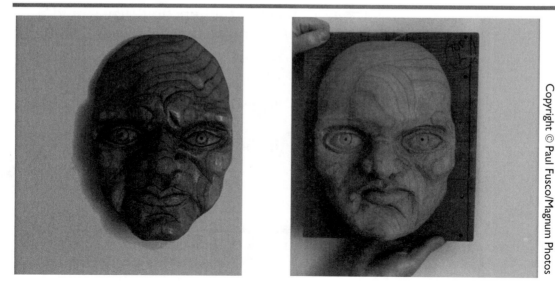

52 Though

ITHER *although* or *even though* is preferable to the unnecessarily truncated *though*.[179]

You should not write *though* when you mean *although*.

UGLY: When she came home just before sunset, Lyn was surprised to find Max sleeping on his head, though he had been complaining of feeling topsy-turvy lately.

GOOD: When he awoke just before sunrise, Max was surprised to find Red sleeping on his head, although it[180] happened every morning.

UGLY: Though the software is fast, the results that it produces are inaccurate.

GOOD: Although the machine is clumsy to use, its graphics capabilities are stupendous.

---

179. Here, and everywhere in this book, I give principles to follow in formal writing. There is nothing objectionable about *though* in casual writing or in poetry.

180. Note how important is the denotation of a pronoun (both *his* and *it*).

UGLY: Mike knew that he wanted to play with DNA, though he was not sure on which strands he wanted to concentrate.

GOOD: Russ knew that he wanted to play in the band, although he was not sure which songs his audience would permit him to sing.

You should write *even though,* rather than simply *though,* when you mean *despite the fact that. (Although* and *even though* have similar meanings, but *even though* emphasizes *despite* more heavily.)

BAD: Larry always wore a hat, though once it blew off into a barbecue fire, causing a massive conflagration.

GOOD: Mark always carried his stethoscope, even though it occasionally got caught in his zipper.

BAD: Lynne just kept doing her job, though everyone tried hard to make her life difficult.

GOOD: Rosalind took the job, even though everyone at the laboratory was clearly several standard deviations off the curve.

BAD: Red allowed Lyn to clean his ears, though he hated the sensation.

GOOD: Lyn allowed Red to burrow into her hair, even though he occasionally got carried away and sank his teeth gently into her scalp.

BAD: I am leaving you, though I love you.

GOOD: I am leaving you, even though I love you.

GOOD: Although I love you, I am leaving you.

SPLENDID: Although I will not leave you, even though you ask it,
I will give you a vacation.

THE PRINCIPLE FOR LUCID WRITING here is that, ~~though~~ even though the word is common, you should not use *though* on its own; you should use *although* or *even though*.

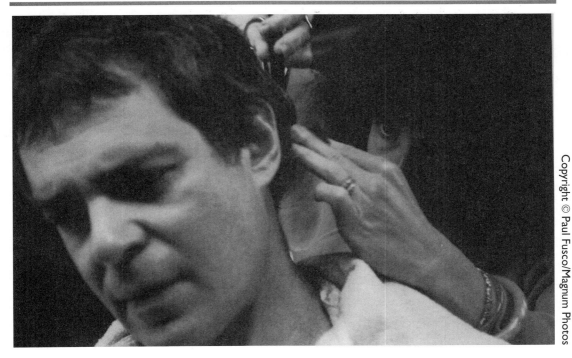

53 References to Parts

WHEN YOU REFER to parts of sequences, of figures, of tables—or to any other components of a whole—use the letter or number only; there is no need to enclose the letter or number in parentheses.[181]

Whenever you refer by number to a part of a *document* (for example, to a chapter, section, figure, table, box, example, exercise, or equation), you should set the word denoting the part with a lead capital letter: Chapter 3, Section 5, Figure 6.3, Table 3, Box 7.2, Example 5, Exercise 4.8, or Equation 6, for example. You should not capitalize words denoting other kinds of parts, however: step 1, phase 3, day 56, for example. You should also not capitalize words denoting parts of a document when they are set without a number: chapter, figure, table, theorem, for example.

UGLY: Read section 7.2, and answer exercises 7.8 and 7.10; then, read the following Section. You can treat each Chapter in the same way.

GOOD: If you understand Equation 4, you will be able to answer Question 3 in Box 8; if you do not understand it, read the preceding chapter. Apply this rule to all chapters in this book.

---

181. Certain publishers do use the parentheses; the style suggested here is my preference because it is simplest. If you wish to follow an alternative style, simply be certain to implement it consistently.

UGLY: Complete Step 3 before you enter Phase II.

GOOD: Sit on step 3, and ponder phase 2.

When you refer to a step or phase, or to any other unit in a sequence, you should write only the number,[182] without parentheses.

UGLY: In phase (II), we will measure the CPU burst times, and will calculate the turnaround time for the different scheduling algorithms.

GOOD: In step 1, we ensure that, if a process P_i is executing in its critical section, then no other process can be executing in its critical section.

GOOD: Cell 3 holds the value for the restaurant bill.

GOOD: Box 7 contains your birthday present.

GOOD: Kiss 8 is for hanging tight for 5 weeks while I refused to answer your calls.

When you refer to a part of a figure, exercise, example, or other component of your document, on its own (that is, usually within the figure caption or exercise), you should write only the letter, without parentheses.

UGLY: Test the algorithm that you developed in part (b) on three different data sets.

GOOD: Reconsider your answer to part d in light of the following: *Mors certa; amor incerta est.*

GOOD: Part c shows the history of various versions of the Unix operating system.

---

182. You should write the letter, of course, if you have a good reason (such as a publisher's or funding agency's mandate) for using letters rather than numbers.

> GOOD: In part d, we described the logical and physical address spaces. Draw a figure that shows the relationship of those components to the CPU, to the page table, and to physical memory.

 When you refer to a part of a numbered figure or exercise (or other component) in text, you usually should give the figure or exercise number followed by the parentheses-enclosed letter; that is the style followed by most publishers. You can, if you prefer, adopt the style of omitting the parentheses, provided that you do so consistently.

> UGLY?: Figure 4a shows why virtual memory can be much larger than physical memory.

> UGLY?: Figure 5.7a shows the page table when a portion of the pages is not in main memory.

> GOOD: Figure 2(b) shows the transfer of paged memory to contiguous disk space.

> GOOD: Figure 6.2(c) shows the steps in handling a page fault.

 When you refer within parentheses to a part of a numbered figure (or other component) in text, you should simply give the number and figure together, without the second set of parentheses.

> UGLY: Tania and Moisés canceled the party because they were worried about flaws in the floor-planning algorithm (see Figure 3.4(a)) that they had used for hot-tub seating.

> UGLY: Tania spent all day recalculating the degree of increase in mass (see Figure 4(a)) that a person of mean weight for the Bay Area population would experience after eating one huge barbecued chicken breast, two hamburgers, a scoop of potato salad, a side of coleslaw, and a handful of tortilla chips with salsa, as well as drinking four bottles of beer.

GOOD: Moisés spent each evening that week using an innovative investigative technique (see Figure 3.4b) to collect data on the volume of the space required in the tub for each of his friends.

GOOD: Tania and Moisés invited everyone to come over to watch the World Cup finals (see Figure 2a for an explanation) amid plenty of beer and barbecued ribs—but not to go near the water.

If you find the distinction between nonparenthetical and parenthetical callouts confusing or awkward, you can (as I mentioned) choose never to use the parentheses, to be consistent with other references to parts. That is why certain of the preceding examples are classified as only questionably (rather than definitely) ugly.

UGLY?: Example 4.9c shows that readers are not always able to deduce more than the simplest rules for styling, and argues that a more practical approach may be to value apparent consistency over actual but overly complicated consistency.

UGLY?: See the preceding example (Example 4.9c) to be sure that you understand why certain people believe that inconsistency is the bugbear of great minds.

THE PRINCIPLE FOR LUCID WRITING here is that, as defined in rule 1, you should generally omit parentheses around letters or numbers that designate parts, unless, as defined in rule 2, you use parentheses for callouts or for parts of numbered manuscript components.

54 Dates and Times of Day

THE FEW simple rules about styling dates and times of day are primarily matters of convention. It is a good idea to follow them to make your writing consistent, both with your own text and with that of other writers.

 You should use a comma after the day of the month when you give a date with the year.

> UGLY: What happened on March 15 1022?
>
> GOOD: August 9, 1991, is a special day for Lyn.

You should not use a comma after the month when you give a date with no day specified but with the year.

> UGLY: By the end of March, 1994, Lyn had not written a single word of her manuscript.
>
> GOOD: By the end of October 1994, Lyn's book had been published.

You should spell out *first, fifteenth,* and so on in a date, as elsewhere.

> BAD: Lyn suffered a serious car accident on August 9th.
>
> GOOD: For years, Lyn has found August ninth to be either wondrously good or appallingly bad.
>
> GOOD: Peter suggested making the ninth of August Lyn's honorary birthday.

BAD: *Computer* was a job description in the 19th century.

GOOD: Homer was born during the nineteenth century.

BAD: Gunnar and Lyn had a few frightening adventures back in the 60s.

BAD: Several of their acquaintances OD'd in the 60's.

GOOD: Overall, the sixties was a marvelous time of growth and discovery for Gunnar, Jessica, Lyn, Bruno, Lisa, and Marilyn, and several other friends.

Note, however, that you should use the numerals with an s and no apostrophe when you give the full year.

BAD: In the nineteen-sixties, the war took many lives.

BAD: In the 1960's, several colleges were closed down temporarily by protests.

GOOD: In the 1960s, Lyn learned go-go dancing!

You should hyphenate *mid-* but not *early* or *late* when you are speaking of a portion of a decade or century.

BAD: This dress was stylish in the mid 1940s.

GOOD: Flappers were abundant in the mid-1920s.

BAD: During the mid twentieth century, numerous species became extinct, due to the lack of habitats.

GOOD: During the mid-eleventh century, certain scholars postulate, many people stank, due to the lack of bathing facilities.

BAD: Where exactly is the cutoff line between the late-1840s and the early-1850s?

GOOD: Eighteen-fifty is the cutoff line between the late 1840s and the early 1850s.

BAD: Many patterns of the late-nineteenth century emerged in the early-twentieth century.

GOOD: The early twenty-first century may bring virtual reality to every household in the country.

GOOD: The late twenty-first century may bring reality back into vogue.

You should set times of day in numbers, and should use only one indicator of morning or evening.

BAD: At two A.M. in the morning, when no one is watching, Red goes out to do his exercises.

BAD: At 23:00 at night, secretly, BB unrolls the bathroom tissue and festoons the living room with streamers.[183]

GOOD: At 9:30 P.M., on the nights that Max works late, Lyn climbs into bed with a good book, two cats, and a cup of hot cocoa.

GOOD: By 21:00, usually, Lyn is peacefully asleep with Red draped over her head and BB providing toe warming.

GOOD: At 6 in the morning, rain or shine, Lyn brings Max a cup of ginseng tea and a kiss, and brings Red and BB their yeast and garlic vitamin pills.

---

183. Contributed by Red.

You should use small capital letters and periods for A.M., P.M., B.C., A.D, B.C.E., and C.E.

BAD: Max did not get home from work until 3 AM.

BAD: Lyn woke up at 2 a.m., and worried when she realized that there were only three creatures in the bed.

GOOD: By 3:30 A.M., all was well again as the four members of the household curled up together and drifted into the arms of Morpheus.

BAD: From 3 PM to 7 p.m., Holly and Misha were cooking up a storm in their gigantic marble-sided kitchen.

GOOD: From 8 P.M. to midnight, Holly and Misha were partying with a huge crowd of convivial and well-fed friends.

BAD: In 4 BC, there was neither mass communication nor mass transit; mass hysteria, however, existed already.

BAD: Confucius was born in 551 b.c.

GOOD: By 2000 B.C., the Chinese had already established a comprehensive model of diseases and treatments.[184]

BAD: In 73 AD, there were no public telephones.

BAD: In 73 a.d., there were, however, public baths.

GOOD: In A.D. 537, which of your relatives was where?
       Note that the A.D. precedes *the date.*

---

184. Joseph Norman suggests that the correct name for that model is SINOMED.

BAD: Lyn wondered whether 0 C.E. was identical to 0 B.C.E.

BAD: How many years before 1 c.e. did 1 b.c.e. occur?

GOOD: The abbreviations C.E. and B.C.E. are becoming common substitutes for A.D. and B.C., respectively.

THE PRINCIPLE FOR LUCID WRITING here is that you should learn the contents of this segment so that you will know how to style dates and times of day consistently. You would not want to write at 2:00 pm on December 19 1993, when your colleagues are writing at 2:00 P.M. on December 19, 1993.

55 Reason Is Because

Y OU SHOULD NOT use the phrase *the reason is because*; the reason for this principle is that the phrase is redundant. *The reason that is because*, and *the reason why is because*, suffer from the same disease; use instead *the reason is that, the reason why,* or simply *because.*

BAD: The reason Max's spanking-new machine jammed is because Red objected unreasonably to having his paws photocopied for the book design.

GOOD: The reason the mail carrier has not come is that she was attacked yesterday by a ferocious field mouse.

GOOD: The mail is early, because the letter carrier wants to rush home to her feline bodyguard.

BAD: The reason why Max was late for the movie was because he had lost his watch.

GOOD: The reason why Lyn was laughing was that Max was early for their dinner date.

GOOD: Max was early, because he had lost his watch, and the woman of whom he asked the time had forgotten to reset her watch for the end of daylight savings.

BAD: The reason that Brian was crying was because the bug list kept growing, but not as fast as did the new-features list.

GOOD: The reason that Suresh was giggling was that he was reexamining the latest bug, which turned out to be chimerical.

GOOD: Boris sighed heavily and wiped his brow, because debugging was not always easy.

BAD: Remember that the reason why Lyn was cranky is because she had only 5 minutes of sleep last night.

GOOD: Lyn was liable to burst into streaming tears because the computer crash lost all the corrections that she and Max and Adrienne and Jan had made over the long weekend.

GOOD: The reason why Lyn eventually became slightly less hysterical is that the four musketeers (plus Joe, who arrived back from the wedding in the middle of the crisis) were able to reconstruct most of the manuscript (bar a few leftover booboos) in just 1 day.

THE PRINCIPLE FOR LUCID WRITING here is that you should use *because* on its own, or should use *the reason that* or *the reason why*, but you should never use *because* with the latter two phrases; the reason is that the *because* would be redundant.

56 *With Terms*

WHEN YOU USE common phrases that include *with*, be careful. Often, the *with* is unnecessary; sometimes, the *with* suffices on its own.

 You should *meet* people, rather than *meeting with* them, when the former will suffice.

>UGLY: I would like to meet with you to discuss pointer types.

>GOOD: I would like to meet you to discuss types pointedly.

 You should *visit* people, rather than *visiting with* them.

>UGLY: Holly visited with Misha while he lay ill in bed.

>GOOD: Sachiko visited Peter while he lay snoring on the brown corduroy couch.

>UGLY: Soren went to the swap meet together with Lois.

>GOOD: Jack went to the concert with Emily.

You should *consult* people, rather than *consulting with* them.

>UGLY: I would like to consult with a decision analyst before I draft the global-warming report.

>GOOD: Lyn would like to consult her personal decision analyst before she commits to writing another book.

UGLY: Experts often consult with one another, but ignore what they hear.

GOOD: Professors often consult one another, but ignore their students.

You should place objects or people *with* each other, rather than *together with* each other.

UGLY: This table contains the addresses, together with the telephone numbers.

GOOD: This table contains the marketing targets, with the telephone numbers.

UGLY: Store the disks together with a few magnets if you want to erase your files.

GOOD: Buy the software with the machine to get a package-deal price.

You should avoid using *along with* to mean *as well as* or simply *with*.[185]

UGLY: Lyn packed in Max's briefcase the relevant disks, along with a banana and a handful of dried apricots.

GOOD: Max brought Lyn chocolate, as well as nectarines and mints, when they were working late together.

---

185. Based on a suggestion by Adrienne Esztergar.

UGLY: Lyn took Red along with her when she went to visit Max at Lumina.

GOOD: Max took Lyn with him on his trip to Chile.

GOOD: Red took BB along when he went to check out the neighbors' new Saint Bernard puppy.

THE PRINCIPLE FOR LUCID WRITING here is that you should double-check all common terms that include *with*, to see whether you can cut one of the words to your advantage; consulting with this book, for example, would be awkward.

57 Equals

THE PHRASES *is equal to, is greater than,* and *is less than* prefer to be left intact; do not drop bits of them.

You should use the term *is equal to,* and should avoid *equals.*

BAD: If *x* equals *y,* then we have no more work to do.

GOOD: When *x* is equal to *y,* we exit the loop.

BAD: We see from Figure 3 that *r* equals $\frac{1}{2} d$.

GOOD: Figure 4 shows that setting *r* equal to *d* gets us into trouble.

BAD: Richard knows well that little sleep plus no coffee equals trouble at his early-morning negotiations for his laboratory.

GOOD: Richard knows well that, when Madeline has not had much sleep, and there are no jelly doughnuts available in the morning, his negotiations with her about an early-morning prework bicycle ride will be equal to a rout.

SPLENDID: Lyn knows well that Max has no equals (Max is a most careful writer), and that he is equal to the task of starting and running several companies.

You should be careful when you use the phrase *greater than or equal to*.

> BAD: We set *a* greater or equal to *b*.
>
> GOOD: We set *z* greater than or equal to *p*.

> BAD: Clearly, *w* cannot be greater than or equal 2.
>
> GOOD: Is it nonsense to think that ε could be greater than or equal to δ?

Similarly, you should be careful when you use the term *less than or equal to*.

> BAD: Let *x* be less or equal to *y*.
>
> GOOD: Whenever *x* is less than or equal to *y*, we ring the alarm.

> BAD: Set 0 less than or equals ω.
>
> GOOD: Use a *p* value of less than or equal to 0.05.

THE PRINCIPLE FOR LUCID WRITING here is that you should not lose any pieces of the phrases *is equal to, is greater than,* and *is less than;* doing so equals an error.

58 *Placement of Adverbs*

I n general, you should place adverbs *after* the verbs that they modify, and usually should place them at the end of the phrase or clause to which they belong.

> GOOD: Once you get to the auditorium, slowly walk through the door, even if your desire to rush in headlong is powerful.
>
> GOOD: Once you get to the podium, talk slowly to your audience, even if your heart is pounding.
>
> SPLENDID: Once you get to the top, climb into heaven slowly.

If you place the adverb *before* the verb, you emphasize the adverb. Thus, in the preceding set of examples, if you want everyone to sit up and pay attention to the *slowly,* then the structure of the first example should be your choice. Usually, you do not wish to convey that emphasis, so you should choose as your default the structure of the second example. The third expresses the same thought (for the purpose under discussion), but does so more gracefully by placing the adverb at the end of the clause; it thus receives the classification of splendid. No example in this set is bad, however, and whether the first is ugly depends on the writer's intention.

There are no simple, absolute rules here; for example, whether your sentence is improved when you separate adverb and verb is a matter of rhythm—of ear. If the clause is short—if your reader will not have to backtrack to make sense of an adverb placed at the end of the clause—then you should place the adverb at the end.

GOOD: Pick up lovingly the hamster.

Emphasizes lovingly.

SPLENDID: Pick up the porcupine gingerly.

Placing the adverb at the end is the form most pleasing to the ear, when the sentence is short.

GOOD: Gently stroke your lover's hair.

Emphasizes gently.

GOOD: Stroke firmly your lover's back.

Good default.

SPLENDID: Stroke your lover's cat sensually.

GOOD: Carefully test the null hypothesis.

Emphasizes carefully.

GOOD: Distribute normally the sample population.

Good default for dishonest researchers.

SPLENDID: Apply the test regressively.

SPLENDID: A computer person is anyone who might plausibly be found wandering around in that section of a bookstore where computer-related books are shelved.

In general, you should leave a term such as might be *in one piece; dropping a word into the middle of the term emphasizes that word; when the word is an adverb that precedes a verb, then the emphasis is doubled (emphasized?). (That is, placing an adverb before a verb emphasizes the adverb.)*

 THE PRINCIPLE FOR LUCID WRITING here is that you should understand <u>fully</u> the nuances of adverb placement. Unless you have an excellent reason to do otherwise, you should place adverbs at the ends of the phrases to which they belong. What constitutes an excellent reason, in this case, is a matter of ear and intent.[186]

Copyright © Lyn Dupré

186. For example, in the preceding sentence, if I wanted to modify *place* by *carefully*, I would not add to the clarity of my sentence by locating *carefully* at the end, after *belong;* instead, the *carefully* should go after the *adverbs.*

59 *U.S. Versus British Spelling*

IT IS SIMPLE common sense to use the language of the country in which you are publishing. You probably would not submit a paper written in Urdu to an English-language journal published in the United States. British English is markedly different from United States English; you should consider it to be a different language. Thus, you should use spelling that is standard in the United States,[187] rather than that standard in Great Britain.

United States		*Great Britain*
acknowledgment	*not*	acknowledgement
afterward	*not*	afterwards
ass	*not*	arse
backward	*not*	backwards
beside	*not*	besides
canceling	*not*	cancelling
catalog	*not*	catalogue
center	*not*	centre
check	*not*	cheque
color	*not*	colour
dialog	*not*	dialogue

---

187. Here, as elsewhere in this book, I am assuming that you are writing for publication in the United States. British and U.S. written language differ in punctuation, spelling, and terminology. All the choices that I recommend in this book are for U.S. written language. If you are publishing in Great Britain or in most other English-speaking countries, you should follow a different—albeit overlapping—set of conventions.

United States		*Great Britain*
favor	*not*	favour
flavor	*not*	flavour
forward	*not*	forwards
gray	*not*	grey
judgment	*not*	judgement
labor	*not*	labour
luster	*not*	lustre
modeled	*not*	modelled
raveled	*not*	ravelled
shoveled	*not*	shovelled
sniveled	*not*	snivelled
theater	*not*	theatre
toward	*not*	towards
traveling	*not*	travelling
yodeling	*not*	yodelling

THE PRINCIPLE FOR LUCID WRITING here is that you should use U.S., rather than British, spelling; when in doubt, check the dictionary,[188] and use the *first* spelling given. You should honor the language of the country in which you are publishing, rather than honouring that of a different country.

---

188.   Use only the *Webster's Collegiate Dictionary* that is published by Merriam-Webster, Springfield, MA; there are numerous other versions of *Webster's Dictionary*. My reason for recommending this dictionary is that most publishers in the United States follow it for style, spelling, and usage. If you use other dictionaries, be cautious. The *Oxford English Dictionary*, for example, is British.

60 *Placement of Prepositions*

IT IS CRUEL to leave prepositions *(with, for, to, from, under, on, in,* and so on) dangling at the end of the clauses that contain them; instead, keep them warm and cozy by placing them within or at the beginning of the clause, snuggled next to the words for which they define relationships.

Dangling prepositions constitute another of those genuine errors that are so common that many people's ears have been mistuned to accept them. Remember, however, that many other people's ears have not been thus mistuned. If placing the preposition correctly sounds odd to you at first, give it time. I am confident that, eventually, you will find pleasing those sentences that are cast correctly, and will grimace when you see a lonely preposition meandering around far from its home.

> BAD: Lyn wondered whom she would be living with in Ladakh, and what she was waiting for.
>
> GOOD: Max wondered for whom he would be cooking in Woodside, and for what he was worrying about such questions when he had several companies to run.

> BAD: Knuth looked for an efficient algorithm to solve the problem with.
>
> GOOD: Hollerith looked for a pointed instrument with which to punch the cards.

BAD: Whom did the Countess of Lovelace write interesting letters to?

GOOD: For what purpose did Wirth invent this language?

BAD: What did you wear a gorilla suit for?

GOOD: For what purpose did you put on your horns and tail?

SPLENDID: Why are you wearing that ridiculous getup?

BAD: Red suddenly began to wonder where the mouse had gone to.

GOOD: Red realized under what the mouse was hiding.

SPLENDID: Red realized that the mouse was hiding beneath Lyn's slipper.[189]

BAD: What did you base your thesis on?

GOOD: In which basket did you put all your eggs?

BAD: What type of nose clip did Lyn squeeze her nostrils into after the famous raccoon–skunk territorial wars in the bedroom ceiling?

GOOD: Under which piece of furniture were Red and BB hiding after the skunk let loose?

---

189. Recasting for specificity avoids the preposition problem all together.

BAD: What were Swix and Spud growling over in the creek at Point Reyes?

GOOD: Through which hole in the fence did Buttercup and Hershey escape, to come charging onto the roof of Marina's house when no one was there to prevent such escapades?

BAD: What sort of container did you bring home your new pet porcupines in?

GOOD: In what part of your anatomy are the quills now?

BAD: What did you pick that up from?

GOOD: From whom did you learn that trick?

BAD: Whom does the bell toll for?

GOOD: For whom is Lyn yearning in the deep of the night?

THE PRINCIPLE FOR LUCID WRITING here is that you should wrap up your prepositions at the beginning of or inside the clauses to which they belong, rather than which they belong to.

61 Different From

THE PHRASE *different than* is never correct; the correct term is *different from*.

BAD: Information is different than knowledge.

GOOD: A database is different from a knowledge base.

BAD: Judy said that Steve's morning shower arias were quite different than what she had been led to expect would awaken her as the sun rose.

GOOD: Steve said that Judy's evening chocolate binges were quite different from what he had hoped would take up her attention in the fading light.

BAD: A one-way analysis of variance is different than a Chi-square test.

GOOD: The difference between the means is different from the meaning of the difference.

BAD: Your writing will be different than what it was if you read this book.

GOOD: Your writing will be different from what it used to be if you follow the principles in this book.

GOOD: Following the principles in this book will make your writing different from most of your colleagues' prose.

THE PRINCIPLE FOR LUCID WRITING here is that *different than* is different from good writing practice.

62 Callouts

Y OU SHOULD *call out* all figures, tables, programs, and boxes, or other numbered blocks in text.[190] A *callout* is a reference to such a numbered block of text.[191] For example,

GOOD: Figure 3.4 shows a camel after the final straw.

GOOD: The relationship between client and server does not need to be complicated (Figure 4.5).

GOOD: The various modes of attack on networked systems are detailed in Table 7.3.

You should assign a number to every figure and to every table. Avoid using unnumbered figures because they complicate layout; an unnumbered figure must fall exactly where you refer to it in the text column, whereas a numbered figure can be as far as, say, three pages away, if necessary. In addition, you can refer back (or even forward) to a numbered figure with precision, because the number is a unique identifier.

If you are writing a journal article, use single numbers.

---

190. *In text* here means within the main text of your document, as opposed to in figures, tables, boxes, exercises, and so on. Note that the term is set as two words, whereas the adjective *intext* is set as one.

191. Note that you should always call out all numbered figures, tables, boxes, and programs. You do not have to call out numbered examples, theorems, lemmas, definitions, and so on, although you may do so if you wish.

GOOD: Figure 3 shows an influence diagram for the decision of when to hold the wedding.

If you are writing a book chapter, use double numbers (the first number being the chapter number).

GOOD: Figure 4.3 shows a belief network for assessing whether rain is likely to spoil the bride's gown.

If you are writing a proposal, report, or other document that you will publish yourself, you may want to use double numbers because doing so keeps the numbering isolated within a section. If you end up adding a Figure 4 minutes before the express-mail deadline, you will need to renumber all figures from the old Figure 4 on; if, in contrast, you add a Figure 4.3, you will need to renumber only those figures that occur after the old Figure 4.3 in *only the relevant section*.

If the outline mandated by the organization to which you are submitting your proposal uses letters rather than numbers for sections, use a letter plus a number for a figure or table.

GOOD: Figure B.5 demonstrates why you must give us this funding.

GOOD: Table B.5 shows the divers talent we can muster for this project.

You should call out every figure and every table. Furthermore, you should call out each one in order: First you call out Figure 3.4, then Figure 3.5, and then Figure 3.6. If you find that you are calling out Figure 3.5 before Figure 3.4, reverse the numbers of the two figures.

If your document has other numbered, in-text entities—such as boxes, examples, or programs—then you should call them out as well, and should call them in order.

Also be sure to *set* your numbered entities in correct sequence.

In rare cases, it may be appropriate to give a forward reference to a figure or table; for example, it may make more sense to set a table in Chapter 4 where you discuss it, but to refer to it briefly in Chapter 2. In such cases, avoid confusing your reader: Make the forward references explicit.

GOOD: We give an example of a visual illusion in Figure 4.3, in Chapter 4.

Note that a forward reference is the only case in which you should use this redundant construction.

SPLENDID: To see an optical illusion, you should turn to Figure 5.6, on page 486.

Giving a page number alerts your reader immediately: She knows that this figure is not the next in the sequence.

 You should use a lead capital letter on the words *Figure* and *Table* when you call out numbered figures and tables.

BAD: By examining figure 3.5, you can see why some people hate the holiday season.

GOOD: By examining Figure 3.5, you can see why certain people hate camping.

BAD: The statistics are given in table 2.1

GOOD: The data are displayed in Table 2.1.

 You should write simply *Figure X.Y* or *Table X.Y* the first time that you call out a figure or table. Thereafter, for subsequent references to the same figure or table, you can write *see Figure X.Y* or *see Table X.Y*; if you choose this style, use it consistently.

GOOD: We discussed this idea in Chapter 1 (see Figure 1.2).

You should not include information about the provenance of a figure or table with the callout; that information goes in the *credit line* that is set with the figure or table.[192]

> BAD: Figure 12.4 shows how the worldwide web operates under stress; this figure is courtesy of Networking Made Commercial, Inc.

> BAD: The different message-passing protocols are outlined in Table 4.5; we are grateful to Patricia Meyer and Steven Miller for providing this information.

> BAD: The network layers interact as shown in Figure 3.7; this figure is adapted from Russell [7].

> BAD: The data for Table 7 were collected by graduate-student slaves; their accuracy is thus guaranteed.

If you wish to refer to a part of a figure, you should use one form of the following:

> GOOD: The complex web of life is shown in Figure 4.5(a).

> GOOD: The male and female perspectives are shown in Figure 5.4, parts a and b.

> GOOD: The child's perspective also is important (Figure 5.4c).

> GOOD: See Figure 9.3; the cat's view is shown in part a.

Thus, you should use parentheses around the part letter when it is set with the figure number but is not itself set in parentheses. If you find this style odd or cumbersome, you can omit the parentheses in all cases; however, certain publishers will argue with you.

---

192. It is not possible to give you an example classified as good here, as the point is that you should not include information. The first three examples would be classified as good if the sentences ended at the semicolons and the words after the semicolons were deleted. The final example belongs in the caption only.

Note that you should not refer to all the parts of a figure together; if you want your reader to look at all the parts, just refer to the figure itself. Thus, if Figure 4.9 has exactly three parts,

BAD: Figure 4.9, parts a, b, and c, prove that BB may well be god.[193,194]

GOOD: Figure 4.9 proves, with just one magnificent picture, that Red may well be god.

SPLENDID: Figure 4.9 proves, in three easy steps, that all felines are manifestations of god.

You should be careful not to refer to parts of a figure that do not exist. That is, if you refer to Figure 13.7(a), be certain that Figure 13.7 has at least two parts, that those parts are labeled a and b, and that each part has a caption.

THE PRINCIPLE FOR LUCID WRITING here is that you should provide a callout (for example, see Table 4.2), in numerical order, for every numbered table, figure, program, or box.

---

193. I do not follow the convention of capitalizing the word *god*, because use of the capital G suggests the God of Judeo–Christian monotheism. The lowercase g includes that God, but also encompasses any other gods, including those from other monotheistic systems.

194. Note that *goddess* is a gender-specific word; I prefer to use *god* to denote all deities, regardless of their genders.

63 Exclamation Point

OVERUSED *exclamation points* detract from, rather than adding to, the punch of your writing. Learn how to use exclamation points wisely, and learn how to set them with other punctuation marks.

You should save exclamation points for emphasizing genuinely surprising or earth-shattering material. You should not sprinkle them about indiscriminately, because doing so dilutes their power and gives the impression of an enthusiastically dumb discourser.

> UGLY: Hi Lyn! How is life?! I have been having my bathroom remodeled (!), and there are 15 (!) people working on it! And every single one of them has made a mistake! We are 4 weeks behind schedule! Everything leaks! And I am tired of having to shower irregularly at my friends' houses!!!

> GOOD: Hi Mom, just a quick note to say I got your postcard about the bathroom saga—it sounds dreadful. Life here is fairly hectic too; there are similarities between bathroom remodeling and book production. In any event, we are hoping to get the book out well before the holiday season: About 1 year ahead of schedule!

> UGLY: "Dammit!" screamed Max, "Why doesn't anyone understand that I just don't have time!?"

GOOD: "Dammit!" roared Lyn, "I hate waiting for people who cannot be bothered to be on time."

UGLY: I am amazed! How can you think that of me?! I have never eaten pretzels in bed with anyone else!

GOOD: I am overwhelmed! How can you be so good to me, when I continually eat pretzels in bed?

The purpose of these examples, in case you are wondering, is to show that the exclamation points fail to add power to the words in most cases.

You should not set commas, semicolons, periods, or question marks[195] with exclamation points.

BAD: "Golly gosh darnit!," exclaimed Chuck.

GOOD: "Well, tiddlywinks to you, too!" retorted Anita.

BAD: My heart was broken!; and then my cat threw up all over my bed.

GOOD: The earth moved! And then the waterbed collapsed.

BAD: Judy forgot to pack the food!—we shall all starve!.

GOOD: Steve found the chocolate cake—we shall all get fat!

GOOD: Steve found the alfalfa-sprout and soy-bean burgers! We shall all remain disgustingly healthy!

---

195. In casual writing, you can set an exclamation point with a question mark: You did *what* with my hair curlers?! The combination is not correct, however, and you should not use it in formal writing.

You should set exclamation points inside or outside of quotation marks, depending on over what you are exclaiming.

> GOOD: Max cried out, "Agh, you are killing me!" as he pried Lyn's ice-cold toes from his belly button.

> GOOD: Lyn always giggles with hysterical delight when Max mumbles "umph"![196]

> GOOD: Lyn sniffled, "What difference does it make? I'll still have to rewrite everything from scratch. I hate slobs who call themselves writers!"

> GOOD: Lyn, crushed by a cloud of black depression, would scream piercingly and hammer her fists whenever she heard the word "computerize"!

 You generally should not set exclamation points with dashes; there are exceptional circumstances, however, in which doing so is the most effective way to convey excited, disjointed speech or thoughts.

> BAD: The toilet leaked and wrecked the new maple flooring!—didn't I already tell you that the only way to get the job done right is to do it yourself?

> *The combined exclamation point and em dash are not necessary — didn't could begin a new sentence.*

> BAD: Max missed his flight! — which he does often!

> *Either the first exclamation point should be left in place, in which case which should begin a new sentence, or the second exclamation point should be left in place, in which case the first one is extraneous.*

---

196. We assume that the event worthy of exclamation is Lyn's behavior, rather than Max's.

GOOD: I don't ever want to speak to you again!—get out of
       my house!!—no, wait a minute!—come back!—I
       love you!—that's better!

THE PRINCIPLES FOR LUCID WRITING here
are that you should use exclamation points
sparingly (except in exceptional circumstances!), and that you should use with them
only quotation marks and, occasionally,
dashes.

64 Deduce Versus Infer

THE WORDS *deduce* and *infer* have usefully different meanings that allow you to distinguish the direction of reasoning. *Deduction* is reasoning from the general to the specific, or from the population to the individual. *Inference* is reasoning from the specific to the general, or from the sample to the population.

UGLY: Lyn knows that all men are pigheaded, and that Max is a man. Therefore, Lyn infers that Max is pigheaded.

Lyn's reasoning is from the general to the specific.

UGLY: Lyn knows that living with Max can be both rewarding and irritating, that Max is a man over 5 feet 10 inches tall, and that the average male height in the United States is 5 feet 10 inches. Therefore, Lyn deduces that living with men who are over average height can be both rewarding and irritating.

Lyn's reasoning is from the specific to the general.

GOOD: Lyn knows that all men are marshmallows, and that Max is a man. Therefore, Lyn deduces that Max is a marshmallow.

Lyn's reasoning is from the general to the specific.

GOOD: Lyn knows that living with Max is a riot, and that Max is a man. Therefore, Lyn infers that living with men is a riot.

Lyn's reasoning is from the specific to the general.

Note that inference is trickier than deduction; you can make mistakes easily when you reason from the specific to the general.

UGLY: When Brian dragged a node from one window to another, it left numerous ghost nodes on the screen; Brian deduced that there was a new bug in the system that displayed intermediate sizing.

Brian is reasoning from the specific instance of a ghost-node occurrence to a general hypothesis about a bug type.

GOOD: After Max had given three demos without a single hitch, he inferred that the bug list had been brought under control.

Max is reasoning from the specific instances of smooth demos to the general hypothesis that most of the bugs have been fixed.

GOOD: Max did know, however, that every time he double-clicked on the startup display, the machine hung; therefore, he deduced that he should instead hit Return when he saw that display.

Max is reasoning from the general knowledge that every double-click brings disaster, to the specific hypothesis that, if he double-clicks on a given startup display, the machine will crash.

UGLY: When Lyn first woke up in the middle of the night to find the bedroom permeated by an overpowering smell of skunk, she deduced the population living in the ceiling above the bed comprised a skunk family.

Lyn's reasoning is from the specific (there is at least one skunk in the area) to the general (all the animals living up there must be skunks).

GOOD: After Lyn had received from Red numerous gifts of dead woodrats, she inferred that the squeaking and scurrying in the ceiling was probably caused by woodrats after all.[197]

Lyn's reasoning is from the general (all rats in the vicinity are woodrats) to the specific (the rats in the ceiling must be woodrats).

GOOD: Lyn knew that woodrats have furry tails, so when she saw that the rat that Red had brought her that morning had a furry tail, she deduced that Red was hunting woodrats, rather than roof rats.

Lyn's reasoning is from the general (all woodrats have furry tails) to the specific (the rat that Red brought in is a woodrat). The example also implies, but does not actually claim, that Lyn inferred from the specific (some of the animals that Red hunts are woodrats) to the general (all the animals that Red hunts are woodrats). It is possible that Red was hunting woodrats that night, but would hunt roofrats the next night.

SPLENDID: You may infer from a few examples in this book that Peter generally falls asleep during meetings. However, Lyn can deduce from this propensity that, despite his open eyes, Peter will certainly sleep through his next meeting with her.[198]

---

197. Lyn formulated the companion hypothesis that the rats, on several occasions, startled a skunk who had invaded their ceiling nest.

198. Contributed by Peter Gordon. Honest.

SPLENDID: Lyn remarked to Max that historical analysis is fraught with uncertainty, because you are inferring from a few known incidents to a general hypothesis. Max pointed out that, in contrast, when you predict the future, you use deduction to reason from general principles to a single incident in the future—and that is why the science of making predictions is so reliably accurate.[199]

Be careful not to confuse *infer* with *imply,* which means to indicate or express indirectly.

BAD: When Richard came to the dinner table that night, Betsy, seeing that Richard's coat was sopping wet, implied that either Richard had walked home in the rain or that Richard had forgotten to disrobe before his evening shower.

GOOD: When Betsy burst into tears, Richard, noting that every moveable object in the house was a component of the tangled heap on the living-room floor, and that Betsy's hair was standing on end, inferred that Betsy had had a normal day at home with the children.

BAD: When served microwaved chicken, Daphne told a long story about how perfectly dreadful had been a meal she had eaten at another house that served microwaved chicken; "Oh dear," she added hastily, "I didn't mean to infer that this chicken won't be delicious, as I am sure it will be wonderful."[200]

---

199. Can you identify precisely the logical flaw in Max's argument?

200. Daphne may indeed have made such an inference, but she did not intend to broadcast it by implication.

GOOD: When told that Dona's marmalade tom Jackie is almost as big as he is, Red stomped off in a huff; "Oh drat, please come back," Lyn called consolingly, "I didn't mean to imply that you are not the most hugely immense cat I have ever met."

THE PRINCIPLE FOR LUCID WRITING here is that you should read the examples in this book to infer how to construct your own sentences. When you then review the literature in your field, you may deduce that many authors do not follow the principles in this book (note the implied criticism of much scientific writing).

65 Citations

I T IS IMPORTANT that you use a consistent style to cite your sources. *Citations* are intext references to sources that are listed in the reference section.[201] The style for citations usually is set by the publisher, and you should identify and follow that style. If no style is set, you can follow any style you wish, provided that your style is consistent. However, there are several considerations that may influence your choice.

If you use *numbers,* rather than *names and dates,* you omit important information. The names of the researchers and the date of the study often allow your reader to weight[202] the results that you are citing. If you omit the names and date, and use a number instead, your reader has to flap back and forth to the reference section to get this information. Thus, unless there is a good reason not to do so, I recommend that you use names and dates, rather than numbers. One such good reason might be that you are writing a review article, and that using names and dates would result in your citations overpow-

---

201. Note the difference between *References* and *Bibliography.* The former lists works that are cited in text; the latter simply provides a list of works, for the reader to use as she pleases. When you cite material, you must provide references; including a bibliography, in contrast, is optional. Many authors use terms such as *Further Readings,* instead of *Bibliography.* The latter term is also correctly applied to a list of your own publications, such as you might attach to your résumé.

202. *To weight* means *to assign weights to;* it is different from *to weigh.*

ering your prose. Another might be that you are submitting your article to a journal that mandates that you use numbers.

If you use names and dates, you should use *square brackets,* rather than parentheses, so that your reader does not confuse parenthetical remarks with citations.

> UGLY: Some time ago, Peter found Helen (Gordon, 1983); she has been organizing him ever since.
>
> *A reader might well think that, in 1983, Peter found someone named Helen Gordon, rather than recognizing that the event of Peter finding Helen was reported by Gordon in 1983.*
>
> GOOD: A previous study [Goldstein & Gordon, 1973] demonstrated that placing trust in authors is a necessary but risky policy.

If you use names and dates, you should use *last names only.* If there is more than one researcher in the field with a given last name, however, then you should include the initials.

> UGLY: The Berknet and SLIP disciplines have notable differences that affect implementation by requiring work such as modifications to the kernel [McKusick MK, 1991].
>
> *We assume that there is only one McKusick in the relevant field, so the initials are extraneous.*
>
> GOOD: One center reported that implementation of the system led to an increase in efficiency of 50 percent [Smith LD, 1954].
>
> *We assume that there is more than one Smith in the field (a safe assumption in most specialties), so the initials provide enhanced accuracy for the identifier.*

If you use names and dates, you should use *et al.* for more than two authors. Note that there is a period after *al;*[203] certain publishers prefer to omit the period. Also note that I prefer to cite only one author before *et al.;* certain publishers prefer to cite two or more. Set *et al.* in roman type (et al.) in your citations.

> UGLY: EBCDIC is an 8-bit code that you should use with all IBM equipment [Fung B, Basevich B, Arnold B, Sterling B, Brown D, Steele M, Cheung T, Chanmugam S, 1986].[204]

> GOOD: Forty percent of the staff complained of headaches after installation of the new interface [Giddy et al., 1967].

If you use names and dates, you can use an *ampersand* (&), an *and,* or a *comma* between the names of two authors.

> GOOD: All authors are prima donnas[205] [Gordon & Goldstein, 1964].

> GOOD: All authors are insecure [Gordon and Goldstein, 1964].

> GOOD: All publishers make mistakes [Gordon, Goldstein, 1964].[206]

If you use names and dates, you should not repeat in the citation the name of an author that is given in text.

---

203. There is no period here, because I am referring to only that portion of the term that is made up of two letters.

204. The example contains a double mistake: The authors initials are included, and all the authors' names are given.

205. Had Gordon and Goldstein published their article in the 1990s, they undoubtedly would have chosen a genderfree term, such as *spoiled brats.*

206. So do all authors.

UGLY: Chaitin [Chaitin, 1994] developed a bureaucratic reminder system that was later discontinued because users reported a high level of annoyance.

GOOD: Chanmugam [1994] developed an innovative bug-tracking system that kept everyone marginally amused in the face of excessive frustration.

Be careful, however: Do not credit only the first author with the entire study.

BAD: Arnold [1994] reported that, in their experience, bringing software to market is hard work.

Arnold obviously had help, but there is no indication in the citation that he was simply first (rather than sole) author.

GOOD: Arnold and Sterling [1994] report that bringing a port to market involves fast automatic conversion and tedious manual debugging.

GOOD: Arnold and colleagues [Arnold et al., 1994] report that bringing little piggies to market is a major hassle.

 You should not use *et al.* in the text line; instead, use *and colleagues* or *and associates*.

UGLY: Dupré et al. [1978] reported that most of the scientists attending the meeting were suffering from marked sleep deprivation.

GOOD: Dupré and associates [Dupré et al., 1978] reported severe snoring during the 1977 meeting.

GOOD: Dupré and colleagues [1978] reported a marked increase in attention after removal of the G-string from the sitar.[207]

---

207. Any erotic implications are in the eye of the reader.

If you use names and dates, you can use a comma between the name and the date, or you can use no punctuation:

GOOD: An unlikely team can produce gorgeous results [Norman and Dupré 1994].

GOOD: A sophisticated design can produce an inviting book [Dupré & Norman, 1994].

If you use names and dates, and you use a comma before the date, then you should use a semicolon between different citations. If you omit the comma before the date, then you can use either a comma or a semicolon to demarcate citations.

UGLY: Two authors [Dupré, 1990, Henrion, 1994] independently discovered that working too hard, regardless of in what field, makes you cranky.

GOOD: Two studies [Dupré, 1990; Henrion, 1994] have noted a shift from left- to right-brain activity that occurs at about 02:00 hours

GOOD: Various researchers [Dupré 1990, Henrion 1994] have reported that eating pretzels in bed presents a severe hazard to health and happiness, in terms of lacerations, sleep deprivation, and marital squabbles.

GOOD: Eating chocolate in bed presents its own array of problems, especially if the sheets are white [Dupré 1990; Henrion 1994].

GOOD: Prenuptial agreements are available (Henrion & Dupré, 1995; Dupré & Henrion, 1995) that allow couples to avoid later divorce on the grounds of cruel and unusual pretzel crumbs in bed.

If you use numbers, you should use square brackets and super-scripting to discourage your reader from confusing citations with powers or other numerical references.

> BAD: Fifty-three percent of cars studied (13) were abused by their families.
>
> *A reader might easily think that 13 cars made up 53 percent of the population under discussion, rather than seeing that the 13 is a citation, and that 53 percent might be 4512 cars.*
>
> BAD: We can see that ω^{13} is indeed an interesting variable.
>
> *A reader might easily think that the 13 is a power, rather than a citation.*
>
> GOOD: Forty-two percent of the married people studied[13] had affairs with their cars.
>
> *The brackets make clear that the 13 is a citation.*

 If you use superscripted numbers, you should place the numbers *after* any punctuation marks.

> BAD: As you will recall[18], the program tended to crash[41].
>
> GOOD: As mentioned in our previous report,[18] the earlier version of the program tended to erase the hard disk.[41]

 If you use numbers, you can number your references by order of citation in text. Alternatively, you can order the references alphabetically by first author, number them, and then use those numbers for citations.

You should give the citations at the appropriate locations in your sentence, rather than always placing them at the end of the sentence. If you place all the references at the end, you may leave out critical information about which researcher reported which result.

> BAD: The reports conflicted: One-half reported no adverse affects, and one-half reported fatalities.[12, 15, 17, 31, 45, 48]

BAD: Several researchers have reported good results, but the two negative reports were from studies done on larger populations. [14, 18, 19, 25, 38, 42]

GOOD: Several researchers [14, 19, 25, 42] have reported good results, but the two negative reports [18, 38] were from highly reputable laboratories.

THE PRINCIPLES FOR LUCID WRITING here are ones that you should know already: Always be consistent when you choose among several citation competing styles, and always give your reader pertinent information [Dupré, 1994].

66 *The Fact That*

THE PHRASE *the fact that* is verbose, clumsy, dry, and overly formal. Substitute for it simpler or more elegant phrases.

UGLY: Shellee and Michael want to live in California, despite the fact that they own a gorgeous house in Brooklyn.

GOOD: Al and Mary want to spend their weekends in Santa Cruz, even though they own a cozy condominium in Mountain View.

UGLY: Have you noticed the fact that collisions are not handled well in this data structure?

GOOD: Have you noticed that various collisions have dented Max's car?

UGLY: The fact that Red is immense is immediately evident when you meet him.

GOOD: That Red is breathtakingly handsome is immediately evident when he gazes into your eyes and purrs.

UGLY: The flowchart does not represent the fact that phase 3 may be bypassed in extraordinary circumstances.

GOOD: The data-flow diagram does not show that data may fall into the bit bucket in extraordinary numbers.

UGLY: Many years ago, Lyn was not sure how she felt about the fact that Max had a beard.

GOOD: Many years ago, Lyn was unsure how she felt about Max being weird.

THE PRINCIPLE FOR LUCID WRITING here is that you should avoid using *the fact that*; ~~you should remember the fact that~~ nothing, simply *that,* or a recast will work.

67 Cross-References

T HE PURPOSE of a cross-reference is to allow your reader to find the relevant passage quickly; if you give a vague cross-reference, you will only frustrate and annoy your reader. Thus, you should provide *explicit* cross-references to related material within your document. You should use cross-references liberally, because they keep your reader oriented and make your document easy to navigate.[208]

You should use the smallest-granularity place indicator available when you give a cross-reference.

> UGLY: We shall discuss wombat mating practices later.

> GOOD: We shall discuss the chemical analysis of crocodile tears in Chapter 7.

> SPLENDID: We shall discuss how to remove a live mouse from a cat's mouth, without inducing a heart attack in any one of you, in Section 7.4.9.

> UGLY: You can see from Table 4.5 that the influenza death rate increases as the winter weather becomes harsher.
>
> *If you write this sentence in Chapter 3; your reader is likely to look first at Table 3.5, or otherwise to become confused.*

---

208. I have provided only sparse cross-references, to keep you disoriented.

GOOD: You can see from Table 4.5 in Section 4.7 that the conception rate increases as the winter weather becomes harsher and the days become shorter.

If you write this sentence in Chapter 3, you alert your reader that you are not talking about a table in this chapter, and also tell her where to find the table.

SPLENDID: You can see from Table 4.5 on page 67 that the lunacy rate increases as spring warms the frozen city.

The most explicit locator is a page number; most text formatters will allow you to provide page numbers automatically with cross-references.

UGLY: We have already described in copious detail the two companies that Max is trying to run.

If you write this sentence on page 879, your reader may have to wade through 878 pages to find that detailed description.

GOOD: In Section 2, we discussed how startlingly handsome is Max's visage.

You can tell by the single section number that this cross-reference is written in an article or proposal, rather than in a book; thus, the section is the least specific numbered (or titled) cross-reference, but the example is definitely preferable to the preceding one.

SPLENDID: Recall that, in Section 2.4.9, we described the infinitely interesting characteristics that constitute Max's personality.

We can assume that the sentence provides the most specific reference, because we can also assume that the description of Max's characteristics occupies more than one page.

You should not give *redundant* information in a cross-reference. For example, do not give both the name and the number of a section; the number suffices and is the more accurate locator. If I know that I am looking for Section 3, I know that it must be after Section 2.7.9, and before 4.1.3, so my search strategy is simpler than it would be if I knew only a name. In the latter case, I might have to scan numerous heads to find the section that I want. Also, you should never use *above* or *below,* whether or not they are redundant.

UGLY: The technique is described in the *Methods* section, Section 3.

GOOD: The statistical analysis that we used is described in Section 3.

SPLENDID: We describe the study population in Section 3.[209]

UGLY: As a result of researcher selection bias, the sample was not representative of the population; see Table 4.5 in Chapter 4.

Where else would Table 4.5 be? This form is valid only if you are giving a forward[210] reference and if a chapter is the smallest grain size available.

GOOD: As a result of bias-cut blouses, the researchers selected the sample (see Table 7 on page 8); we give this example to show how results may be skewed.

---

209. The recast from passive to active voice is the basis for the upgrade in the classification, from good to splendid.

210. Conceivably, if you were giving a backward reference—from Chapter 40 to Chapter 4, say—the redundant information might be justified.

UGLY: The correlation between blood-glucose level and belligerence is not necessarily linear (see Figure 8, below).

UGLY: The above table (Table 12.5) shows the data that Red collected by sampling the local delicacies.

GOOD: The graphic in Figure 14.7 indicates that Lyn has not had any lunch.

UGLY: See the review articles cited in the *Further Readings for the Chapter* section at the end of this Chapter.[211]

GOOD: See the articles on the inverse relationship between severity of inebriation and frequency of sexual congress, cited in Further Readings.

THE PRINCIPLE FOR LUCID WRITING here is that you should use liberal, specific, and nonredundant cross-references (see, for example, the top-right corner of page 1190).

---

211. The capitalization of the word *chapter* is incorrect; you should capitalize names of document parts only when those names are accompanied by a number (or letter). The information on the location of the section also is unnecessary, given the title of the section. In addition, if you write such sections (Further Readings, References, Bibliography, and so on), you should provide them in all relevant chapters (except perhaps in Chapter 1), so your reader should already know their location.

68 Proposals

A PROPOSAL has as its goal not so much to inform your reader as to convince—or to sell—your reader. Whether you are writing a white paper, a request for funding, or a business proposition, you should make it easy for your reader to cooperate with you.

You should *write an abstract* that informs your reader what you are proposing to do and why you are proposing to do just that. Assume that your reader will look at the abstract and the budget; if those two components meet her criteria for goodness in a proposal, then (and perhaps only then), she will read the remainder of what you have written. If the components are mandated by the organization to which you are submitting the proposal, and they do not include an abstract, then use the first paragraph of your beginning text as an abstract.

You should *say why your work is important*, and *why your reader should fund it*. Oddly enough, many authors forget to include this vital information. First, you must demonstrate that there is indeed a need (problem, question, void, whatever). Second, you must argue that your reader is the appropriate person to fund the work (or to enter into an alternative contractual relationship with you). I may be convinced that you are right that someone ought to investigate whether clipping nose hair leads to increased sneezing during bouts of influenza; that conviction, however, in no way guarantees that I am convinced that such research is relevant to my goals. Third, you must show that you can meet the need. Fourth, you must show that you and your proposed approach are, on an identified measure, more qualified to meet the need than are your (potential or actual) competitors.

You should *be explicit about what you are proposing to do*. Most funding agencies specify a structure that tells you where to put what information. Nonetheless, remember that your reader is primarily interested in knowing what you will do. I have read grant proposals where, on page 45, I still have not figured out what is being proposed. Be clear about what you have done already, what you have only imagined, what you know you can do, and what you are merely postulating may work. Do not get into the details of your proposal until your reader is fully conversant with the big picture.

You should *not make commitments that you cannot keep*. You should especially not make commitments that other people will know you cannot keep.

You should give the *background* to your work in terms of *work done by other researchers or organizations*, and of *work done by you or your organization or laboratory*. In the first case, you are telling your reader what progress has already been made, to show how your work will contribute. You want to say that other people have done work on designing a cat brush, but no one has yet designed a *self-cleaning, automatic* cat brush.

You should *talk intelligently about why you* (and your laboratory or organization, if applicable) *are particularly well qualified to carry out the proposed work*. You might have done work in a related area, or work that can serve as a basis for the proposed project. You might have access to specialized equipment. You might have established relationships with people who or organizations that are suited to serve as coinvestigators or subcontractors. You might have a geographical advantage. Give your reader an excuse to choose you above your competitors.

 You should *talk intelligently about why your proposed work is the best approach* by a relevant measure—such as the most feasible, least expensive, or quickest—to providing what your reader needs, whether that work consists of answering a research question, designing a system, conducting a survey, mounting a conference, or building an outhouse. Do not fudge your discussion of what the alternatives are; it is considerably more useful to mention the alternatives and to argue cogently against them than to pretend that they do not exist. (If you fail to mention another method, your reader may therefore assume that it must be the preferable option, even if she knows little else about it.) If you do not believe in your proposal, neither will your reader.

You should *specify precisely what you will deliver.* Separate from the work that you propose to do, you may be proposing to deliver a system, a written report, a manual, a revamped public image, or an indestructible outhouse. What will your reader have in her hands if she awards you the contract and waits for the specified time?

You should *indicate what measures you will use to determine your success.* You may need to validate a system, for example, which will require specifying the system goals, arguing that the measures that you propose are directly related to those goals, and arguing that your proposed validation study will measure what you claim that it will measure. You may need to determine whether you have conducted a satisfactory advertising campaign; again, you need to design a study to measure your success. The goals and the validation of your project are intertwined; when you discuss the one, you should also discuss the other.

Your reader wants to know how you propose to demonstrate that you have indeed done what you said you would do. In addition, she wants to know how you will determine the success of your work. That is, you may propose to implement an algorithm, and to test its use in a specific domain. Whether you have indeed implemented the algorithm is a question of whether you have fulfilled the terms of your grant or contract, and only an affirmative is acceptable, barring natural disasters and other good excuses; whether the algorithm turns out to be useful in the domain might be an *answer* that you propose to deliver, and a negative is acceptable in that case.

You should *write authoritatively*, so as to inspire confidence in your reader. Never say that you will try to do a task or—even more wimpily—that you hope that you will be able to do it; say rather that you will do it, or that you will investigate whether doing it is feasible. Never apologize for past less-than-perfect work; instead, explain it. Use *will* rather than *shall* when you describe what you will do. Even more stringently than in your other writing, use only active voice and make clear who will do (or has done) what.

You should *provide a carefully thought-out budget* that allows your reader to see at a glance how funds will be allotted. The budget should be sufficiently detailed that she can suggest areas to cut if she wants to fund you at a level lower than that you have requested, or at least can see whether there is room for you to make cuts. Further, she should be able to tell what she will give up if she reduces funding.

In cases where you are operating without the constraints of established funding agencies or large corporations—if, for example, you are proposing that you serve as a consultant to a relatively small company—you might consider reflecting contingencies in your budget. The notion of including uncertainty in a budget is a novel one, and yet most budgets are subject to the influences of uncertain variables. That is, most people write in explicit numbers, rather than providing a range or probability for a given value. Instead of saying that you will need 100 screws at 4 cents per unit, it might be considerably more honest to say that you will need 50 to 150 screws at 3 to 5 cents per unit, and that if the module turns out to need reengineering—an event with a 40 percent chance of occurring—then you will need 100 to 200 additional screws.

I do not advise you to submit to the National Science Foundation, for example, a probabilistic budget. Depending on the recipient of your proposal, however, you might impress your reader by including ranges and contingencies to the extent practical.

You should *follow precisely the structure mandated by the funding agency or other organization* to which you are submitting your proposal. You should also follow any mandates on page counts, layout, type sizes, reference style, number and binding or stapling (or lack thereof) of copies, and so on. You may submit a well-written, carefully conceived, highly deserving proposal, and lose your good chance at a grant because the clerk in the receiving office has been directed to toss into the wastebasket any proposal that is not in proper format. It happens; do not risk it.

You should *invite your reader to ask questions* or otherwise to seek further information. You should tell your reader how to reach you (and, if you are plural, specify to whom each type of inquiry should be directed). Give your reader as many alternative forms of communication as possible: electronic mail, surface mail, telephone, voice mail, facsimile mail, satellite hookup, or semaphore.

Before you begin to write a proposal that involves *multiple actors*—coinvestigators, or subcontractors, for example—you should *clarify what roles everyone will play* if the contract or grant is awarded.

Then, you should write about those roles in the proposal. This step is important both for obtaining the grant or contract, and for carrying out the work. Naturally, you must determine who will receive what funding, and what each person will do and deliver, by when, in return. One person failing to meet the terms can botch up an entire project, so you should not only specify what each person is supposed to do, but also *agree on what the default will be.* For example, if Max fails to write his portion of the code by Monday noon, Brian will write it instead, and Brian will bill his hours against Max's portion of the budget.

For a *multiauthored proposal,* you should *plan an outline to which everyone agrees.* A common problem with multiauthored proposals is that five different people are writing five different proposals, and it shows. One person—the principal investigator, president of the company, or anyone designated—must take responsibility for the document, articulate its structure, and outline its contents. Even when you have all agreed on what work you propose to carry out, different members of the team may have various ideas about what to emphasize. One person may be more interested in the theory behind a project; another may think that the main point is to build a practical system. Get everyone to agree on such points. To the extent practical, you should also *agree on the terminology* that you will use in the document; if one section discusses the history of apples and the next describes the canning of pears, your reader will be confused (if you intend her to think about a single fruit). Time invested up front will save you time and protect you from frayed nerves later.

For a *multiauthored proposal,* you should *make up and agree to a page budget,* which specifies who will write what sections, and in how many pages. The purpose of the page budget is twofold. First, many funding agencies and organizations put an overall page limit on proposals, but allow you latitude in the length of any given section. If your overall limit is 50 pages, and one person writing one of 14 sections writes 40 pages, you have a major cutting headache ahead of you. Second, the lengths of the various sections should correspond to those sections' importance. If work by other laboratories, for example, is only vaguely relevant, it should not take up 90 percent of your proposal.

For a *multiauthored proposal,* you should *set explicit deadlines* for each person. Allow time for the substantial editing that may be necessary, even if you have all agreed to an outline and page budget. You will definitely have to edit for consistency of style and usage, and you will definitely find that certain writers are more skilled than are others. *Set defaults for nondelivery.* If Len does not write the budget by Friday noon, then Bob will write it, but then Bob rather than Len will be funded by the grant, if funding is awarded.

More than 2 minutes before the deadline, you should *make sure that your printer, copier, stapler, car, and other necessary pieces of equipment are working.* You should also know what fallback options you have if, for example, your copier develops a pharyngeal obstruction and you miss the deadline of your express-mail carrier. Certain organizations will accept proposals by facsimile, followed up by signed hardcopy; others will not. There are usually airport delivery services of various types, including couriers, available at a price. Know where the services are located (and the routes to them), what the prices are, and what the deadlines are, well in advance of any last-minute crisis, so that you do not have to waste precious time on the telephone or, more frustrating still, do not discover, until 5 days after you missed the deadline, that you did have a workable alternative.

After you submit the proposal, you should do your best to *forget about it until you get a response.* (But remember that certain organizations may ask you to defend your proposal in a second round.)

THE PRINCIPLE FOR LUCID WRITING here is that you should write a proposal such that your reader understands immediately what you intend to do, why you are especially qualified to do it, why it is important, and why she should fund it.

69 Better, Best, Worst

Words such as *better, worse, best, worst* are vague and can be ambiguous. You should instead use words that specify the measure of goodness or badness that you are using in your evaluation.

UGLY: This algorithm is better.

GOOD: This algorithm is faster.

GOOD: This algorithm is cheaper to run.

GOOD: This algorithm is more efficient.

GOOD: This algorithm is more accurate.

GOOD: This algorithm is less costly.

GOOD: This algorithm is more effective.

GOOD: This algorithm is less confusing.

GOOD: This algorithm is more awkward.

GOOD: This algorithm is more elegant.

GOOD: This algorithm is less clumsy.

GOOD: This algorithm is more appropriate.

GOOD: This algorithm is less likely to cause trouble.

GOOD: This algorithm is less difficult to implement.

GOOD: This algorithm is more to my liking.

As you can see from this example, *better* can cover substantial acreage. You should instead say what you mean. Let us consider a few more examples.

UGLY: This purported hand-writing–recognition system is the worst that I have ever seen.

GOOD: This speech-recognition system produces the most inaccurate examination-note translations of all the medical-dictation systems that I have used.

UGLY: Bob's cold was worse than any he could remember.

GOOD: Lyn's struggle with toxic shock syndrome, subsequent to her knee surgery, was more debilitating than had been any bacterial or other invasion that she had experienced previously.

UGLY: Using this debugger will make your program worse.

GOOD: Using the new debugger will make your program's behavior still more unexpected.

UGLY: The best way to convince your boss is to show her the decision model.

GOOD: The most effective way to convince your cat is to show him the food dish.

UGLY: The best grass is across the fence.

GOOD: The greenest pasture is in your own backyard.

UGLY: Neither paging nor segmentation is better, so combining them is best.

GOOD: Both paging and segmentation schemes have advantages and disadvantages; certain systems, such as MULTICS, combine the two schemes and thus improve each one.

UGLY: Red is the best cat in the world.

GOOD: Red is the best cat in the world, on every conceivable measure of goodness.

SPLENDID: Red is the cat's whiskers.

Note that there are phrases in which *best* is sufficiently precise.

GOOD: At best, Max will be rewarded for doing his best by the development of a huge, successful company and by the possibility of retirement at an early age; at worst, he will have had fun giving a new venture his best shot, and doing his best by his employees.

THE PRINCIPLE FOR LUCID WRITING here is that you should avoid using valuative[212] terms that fail to specify the measure that you are applying. Rather than writing better, for example, you should write more concisely.

---

212. *Valuative terms* assign a value, or worth; *good, better,* and *best* are thus valuative terms. Other valuative terms are more subtle: *blue-haired little old lady,* for example, carries valuation that may be insulting to certain people.

70 Missing Words

 THERE ARE several useful techniques that you can use to signify missing words.

In quoted material, you can use points of ellipsis[213] to indicate that text is left out.

> GOOD: Doug reported that, "The committee on physicians' guidelines met for 94 hours straight. The members discussed many items, including the following:.... The meeting was then adjourned. Several scientists suffered minor injuries in the race for the door."

 You can use a 3-em dash (———) to indicate a missing word. This technique is a good one to use when, for whatever reason, you do not wish to set the word—for example, if the word would offend readers who have delicate sensibilities. To indicate missing letters, you can use a 2-em dash (——).

> GOOD: Max expostulated, "What the ——— happened here?" when he saw what BB had done to his most critically important papers.

> GOOD: If the gods are just, Anthony F—— will be a famous actor one fine day.

---

213. We shall not delve into the details of correct use of points of ellipsis in this segment; see Segment 86 for such details.

When you have two hyphenated terms in which the second word is the same in both, you can simply omit the second word in the first term. Note that it is not incorrect to repeat the word; you should, however, choose which style you want to use, and should use that style consistently throughout your document in all similar situations.

> GOOD: You can set up the program to calculate the value using either run-time[214] or compile-time techniques.

> GOOD: Run- or compile-time techniques will give the same value in this case.

 Note that you should not leave out the first word in the first term when the first words in both terms are the same.

> BAD: The HIS included patient-admitting, -monitoring, and -billing modules.

> GOOD: The LIS included test-ordering, test-analyzing, and test-reporting functions.

 THE PRINCIPLE FOR LUCID WRITING here is that you should use the various ——— that indicate missing words and l——ters.

---

214. Note that the adjective–noun pair *run time* is not hyphenated; only when the pair is used as a compound adjective, as it is in the example, should you use the hyphen.

71 Aggravate

AGGRAVATE and *irritate* are not synonyms; *aggravate* (which is derived from the Latin *aggravare,* to make heavier) is commonly misused to mean *irritate*. Learn instead to use each word correctly.

Aggravate means to increase in severity.

> GOOD: Drinking wine will aggravate your headache.
>
> GOOD: Everyone started screaming at once, which only aggravated the confusion.
>
> GOOD: Smoking a Cuban cigar is likely to aggravate a cough.

Irritate means to annoy or provoke when the object irked is animate.

> GOOD: Max became irritated when no one worked as hard as he did.
>
> GOOD: Max was irritated to find that he had lost his keys again.
>
> GOOD: Red was irritated when BB leapt on him from the top shelf of the closet.

Irritate can also mean to cause irritation, as in redness, swelling, or inflammation.

> GOOD: If you keep picking at that scab, you will irritate the skin.
>
> GOOD: That toe looks irritated; you had best wear sandals.

GOOD: Lyn's eyes were red and irritated, because she had been staring at the computer screen for 6 months.

You should not use *aggravate* when you mean *irritate*.

BAD: Lyn became aggravated when she could not figure out a use for the `offset` function.

GOOD: Lyn became irritated when she forgot to put the global declaration in the handler when she tried to do field decomposition.

BAD: If you try to derive an analytical description for the optimal discriminant surface, in a two-category classification problem, without knowing the class conditional probability-distribution densities or the a priori probabilities, you will just cause aggravation.

GOOD: If you fail to distinguish criticisms of Bayesian pattern recognition that are concerned with Bayesian statistical measures from those concerned with only the appropriateness of a Bayesian approach, you will cause only annoyance.

BAD: It is exceedingly aggravating to lose all your work because of a power outage, especially if you forgot to save your file.

GOOD: It is decidedly irritating when the person to whom you are speaking on the telephone continues to type, especially if you can hear the key-depression clicks.

BAD: Do not aggravate your spouse, or you will simply cause the situation to deteriorate.

SPLENDID: Do not irritate your spouse, or you will simply aggravate the situation.

THE PRINCIPLE FOR LUCID WRITING here is that *aggravate* does not mean *irritate;* you should use *aggravate* to mean *to increase in severity.* Otherwise, you may irritate your reader.

72 *Upon*

Upon (which means *up on*) is only rarely the correct term to use; in most cases, the more elegant and simple *on* suffices.

UGLY: I shall discourse upon Silicon Valley formalisms in Chapter 5.

GOOD: We shall comment on the spit-happens phenomenon in Chapter 2.

UGLY: Upon glancing down at his lap, after working with great concentration for several hours, Max was appalled to find that Red had left upon it yet another small carcass.

GOOD: On coming home, Lyn was distressed to find BB stuck in the clothes-dryer duct.

SPLENDID: When they got into bed, Max and Lyn were nonplussed to find themselves covered in white fur.

The construction on doing X *is awkward; you can often avoid it with more elegant phrasing.*

UGLY: Maria and Geoff decided upon a control technique that used agendas in a separate element class.

GOOD: Malcolm and Gelareh agreed on a new control WME.

UGLY: This principle is based upon Occam's razor.

GOOD: I would bet all my money on Schrödinger's cat.

UGLY: Marina gazed sadly upon the remains of the petunias.

GOOD: Pat gazed gladly on[215] the rains in Petaluma.

UGLY: To prepare yourself for this lesson, put your finger upon the Enter key.

GOOD: To prepare yourself for lassitude, put your head on my shoulder.

UGLY: Nick rested upon his laurels, and Laurel Ann chastised him for his laziness.

GOOD: Laurel Ann banked on the ties of friendship to protect her when she expressed her opinion that Nick should write additional books.

UGLY: Moisés and Tania were practicing the tango upon the dance floor.

GOOD: Max and BB were practicing *one-legged stork fishing in small pond* on the living-room floor.

---

215. Pat would be more likely to gaze *at* the rains, whereas Marina would be more likely to gaze *on* the petunias. English is an endlessly fascinating and complex language.

Note that, occasionally, *upon* is the term of choice.

> GOOD: Max leapt upon the chandelier and swung across the room to transport Lyn to safety.

> GOOD: Lyn leapt upon her white horse and rescued Max from his distress.

In both cases, *on* would be correct, but if you want your reader to envision vertical ascension, use *upon*.

You can also use *upon* in its tried and true—and worn out—place.

> GOOD: Once upon a time, a rustic woman skilled at wood working and plumbing lived with a peasant who was an exceptional omelet chef in the woods with their two feline children and the rats and the skunks; after consulting with her guardian angel, she decided to write a book about magic and mythology.

> GOOD: Lyn wished upon a star, and soon thereafter met Max.[216]

THE PRINCIPLE FOR LUCID WRITING here is that *on* does not usually require any help from *up*, and you should never waste extra letters by putting them ~~up~~on pages where they are not needed.

---

216. Lyn later presented to Max this irrefutable evidence of the ability of an astral body to respond to gentle supplication. Max remained unconvinced, however, maintaining that his volition had influenced the stars.

73 Whether Versus If

THE WORDS *whether* and *if* have different meanings. So that you can write precisely, you should distinguish the two terms.

You should use *whether* when you could substitute *whether or not* without destroying the meaning of your sentence (although you should not generally add the *or not*—see the next point).

UGLY: Moisés mumbled, "I don't know if this undertaking is a good idea."

GOOD: Tania sighed, "I don't know whether Moisés has any ideas left."

SPLENDID: Max suggested that whether you thought Moisés had ideas was immaterial if you knew the man.

UGLY: Ted did not know if the variable n was an integer or a real number.

GOOD: Darlene did not know whether Bachmann's O notation could be used with $S(x)$, where x is a real number.

SPLENDID: Ted asked Darlene whether she planned to draw up the accounts for the grant, if she had time that evening after dinner.

UGLY: Sally wondered if, now that she had tenure, she could stop working long enough to go birding in Alviso with Max and Lyn.

GOOD: Gordon wondered whether, given how busy Sally was these days, he should plan a huge party for her.

SPLENDID: Gordon decided that, regardless of whether Sally was busy, when she got her tenure appointment, he would take her to Inverness for a romantic adventure.

In most cases where *whether or not* is correct, you should use simply *whether*. Use *whether or not* when you mean *regardless (or irrespective)*[217] *of whether*. Keep intact the phrase *whether or not;* do not spread it out over your sentence.

UGLY: You can use the ID3 approach, whether you know what discrimination trees and features values are or not.

GOOD: Whether or not you understand why, you might accept that ID3 is valid when the data consist of many patterns, each of which comprises an extensive list of feature–attribute values.

GOOD: ID3 will not let you update the decision tree easily; you have to rebuild the entire tree, irrespective of the number of patterns classified incorrectly.

UGLY: Lyn asked Doug to review her manuscript, whether he approved of it or not.

GOOD: Doug asked Lyn to include an example about him, irrespective of whether he had completed his review.

---

217. Note that your options are *regardless* and *irrespective; irregardless* is not a word.

GOOD: Lyn asked Doug to supply the example, regardless of whether he was so busy that he did not have time for frivolities such as breathing or sleeping.

You should use *if* when you are simply placing a constraint.

GOOD: "If it *is* a good idea," Moisés continued to ruminate, "then I need to consider the ramifications."[218]

GOOD: If you are given the mean, range, and S value for a test distribution, and a single person's score on the test, can you calculate that person's z score?

GOOD: Try to understand the concept of relationships in logical data models, if you want to give your brain a workout.

GOOD: If either of the pen dimensions is set to a negative value, the pen assumes the dimensions (0,0), and the system performs no drawing.

GOOD: If $a = b$, then go to step 4.

GOOD: Do you know what you should do if your cat brings home a squeaking woodrat?

THE PRINCIPLE FOR LUCID WRITING here is that, if you mean *whether or not*, you should use *whether*, so whether you should use *if* or *whether* depends on what you intend to say.

218. Based on a suggestion by Richard Rubinstein.

74 Sections and Heads

IT IS good practice to subdivide your writing into as many section levels as is appropriate, so that you guide your reader. Subsections keep your reader oriented; give your reader small, digestible chunks to absorb; and provide heads that summarize contents.

You should *use numerous heads* to keep your reader oriented. If you write for pages and pages without a head, your reader is likely to get lost—to forget what the subject of the section is.

Heads can go to any level, although usually the lowest level used is 5 or 6. (Within a chapter or article, the top level is a level 1 head, the next is a level 2 head, and so on.)

Chapter 4 Varieties of Domestic Cats	[chapter title]
Section 4.1 Manx	[1 head]
Section 4.2 Domestic Shorthair[219]	[1 head]
Section 4.2.1 Tabby	[2 head]
Section 4.2.2 Calico	[2 head]
Section 4.3 Siamese	[1 head]
Section 4.3.1 Points	[2 head]
Section 4.3.1.1 Blue	[3 head]
Section 4.3.1.2 Seal	[3 head]
Section 4.3.1.3 Red	[3 head]

---

219. Marina Nims points out that all the cats listed are domestic; *domestic shorthair,* however, is a breed.

Section 4.3.2 Varieties	[2 head]
Section 4.4 Abyssinian	[1 head]
Chapter 5 Big Cats	[chapter title]
Section 5.1 Jaguar	[1 head]

You should always use zero, or two or more, heads at any given level. If a portion of your text can be subdivided, then logically there should be at least two such divisions.

BAD: Section 4.4 Persian
Section 4.4.1 Himalayan
Section 4.5 British Blue

In such a case, you could either subdivide 4.4 into, for example,

GOOD: Section 4.4 Persian
Section 4.4.1 Common
Section 4.4.2 Himalayan

or you could simply drop the head *Himalayan.*

You should *number* your heads. Numbers tell your reader what the level of a head is, make it easy to find a given section, and allow you to give explicit cross-references. In a chapter, you should use double numbers (<chapter number>.<section number>; for example, 4.7) for level 1 heads. In a journal article, you should use single numbers (for example, 4) for level 1 heads. In a proposal, manual, or other mid-sized document, you can use either; I recommend that you use double numbers, because that numbering scheme lets you number your figures and tables by section, making renumbering easier.

You should use *informative* heads that summarize the material in the section. You should not, for example, entitle the leading text of your manuscript *Introduction.* Everyone knows that the leading text is, by definition, introductory, so such a head adds no useful information and just takes up space. You can, if you wish, leave out the initial head all together. Or, preferably, you can use a head that, for example, sums up the research problem that you will address in your article. Heads such as *Summary* and *Conclusions* are reasonable (because they occur after an amount of other text that is not prespecified, and thus are informative), but always think about whether you can improve them by substituting a more specific phrase.

UGLY: Section 1 Introduction

GOOD: Section 1 Assessment of the Benefits of Qualitative
                 and Quantitative Reasoning

GOOD: Section 7 Conclusions

SPLENDID: Section 7 Implications of This Work for Public-Policy
                    Assessment in the 1990s

GOOD: Section 7.2.9 Summary

SPLENDID: The principle for lucid writing here is that you should
          use the most informative title that you can invent.

You should not repeat heads. That is, you should not use the same head twice, whether at the same level or within different levels, and you should not break down an upper-level head into two or three lower-level heads. You should not squirm out of this rule by using synonyms. (For an example that is correct, see the cat example at the beginning of the segment; no head is repeated.)

BAD: Section 1.1 Basis for the Research

     Section 1.2 Reasons for the Research

     Section 1.3 Motivation for the Research

BAD: Section 8 Summary and Conclusions
          Section 8.1 Summary
          Section 8.2 Conclusions

BAD: Section 3 Systems with Graphical Interfaces
          Section 3.1 The WEIRDO System
          Section 3.2 The ALBEN–FERRIS System
      Section 4 Systems with Unusual Interfaces
          Section 4.1 The WEIRDO System
          Section 4.2 The HENRION–DUPRÉ System

You should keep heads parallel. That is, all your heads should be in the same form.

BAD: Section 5 Whether to Have Sex
      Section 6 Sex Can Be Fun
      Section 7 Is Sex Dangerous?
      Section 8 You Should Be Careful with Sex

GOOD: Section 5 Whether to Fall in Love
       Section 6 Why Falling in Love Can Be Fun
       Section 7 How to Avoid the Dangers of Falling
                  in Love
       Section 8 What Signs You Can Recognize

GOOD: Section 5 Assessment of Cohabitation
       Section 6 Enjoyment in Cohabitation
       Section 7 Domestic Violence and Cohabitation
       Section 8 Cohabitation and the Open-Tube-of-
                  Toothpaste Syndrome

UGLY?: Section 5 Deciding Whether to Wed
        Section 6 Creating a Memorable Wedding
        Section 7 Avoiding Homicide on Your Honeymoon
        Section 8 Deciding Whether to Divorce

The final set of examples is questionably ugly because most people prefer to avoid using *ing* words in heads, titles, and captions.

You should style heads with capital and lowercase letters (called *cap/ lc* in publishing). Most publishers will set the style for the different head levels, so your heads may be changed later. However, if you set your heads with lead capital letters, or with all capital letters, and the style calls for capital and lowercase letters, then someone other than you may decide which words to capitalize. Even more alarmingly, your heads may be rekeyed, with new errors introduced into your carefully proofed copy. If you style your heads (and chapter titles) using capital and lowercase letters, then someone other than you may change them to all capital letters or to lead capital letters, but this latter task requires almost no judgment (there are exceptions!), and leaves little room for problems (it does not even require rekeying). All the section (or, in this case, segment) titles—both example and real—in this book are styled with capital and lower-case letters.

> GOOD: This Sentence Is Styled Cap/lc.
>
> GOOD: Max Runs Energetically to Skyline Every Morning.
>
> GOOD: Lyn Walks Sedately to the Mailbox Every Afternoon.

Of course, if you are designing your own document, then you can set your heads in whatever style you choose. Be consistent, however!

Based on similar considerations, you should leave your heads left justified; if the design calls for another style, your publisher will have no trouble converting from normal, left-justified text.

If you are creating your own design, you should differentiate head levels typographically, as well as by numbering. The style that you choose (typeface and size, spacing before and after, boldface versus normal type, and so on) should reflect the level: Higher-level heads should be bigger, or otherwise easier to see. If you are using a full-page width (rather than setting type in columns), you should left-justify the heads.

THE PRINCIPLE FOR LUCID WRITING here is that you should break up your writing by sections and subsections, each of which you should number. In Section 1, for example, you might set subheads for Sections 1.1 and 1.2, and perhaps for Sections 1.2.1, 1.2.2, and 1.2.3.

75 Comprise

THE WHOLE comprises the parts.

You should not use *comprise* to indicate that the parts constitute (or make up) the whole.

> BAD: Mind and body are sometimes thought to be the two parts comprising a person.

> BAD: A person is comprised of one or many parts.

> GOOD: A person comprises heart, soul, and mind.

> GOOD: Analytic and synthetic together constitute a mind.

> GOOD: Creative and scientific skills contribute equally to a person's insight.

> GOOD: A person is made up of her past experiences and yesterday's dinner.

> BAD: By this perspective, numerical probabilities, if–then default rules, plain beliefs, and logical constraints comprise an information-representation scheme.

> GOOD: In our view, the proposed framework is a single network representation that comprises probabilities, defaults (plain beliefs), and logical statements.

> GOOD: We believe that situation assessment, course-of-action evaluation, and planning together constitute a major challenge for artificial-intelligence research.

GOOD: Moisés argued that explicit uncertainty, hard functional constraints, and ill-formed beliefs make up the inherent problems in these applications.

GOOD: The single computational framework is made up of tools for mixing, managing, and analyzing probabilistic information and logical expressions.

You should *never* use the phrase *is comprised of*.

BAD: This book is comprised of 150 principles of lucid writing, numerous examples, and a dollop of lunacy.

GOOD: Red's world often comprises the view from the top of the garage roof and the sounds of the wild woozles in the distant hills.

GOOD: BB's world often is made up of the view from the top of Lyn's computer monitor and the sounds of Lyn's fingers tapping gently on the keyboard.

BAD: The software package is comprised of numerous lines of buggy code.

GOOD: The software package comprises a number of useful tools.

GOOD: Many independent modules constitute the software package.

GOOD: Not much other than the interface makes up the software package.

GOOD: The software package is made up of bits and pieces that you can buy separately if you prefer.

BAD: A difficult marriage is comprised of tense partners.

GOOD: A peaceful marriage comprises patient, supportive partners.

GOOD: Max and Lyn constitute a halcyon partnership at least once per month.

GOOD: Richard and Betsy make up an unexpected marriage.

GOOD: The company's partnership is made up of a demo giver and a finance finder.

THE PRINCIPLE FOR LUCID WRITING here is that the whole comprises the parts, whereas the parts constitute or make up the whole. If you ever find yourself using the phrase *is comprised of,* give yourself three lashes with a red pencil and recast your sentence.

76 *In Order To*

THE PHRASE *in order to* is fusty and verbose; in almost all situations, *to* will suffice. If *to* is insufficient, I prefer the less klutzy *so as to,* rather than *in order to.*

UGLY: Max's new colleague pontificated, "In order to get ahead, you should learn to laugh when insulted."

GOOD: Max replied, "To get ahead, you should learn not to care about the fools who insult you, and not to insult the cares that fool you."

BAD: Use the following equation in order to test whether there exists a linear dependence.

GOOD: To show that this algorithm determines correctly the lexicographic basis in a finite number of steps, you will need to read the entire book.

UGLY: Judy bought Steve new shoes, in order to protect the white linoleum.

GOOD: After Steve had stained the linoleum irretrievably, he wrapped Judy in piles of fragrant rose petals, so as to comfort her.[220]

---

220. *To* alone would suffice; I am showing you a good use of *so as to.*

UGLY: Sally used confidence intervals in order to estimate the population mean, μ.

GOOD: Gordon used the notion of levels of significance to determine the probability of a result being caused by a sampling error.

UGLY: Red fought with the neighbors' tomcat, in order to avoid losing his self-esteem.

GOOD: Lyn climbed a tree, to avoid fighting with Max.

SPLENDID: Red climbed the tree after Lyn, so that he could help Lyn to climb down.[221]

SPLENDID: "I left the house in order to please my host," said Peter after his departure; he had made the bed, folded the towels, and washed the dishes.

THE PRINCIPLE FOR LUCID WRITING here is that *in order to* is a clumsy phrase that you should avoid using, ~~in order~~ to improve your writing, ~~in order~~ so as to communicate effectively.

---

221. *So that* is kinder to the ear than is *so as to* or *to*.

77 Em Dashes

An em dash—the punctuation-mark dash—delineates a thought that is not a critical part of your sentence. The preceding sentence provides a good example of this principle. The em dash can function as a strong comma, or as a semicolon; it is, in essence, a loosely defined break or pause in a sentence that calls attention to the material that it sets off.

Remember that there are three kinds of dashes:

- - hyphen
- – en dash
- — em dash

 You should use two em dashes if you wish to include—in the middle of a sentence—a thought that is slightly tangential to your subject, but to which you wish to call attention. (If you want instead to downplay the tangential thought, you should use two parentheses.) When you use dashes to set off a thought in this way, do not forget the second dash and substitute a comma instead.

> BAD: Max had many worries—his company, his house, his car, and his socks, so he tended to frown often.

> GOOD: Max had many sources of joy—his company, his research institute, his house, his gourmet dinners, and his three felines—so he managed to survive from one day to the next.

GOOD: Max had many documents to write (journal articles, white papers, proposals, business plans, company policies—all the usual nonsense), so he decided to take off the afternoon to go to the beach. Lyn fainted—from delight.

BAD: Turing decided—after many tests, that the machine was not behaving sufficiently irrationally.

GOOD: Minsky, Simon, Newell, and McCarthy—all grand men, as we know—decided to play a genuinely intelligent game.

GOOD: Von Neuman (who has one characteristic in common with Wright, and one in common with Archimedes) took a long hot bath.

Note that you should use em dashes, rather than commas, to insert a clause that begins with *for example* or *that is*.

BAD: Howard, that is, the man known for his decisiveness, is sitting at that table.

GOOD: Shannon—that is, the man asking for information— has been cooking up a new theory.

BAD: Various birds, for example penguins and roast chickens, have had an illustrious history in the field.

GOOD: Various towers—for example, those of Hanoi—have featured largely as well.

 You should use one em dash if you wish to tack on an extra thought at the end of a sentence. Note that a colon or semicolon serves the same function as a period in such cases, so it is acceptable to place a clause delineated by dashes before any of these punctuation marks, and thus to leave out the second dash.

> GOOD: Spencer also has many business concerns—his coffee company being the main one; nonetheless, he takes off every Sunday to be with Thérèse.

> GOOD: Thérèse is a thoroughly delightful woman; she is warm, talented, gorgeous, and creative—and she is one of the greatest cooks in California.

> GOOD: To get the Rose Café off to a good start, the Stars sweated over every aspect—including the amazing menu: 274 different kinds of coffee drink, and 189 different kinds of pizza.

> GOOD: Peter has a tendency to fall asleep at the first opportunity—or, occasionally, when there is no opportunity visible.

> GOOD: Peter once spilled Rocky Road ice cream on his trousers—and made a huge stain; he was shy about removing them to have them cleaned.

> GOOD: At one Japanese restaurant, Peter ordered in fluent Japanese—and the waiter brought so much sushi that the table almost fell through the floor: tuna, mackerel, eel, shrimp, lobster, and on and on.

Note that you should use an em dash, rather than a comma, when you add a clause that is not a full sentence and that begins with *that is* or *for example.*

> BAD: The gap between the read–write head and the hard disk is so small that you can cause a crash by inserting into the gap various particles, for example, hair, dust, or smoke.

GOOD: To ensure that the hard disk and drive assembly remain dust free, manufacturers adopt clean-room standards—for example, requiring employees to wear what often resemble Halloween ghost costumes.

BAD: Max's to-do list for that day was about as long as usual, that is, absurdly long.

GOOD: Max had all the necessities of life—that is, food, shelter, warmth, and on-demand snuggles.

You should not use more than two em dashes in any one sentence. If you use more than two, your reader becomes confused about what is the baseline sentence and what is the addition. You can use parentheses or commas instead, in such cases; you can break your sentence with a semicolon and use one dash in each of the portions; or you can break your thoughts into separate sentences, each with only one pair of em dashes.

BAD: Mark kept acting like a doctor—running around with his stethoscope swinging and his white coattails flying—yet he also did well as a scientist—flying around the country reviewing goodness knows what—and managed to smile most of the time.

GOOD: Elysse had done interesting work for her doctorate (getting chased through the jungle by someone wielding a machete), and then had to write her dissertation in Palo Alto (which was tame by comparison).

GOOD: Max kept acting like a CEO, running around with his briefcase swinging and his black hair flying; yet he also did superb work as a consultant, zooming around the country reviewing myriad intricate decisions—and managed to grimace most of the time.

GOOD: Lyn had done interesting work for her doctorate—punting on the Cam and sampling the wares of English pubs. Then, she decided instead to write a health-services annual report in Princeton—which was boring by comparison.

GOOD: Lyn was finding it uncommonly difficult to spend all morning editing someone else's work—especially when that work was not particularly enthralling—and then to spend all afternoon doing her own work (especially when that work consisted of complex scheduling of numerous book-production tasks).

The exception to the preceding advice occurs when you use em dashes to indicate disjointed quoted speech. You can also use an em dash to indicate interrupted quoted speech.

GOOD: Richard was teaching Madeline to ride a bicycle: "Put your foot there—other side now—pedal fast—lean left!—steer!—use the brakes—be careful of the rock—hey, slow down—wait for me!"

GOOD: Adrienne called Lyn to explain the latest problem: "We'll need to get those icons up to a higher resolution as EPS files, and I don't have the software to convert them, and Joe is leaving in an hour on a flight to the East Coast so that he can attend his friend's wedding, so ask Joe—" and then the line went dead.

GOOD: Joe called Lyn from the airport in Boston, where he had flown that day to attend the wedding, even though the wedding was on Long Island, and started to explain to her, "You can deal with those files easily. Just tell Adrienne to open them up, to cut and paste the separate icons, and then the tricky bit is that she has to—" and then that line went dead as well, and Lyn burst into tears.

GOOD: To cheer Lyn up, Max held Lyn tight and exclaimed, "You are so gorgeous, and talented, and funny, and —";[222] Lyn never heard the rest, because she could not resist planting a wet sloppy kiss on his external speaking apparatus.

THE PRINCIPLES FOR LUCID WRITING here are that you should use em dashes—singly or in pairs—to add to your sentences information to which you wish to call attention, and that you can also use em dashes to indicate disjointed or interrupted quoted speech.

222. Note that the em dash here signifies that Max stopped speaking; it does not indicate that he spoke words that the author has failed to print. For the latter situation, the correct punctuation would be a three-em dash, ———.

78 Eminent, Emanate, Imminent, Immanent

THE WORDS *eminent*, *emanate*, *imminent*, and *immanent*, although they sound and look similar, have different meanings. All the words are correct, so the example sentences formed from them are all classified as good. Do not embarrass yourself, however, by substituting one for another.

You should write *eminent* when you mean *outstanding*.

> GOOD: Lyn asked Max, an eminent decision analyst, what he wanted to do about the unpleasant odor that permeated the bedroom.

You should write *emanate* when you mean *to issue forth from*, or to *spring from*.

> GOOD: When Max and Lyn tracked down the offensive odor that emanated from under the chest of drawers, they found a decomposing lizard.

You should write *imminent* when you mean *about to happen*, or *impending*.

> GOOD: When Max moved aside and Lyn saw what was left of the lizard, she was in imminent danger of losing her breakfast.

You should write *immanent* when you mean inherent, or existing only in the mind.

> GOOD: An immanent god is an abstract principle that exists in the mind; the smell of the lizard, in contrast, was verging on transcendent.[223]

You should not substitute one of these words for another. Consider the differences among the following sentences:

> GOOD: Max had an eminent idea; that is, Max had a brilliant idea.

> GOOD: Max had an idea that emanated[224] from his past: that is, the idea leaped from Max's history.

> GOOD: Max had an imminent idea; that is, Max was about to have an idea, or Max had the beginnings of an as-yet-unformed idea.

> GOOD: Max had an immanent idea; Max's idea had existed in his mind only.

> SPLENDID: Lyn is an imminent eminent emanator of hitherto immanent ideas.

> GOOD: Carver is an eminent scientist.

> GOOD: Carver emanates charm and brilliance.

> GOOD: Carver's imminent arrival sent Lyn dashing around the house.

> GOOD: Carver was not sure what it would mean for him to be an immanent guest in Lyn's house.

---

223. An immanent god may appear closer than a transcendent one, however.

224. Unfortunately, *emanant* is not considered to be a word. Perhaps we should coin it: An *emanant idea* should be one that leaps from your mind.

GOOD: His Eminence waited for the smoke to emanate from the chimney, and relished the imminent reaffirmation of divine immanence.

THE PRINCIPLE FOR LUCID WRITING here is that it is eminently reasonable to emanate authority and knowledge when you write. Maintaining immanent knowledge carries the imminent risk of forgetting it.

79 *Expected but Nonarriving Agents*

WHEN YOU CREATE in your reader the expectation that you are about to name an object or creature about which or whom you have been speaking, be careful to name the one that you mean.

> BAD: Galloping down the road in a frenzy, Lyn saw a horse that had thrown its rider.

After reading the first clause, you are primed to find out precisely who or what was galloping down the road in a state of distress. When Lyn turns up, you immediately form a mental image. There goes Lyn, flying down the road, as she notes over her shoulder in a field a horse, which has recently thrown its rider, grazing peacefully on the daisies. What is wrong with this picture?

> GOOD: Galloping down the road in a frenzy, the horse that had thrown its rider sped past Lyn without noticing her feeble attempts to flag it down.

> BAD: Heaving her backpack over her shoulder, Lyn's vacation started as she ran to catch the plane.

> GOOD: Throwing down her backpack, Lyn started her visit with her mother with a hug that nearly suffocated both women.

BAD: Given that she was throwing her favorite teddy bear across the room, Peter deduced that he had crossed one of Sachiko's boundaries.

GOOD: Given that she was singing and waltzing around the living room with a feather duster, Sachiko, Peter deduced, was in a mood more conducive to productive conversation.

BAD: Giving up her job and traveling for a year or two, Greg surmised might be a good way for Nicola to relax after an arduous first year teaching.

GOOD: Giving up his own job and traveling, Greg surmised, might be a good way for him to discover something he knew not what.

BAD: Sneezing and coughing and sleeping all day, and never purring, Lyn was getting worried about Red.

GOOD: Purring all day and hunting all night, Red demonstrated to Lyn that a trip to The Dreaded Vet was not required.

Note that a special case of this problem occurs when you follow the comma with a clause in passive voice:

BAD: To solve the problem of derangements, generating functions were used by Max.

GOOD: To introduce power series, Max defined the general binomial series $B_t(z)$ and the generalized exponential series $E_t(z)$.

THE PRINCIPLES FOR LUCID WRITING here are that, to be fair to your reader, <u>you</u> should name whom you intend to name, and should deliver on any promises that you make about the forthcoming identification of an agent.

80 *Its Versus It's*

Y OU SHOULD distinguish between *its,* which means *belonging to it,* and *it's,* which means *it is.* Using the incorrect term is likely to embarrass you. (Using *its'* will humiliate you.)

Be aware that, in formal writing, you should avoid *it's* because you should avoid contractions. In casual writing, you can use *it's,* provided that you use it correctly.

BAD: This approach will not work for a complex project, because, if the rule base is large, its inefficient to edit, compile, and run repeatedly.

BAD: The final result was unexpected, because the rule base added a new rule to its' own contents.

GOOD: We should proceed carefully, because, if a rule base is divided into more than one file, it's a good idea to place the attribute declarations in a separate file that is compiled first.

GOOD: Lyn was puzzled, because, if a rule's specificity is the number of value tests performed on the LHS, then what is its sensitivity?

BAD: Its been a long time since Max and Lyn spent a few days doing nothing of import but enjoying each other.

GOOD: It's been a great year for workaholism in the Dupré–Henrion household.

GOOD: One of the delicate spotted fawn twins was prancing along behind its mother, outside of Lyn's study.

BAD: Raquel believes that its acceptable to wake up people by telephoning them at 2 in the morning.

GOOD: Max believes that one way to answer the telephone is to drop its handset into the wastepaper basket.

GOOD: Lyn believes that it's delicious to turn off the telephone in the evening to ensure a peaceful night.

BAD: When you have to choose between disturbing a cat or remaining trapped in an armchair while that cat snoozes on your lap, its a tough call.

GOOD: When Lyn accidentally knocks BB off the bed, BB thinks it's tough to be a loyal cat.

GOOD: BB wondered whether the banana slug knew how silly its underbelly looked as it traveled slowly across the picture window in the bedroom.

THE PRINCIPLE FOR LUCID WRITING here is that it's a good idea to use *its* to mean *belonging to it,* and *it's* to mean *it is.* It is also a good idea to avoid contractions in formal writing.

81 Adverbs Versus Adjectives

WORDS THAT modify the way that an activity is performed *(adverbs—words that modify a verb)* are different from words that modify a result or product or object *(adjectives)*. Often, the difference is signified by whether the word ends in *ly* (if it does, it is an adverb); in other cases, the adverb and the adjective are different words.

Consider the following examples:

> BAD?: Darlene sliced the onions thinly.
>
> GOOD?: Brian chopped the parsley fine.

The first sentence is bad if and only if you mean that Darlene made thin slices; if you mean that she went about slicing in a thin manner—we imagine the wrist-motion equivalent of a thin voice—then the first sentence is good. Similarly, the second sentence is good if and only if you mean that the result of Brian's labor was parsley chopped to fine granularity; if you mean that Brian performed his job brilliantly, then you should instead say

> GOOD?: Brian chopped the fresh parsley well (that is, Brian chopped the parsley with élan).

Here is another example:

> BAD?: Garrett did not like to drive slowly.

If what Garrett did not like was to undertake in a slow manner the captaining of his car (perhaps reaching lazily for the gear shift, or dawdling about dimming his brights), then and only then the sentence is good, not bad. If, however, Garrett was a speed freak, then the sentence should be

GOOD: Garrett did not like to drive slow.

Here are more examples:

GOOD: Sachiko thought that Peter smelled badly.
       Perhaps Peter had a cold.

GOOD: Sachiko thought that Peter smelled bad.
       Perhaps Peter had not bathed recently.

SPLENDID: People smell; life stinks.[225]

GOOD: If we plot $V_1 - V_2$ horizontally, and *I* vertically, then the transconductance amplifier will work in only the first and second quadrants, and will function as a two-quadrant multiplier.

       This example may be misleading, because we might imagine ourselves lying on our backs to plot $V_1 - V_2$, and standing up to plot I. However, it is correct to say that you plot a function horizontally, meaning that the plotting is carried out in the horizontal direction.

GOOD: In the graph of the response of the circuit, the horizontal axis represents time, and the vertical axis represents ΔV_{out}.

       Here, it is the axis, rather than the plotting, that is horizontal or vertical.

---

225. Contributed by Richard Adamo.

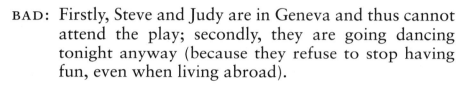

GOOD: Max mixed the milkshake smoothly.

Max was his usual urbane self while mixing the milk-shake; that is, he carried out the action of mixing without an awkward gesture.

GOOD: Lyn whirled the yogurt and wheat bran smooth.

Lyn stressed out the blender until the drink had no lumps in it.

BAD: Firstly, Steve and Judy are in Geneva and thus cannot attend the play; secondly, they are going dancing tonight anyway (because they refuse to stop having fun, even when living abroad).

GOOD: First, Max and Lyn are in Woodside and thus cannot hear the traffic; second, they have on their earmuffs tonight anyway (because the house has an average temperature of 40°F in winter).

BAD: Cheap car, runs good.

GOOD: Expensive copy editor, performs well.

GOOD: When Lyn bounces out the door to hurl herself into his arms in greeting, Max is reassured that Lyn feels both well and good.

SPLENDID: When asked how he was, Joe replied, after his usual pause for careful thought, "It depends on whether you mean how good I am at existing, in which case I am well, or what value I put on myself, in which case I am good."

THE PRINCIPLE FOR LUCID WRITING here is that you should be careful about what you are modifying when you use a modifier. You should mark this principle well so that you can write good prose.

82   *Persons Versus People*

PEOPLE IS the plural of that creature denoted by *person*. *Persons* is the correct, albeit stilted, word to use when you wish to denote multiple bodies.

UGLY: The system is now being used by 145 persons.

GOOD: The installed base comprises 37 people.

GOOD: The users carried their palmtop machines on their persons.

UGLY: The jazz club had so many persons in it, all generating so much smoke, that Tom could not see more than a foot from the stage.

GOOD: Tom was making love to the ivories, and all the people were going crazy.

GOOD: Tom was carrying a fountain pen on his person, just in case the fans asked for his signature.

UGLY: Max found most persons unreasonable.

GOOD: Lyn found most people riotously funny.

SPLENDID: After eating the expensive meal, Max and Lyn found that they had no money on their persons, so they had to ask people for help.

UGLY: No matter how many persons you please, there will always be someone who has a complaint.

GOOD: Although they may not always acknowledge it, people prefer to be happy, if given the chance and the requisite skills.

SPLENDID: Although they may not always consider it, people prefer not to have pigeon droppings on their persons.

Note that *peoples* is a plural denoting multiple sets of people who are grouped by their culture, country, tradition, or kinship.

GOOD: After their leaders signed the peace treaty, the peoples of all sides in the conflict celebrated for many days.

GOOD: Peoples of the world, unite!

SPLENDID: Various peoples peopled the new world, to many people's delight.

THE PRINCIPLE FOR LUCID WRITING here is that you should almost never use *persons*; you should use *people*[226] when you intend to denote the plural of *person*, and *peoples* when you intend to denote multiple groups of people. People may wish to keep this book near their persons.

226. I am not suggesting that you should use people. The principle provides an excellent example of why you must differentiate between a word and that word's denotation.

83 Cap/lc

Y OU WILL frequently want to style lines in capital and lowercase letters—called *cap/lc* in publishing. You probably will want to use cap/lc for chapter titles, section headings, or publication titles.

In general, whenever your document is being published with a design other than your own, you should set these items cap/lc. If the design calls for all capital letters, called *all caps,* for all lowercase letters, called *lc,* or for only the first letter to be capitalized, called *lead cap,* then the conversion is simple; if, however, you set your heads in all caps, lc, or lead caps, and the design calls for cap/lc, someone other than you will have to make judgments about what to capitalize; in addition, your publisher may rekey your heads, with the risk of introducing errors into your carefully proofread manuscript.

Do not use cap/lc in running text where lowercase letters are serviceable. Cap/lc is visually distracting in running text, and adds false emphasis to your terms—it is usually unnecessary. Use it only when you have to, such as when you are setting titles of books or movies.

In general, you should not capitalize any words in your text unless you have a strict reason for doing so. A few writers sprinkle capital letters all over the page, for no good reason. Reserve capitals for titles, names, and other needful occasions.

The nine rules that govern how you set phrases in cap/lc are simple; they are listed on the following page *in order of precedence,* each subsequent rule being governed by those that precede it:

1. Always capitalize the first word (that is, begin the first word with a capital letter), unless, for another reason, the first letter would never be set as a capital. For example, if the first letter is the lowercase name of a variable—say, *n*—then you should not capitalize it.[227]

2. Always capitalize the final word (unless, again, for another reason, the first letter would never be set as a capital).[228]

3. Always capitalize words containing five or more letters (unless the first letter would never be capitalized).

4. Do not capitalize prepositions *(on, in, over, from, to, with,* and so on; and, in people's names *von, del, de,* and so on, if the person who owns the name sets it lowercase).

5. Do not capitalize articles *(a, an, the).*

6. Do not capitalize the *to* in an infinitive; do capitalize the other part of the infinitive—for example, *to Be.*

7. Do not capitalize connectives *(and, or,* and so on).

8. Capitalize every word.

9. Consider each word in a hyphenated word to stand alone, and capitalize it or do not capitalize it according to the preceding rules.

Thus, you should set the first letter as a capital (with exceptions), and should set every other word lead cap, *unless* that word is not the

---

227. In disciplines that use numerous such variables, you may wish to adopt the convention of capitalizing the *second* word of a sentence that begins with a word that must begin with (or is) a lowercase letter. An example follows:

d Here is used to denote the decision under consideration.

228. Rule 2 can lead to cap/lc lines that look inconsistent; for that reason, I consider adoption of the rule optional. If you do use it, however, then use it consistently throughout your document.

final word and is a *preposition, connective,* or *article* of less than five letters, or is the *to* in an infinitive.

GOOD: Now Is the Time for All Good Men to Come to the Aid of Their Karakuls

GOOD: Graphics Operations on Rounded-Corner Rectangles

GOOD: Max and Lyn Practicing Footling Nonsense Under the von Neumann Architectural Model

GOOD: The Case of *Max v. Lyn* to Be Heard in Cosmic Court Tomorrow

GOOD: All of Versus Some Of[229]

GOOD: Richard Tickles Madeline While Sophia Bites Betsy

GOOD: Chart of Correlation of Years of Breast Feeding and Number and Severity of Toothmarks

GOOD: How to Get More Memory Crammed into Less Space

GOOD: Screen Dump from Dr. Henrion's Brainchild

GOOD: BB Has Kittens; Household Joyously Disrupted; New Relationships Formed!

THE PRINCIPLE FOR LUCID WRITING here is that, in cap/lc lines, You Should Set the First Letter as a Capital (with Exceptions), and Should Set Every Other Word Lead Cap *Unless* that Word Is Not the Final Word and Is a *Preposition, Connective,* or *Article* of Fewer than Five Letters, or Is the *to* in an Infinitive.

---

229. This line provides an example of how application of rule 2 can lead to the appearance of inconsistency.

84 Feel Versus Think

F EELINGS ARE not the same as *thoughts*, *perceptions*, or *beliefs*. Failure to distinguish *feel* from *think* will lead you to make false claims about the emotional or perceptual capacities of your actors.

You should write *feel* only when the creature under discussion is experiencing an emotion, is responding to a perceptual stimulus, or is in a given overall state.

> GOOD: Jade feels slightly soapy when you rub it.
>
> GOOD: The covering used on the wrist-support pad and on the mousepad feels scratchy.
>
> GOOD: After climbing Windy Hill, Lyn usually feels pleasantly exhausted.
>
> GOOD: Shellee felt delighted to be back in California after all those months in Brooklyn.
>
> GOOD: Michael felt a slight tug as his shorts were caught by a thorn during the hike.
>
> GOOD: Max felt guilty because he had not taken Lyn dancing for more than 3 months.
>
> GOOD: When Max rubbed her back, Lyn felt deliciously sinuous and sensual; she also felt good.

You should not write *feel* when you wish to denote an intellectual endeavor. You should use the most appropriate word to denote the thought process that you intend.

BAD: Max and Lyn felt that the giant lizard in the bedroom was another of Red's involuntary houseguests.

GOOD: Max and Lyn felt frightened out of their wits when, late one night, Red dropped a rattlesnake into the bed between them.

GOOD: Max argued that the scurrying in the ceiling was probably caused by rats, whereas Lyn analyzed the noises and decided that they were unquestionably made by a porcupine running under a copper pipe.

GOOD: Max and Lyn thought that enough was enough; they believed that the time had come to limit the household census.

BAD: Helen feels that computers are not always simpatico companions.

GOOD: Helen thinks that computers can be extremely uncooperative.

GOOD: Helen feels exhausted when she thinks of all the electronic mail she has to answer.

BAD: The researchers felt that their data indicated a negative trend toward the wrong direction.[230]

GOOD: The investigators believed that their results were remarkably reliable.

GOOD: The investigators felt ashamed of their sloppy data-collection methods.

---

230. It is a common error to speak about *trends toward directions* (for example, *a trend toward the positive direction*); the error is compounded when the writer uses the British spelling *towards*.

BAD: Peter felt that Lyn was not working as hard as she should have been.

GOOD: Peter felt lousy after eating cookies that Lyn baked.

GOOD: Peter thought that perhaps Lyn had added crow to the dough.

THE PRINCIPLE FOR LUCID WRITING here is that you should limit your use of *feel* to those situations in which a creature is genuinely experiencing a feeling; you should use *think, believe, argue,* and so on to denote intellectual activity. You may feel confused, but you should not feel that this segment is confusing.

85 *Parallelism*

THE NOTION of *parallelism* is critical to good writing, yet it is a difficult one to convey without use of technical language. Once you tune your ear, however, parallelism is obvious, and lack of parallelism is not only obvious, but also irritating.

Sentences, and parts of sentences, have form and structure. Basically, you maintain parallelism by ensuring that the various parts of your writing have the same structure and form.

To understand what structure and form are, consider that a question

> GOOD: Will you sleep well tonight?

is different from an assertion

> GOOD: You will sleep well tonight.

and from an imperative

> GOOD: Sleep well tonight!

Similarly, a head with an article (such as *the* or *an)*

> GOOD: The Fattest Cat in Woodside

is different from a head without an article

> GOOD: Cutest Fluffball in California

and from a head that contains a verb:

> GOOD: BB Mated Successfully!

The following examples all are parallel. (Note that content does not affect parallelism: Three entries mention more than one creature.[231])

GOOD: What Red caught.

GOOD: How Red caught the king rat.

GOOD: When Red dropped the rat.

GOOD: Where Lyn found the rat.

GOOD: Why Lyn screamed.

The next sentences, however, are different from the preceding set and from each other:

GOOD: Lyn screamed because the rat dropped out of her bedroom closet onto her shoulder.

GOOD: Cats hunt.

 You should write *parallel chapter titles and heads*. Ideally, all your heads, on all levels, should be parallel with one another. If circumstances dictate that global parallelism would be awkward, however, you can get away with maintaining parallelism within a set of heads on a given level.

BAD: Chapter 1   The Problem

Chapter 2   Solving the Problem

Chapter 3   A First Solution of the Problem

Chapter 4   Second Problem Solution

Chapter 5   Why Solve the Problem?

Chapter 6   How the Problem is Solved

Chapter 7   Have You Solved the Problem?

Chapter 8   You Should Do Problem Solving!

---

231. I assume that even a past creature deserves to be thought of and referred to as a creature. I also assume that any creature caught by Red is a past creature.

GOOD: Chapter 1 Problem

Chapter 2 Method for Solution of the Problem

Chapter 3 First Solution

Chapter 4 Second Solution

Chapter 5 Rationale for Solutions

Chapter 6 History of Solutions

UGLY?: Section 1 Designing[232]

Section 2 Prototyping

Section 3 User Testing

Section 4 Refining

Section 5 Implementing

GOOD: Section 1 Design

Section 1.1 Initial Meeting

Section 1.2 Project Objectives

Section 1.3 Specifications Documents

Section 1.4 Time Line

Section 2 Prototype

Section 2.1 Why Should You Build It?[233]

Section 2.2 How Can You Build It?

Section 2.3 With What Can You Build It?

---

232. This example receives the questionably ugly classification because many people prefer to avoid using the form *Xing* in heads. The example is parallel, however, and there might be circumstances in which it is the most informative and least awkward choice.

233. I do not recommend that you set heads as questions; if you do so, however, then be sure that at least all heads in a set on a given level are cast as questions.

You should write *parallel entries* in a list or table column.

BAD: Homer built the kitchen cabinets using several wood-working techniques:

- Wood carving
- Sawing of wood
- Planing the wood
- The drill
- Sanded it smooth

GOOD: Lyn carves wooden masks in eight steps:

1. Select a block of hardwood
2. Create a form
3. Attach the block to the form
4. Rough out the facial structures
5. Carve the details
6. Scrape the wood smooth
7. Saturate the wood with thin shellac
8. Sand and oil the wood

BAD: In this talk, I shall discuss

- The QMR-DT Project
- Why We Decided to Submit the Grant
- How Did We Handle Import Values?
- Assessing Prevalence
- Analysis of Trial Runs

GOOD: In this paper, I shall discuss why we undertook the research, how we carried out the research, what our results were, what our findings mean, and where future researchers might concentrate.

BAD: The bag of a well-known cat food is printed with this text, with accompanying pictures of healthy cats:

- Health & Vitality
- Bright, Clear Eyes
- Shiny, Full Coat
- Low in Magnesium

GOOD: As no self-respecting cat owner wants to induce a magnesium deficiency in the cat, although she might be willing to buy catfood in a bag that had not been subjected to an editor, the following phrasing would be preferable:

- Good Health and Vitality
- Bright, Clear Eyes
- Shiny, Full Coat
- No Problems from Excess Magnesium

SPLENDID: A cat who eats this food will have all the preceding characteristics; the food is low in magnesium, high in fat, and rich in vitamins.

You should write *parallel clauses* within a sentence

BAD: Adrienne was so worried about work that she contemplated canceling the final day of her honeymoon, had driven home quickly, and to jump out of the car and turned on her terminal.

GOOD: Brian practiced being a good husband by soothing Adrienne, saying not a word in protest, considering carefully whether he would rather be stuck in an airless office in front of his computer, and whispering about opulent sunsets and strolls through charming landscapes.

BAD: Once home, Max whipped out his papers, spread them on his desk, read them, and never even noticing Lyn who was waiting in her silk gown.

GOOD: Once home, Lyn stripped off her clothes, ran a hot bath, soaked herself, and never even noticed Max.

BAD: Max feels pressured because Lyn is talking about beautiful places to go for vacations; when Daphne wanted him to visit England more than once per century; by Devon, who wants checks signed; his in box continues to be full; and Lyn whispering about going to a play, why not attend a movie, and Al and Mary's hot tub.

SPLENDID: Lyn loves to go for long hikes ending at sunset; to sit on the deck watching the fawns graze; and to wake up Max every morning by playing music, brewing ginseng tea, and bestowing a kiss.

BAD: Marina wanted to take the photograph for the book cover, but she was a perfectionist. Thus, she was worried about the quality of the light, about posing the models, about would the children interfere, the problem of props, and could she get the film to the developer on time?

GOOD: Reno and Cosmo wanted to paint a portrait of Swix, but they were inexperienced artists. Thus, they took out their paintbox, spread paper all over the floor, poured paint all over Swix, told Swix to lie down on the paper, told Swix to roll over, let Swix return to snoring on the white living room carpet, and then presented Marina with the product of their labor.

You should generally write *parallel sentences or fragments* when you are describing *a series of steps*.

BAD: The specified code is compared against the values in the left column of the table. We consider as a match the greatest value in the left column that is not greater than the code. You should use offset to determine which value in the row that contains the matching table cell will be returned. A code with a value less than the first value in the left column of the table will be returned as ERR.

GOOD: Enter the query criteria on your worksheet. Enter the field names that you will want to copy in the output area of your worksheet. Enter `\Data Query Input`, and specify the range for your database, including field names. Choose `Criteria` from the `Query` menu, and specify the location of your criteria.

GOOD: Here are three ways to enjoy yourself: eat, hike, and be in love.

GOOD: In David's class on mysticism, pay attention, do not take notes, and argue as much as you like.

THE PRINCIPLES FOR LUCID WRITING here are that you should enforce parallelism throughout your writing, should pay attention to parallelism at each level, and should cast all like entities in like form. The three parts of the preceding sentence are parallel.

86 Points of Ellipsis

POINTS OF ELLIPSIS indicate missing words in a quoted passage, or indicate missing entries in a series. Points of ellipsis are three dots (...) set with thin spaces between them. In many text-editing systems, the three points of ellipsis are available as a single character.

You should not use points of ellipsis to indicate missing lead words in what is obviously a fragment of a sentence.

> BAD: Max had plenty to say about stubbing his toe, including that he had "...always hated that chair anyway," and thought that "...it should be taken out and shot."

> GOOD: Lyn motor-mouthed about her dinner with Egar, expressing thoughts such as that she had "always thought he has a unique perspective," or "could not understand why he didn't pursue a career in art."

> BAD: Lyn was reading to Max at breakfast, "...most famous device to utilize the junction between n- and p-type semiconductors is the light-emitting diode."

> GOOD: When she had confirmed that Max was indeed stunned by this earth-shattering news, Lyn continued, explaining enthusiastically that the "charge carriers on the n and p sides of the junctions are electrons and holes, respectively."

BAD: *The New Yorker Times* [16, p. B5] reported that the 140 teen-age fans "...tore the star's clothes off," and "...leapt on his naked body."

GOOD: The *Herald Tribunal* [35, p. 21] reported that the reception held for the book signing had been "a spectacular success."

 You should use points of ellipsis to indicate where words are missing in the middle of a sentence.

BAD: Anthony noticed that, if he left out certain words, he could turn around the review that panned his work, making it read "This is the best play ever written! I recommend [that you] see it!"[234]

GOOD: Martha said, let's edit the review as follows: "This is... the best... play ever... written.... I recommend... see it."

BAD: Peter said, "This manuscript needs repagination."

As it stands, the example looks correct, but that is the problem: Words have been left out, and you have no way of knowing that there is a gap.

---

234. I assume that you can use your imagination to fill in what the review might actually have said. Here is one possibility: "This is the worst play I have ever seen, and it is being produced at the best time of year. The only play ever put on (or written) that was more boring is that mounted at the Athenaeum last year. I recommend shooting producers who choose such garbage. If you think that's extreme, I can only say I'd like to see it done." You might try this exercise whenever you see broken quoted material praising a book, play, concert, or other subject that is under review.

GOOD: Peter scolded, "This manuscript needs plenty of work, such as...; it also needs repagination."

Now you know part of the story, and you know that you do not know the whole story; your state of enlightenment has improved dramatically.

You should use points of ellipsis followed by a period,[235] exclamation point, or question mark to indicate that a fragment of quoted material was taken from a complete sentence.[236]

BAD: Max wondered to himself, "When will I."

As it stands, Max has not spoken a sentence, so he has no right to the period set at the end of his speech.

GOOD: Lyn cried, "I cannot do...."

BAD: Helen, when she read the manuscript that Lyn and had cooked up over the course of week, thought "What on earth is she."

GOOD: Helen, having sacrificed her weekend to read the manuscript at Lyn's behest, now pondered, "What on earth is this author...?"

You should use a period, exclamation point, or question mark followed by points of ellipsis to indicate that you have left out material between quoted sentences.

---

235. You can distinguish a period from the initial point by checking whether the dot is closed up to the preceding letter; if it is, it is a period. In certain typefaces, the period is also lower and heavier.

236. Certain experts prefer to set the punctuation mark before the points of ellipsis. I find that notation misleading, because it implies that the omission occurs after the punctuation mark, when in fact the omission occurs before the punctuation mark.

GOOD: Red told Lyn, "All I want is a sunny spot in which to lie, and a few hours per day of petting. And I'd like my bowl to be full at all times. ... And a nice crunchy mouse would come in handy now and then." (Purr.)

GOOD: BB was thinking, "He's such an old grump. I'm just trying to groom him!... And all he does is try to bite me. " (But then Red put his arms around her neck.)

To indicate that you have left out one or more full paragraphs between others, you should use points of ellipsis after the period of the sentence preceding the omission. If the next quoted paragraph also is missing one or more leading sentences, then you should place points of ellipsis at its beginning.

GOOD: The following is from Dupré's *Morning Climb* [1995]:

> Climbing Windy Hill can be a spiritual experience, which is one of the reasons why I do it every morning. Another reason is that it's a good 8-mile hike. Often, when I start out in the valley, the sun is shining bright from an azure California sky, with no clouds in sight....
>
> The fog comes swirling over the top, racing down the valleys, reaching out its fingers to greet me, envelop me, wrap me in a cold white blanket. All my senses become muted: sounds are dulled, I cannot see, I can taste only the fog. I am alone....
>
> ...I crest the hill suddenly, panting and sweating, and below me the vista opens: many green rolling hills, with the ocean behind them. Sometimes, all is clear; more often, the gray fog covers and mimics the water, affording me only chance glimpses of the scene below....
>
> ...And returning I feel my limbs heavy, the morning's tensions washed away, leaving me at peace with the world and myself. I am replete. [Dupré, 1994, p. 88]

You should use points of ellipsis to indicate missing entries in a series. In equations, the points should be centered vertically if they are set between operators (that is, if you place the points of ellipsis between plus signs, minus signs, times signs, and so on, then you should raise them so that they are off the base line, to one-half the height of, say, an o; see the second example that follows).

GOOD: $(x + 1), (x + 2),\ldots, (x + n)$

GOOD: $a + b + c + \cdots + z.$

GOOD: Bring along a sandwich for Mary, one for Al, one for Jeff, one for Elizabeth, one for Misha, one for Jean, one for Carlo,..., and one for Lyn.

You should use points of ellipsis with nothing after them (except the closing period) to indicate missing entries in an infinite series. To indicate a finite series, you should specify the endpoint.

GOOD: $(x + y + 0), (x + y + 1), (x + y + 2), (x + y + 3)\ldots.$
        No endpoint indicates an infinite series.

GOOD: $(x + y + 0), (x + y + 1), (x + y + 2),\ldots, (x + y + n).$
        The endpoint $(x + y + n)$ *delimits a finite series.*

You should not use points of ellipsis to indicate that you are not finishing your own sentence or thought. In such cases, you should use *and so on* or (in parentheses only, and then only if that is the style that you have adopted) *etc.*

BAD: I just wish that you would take a vacation with me, laugh more often, want to dance with me,....

GOOD: I just hope that you will finish your dissertation, get a good job, buy a house, and so on.

There is an exception to this principle, however: When you want to emphasize that your reader should use her imagination—that is, when, in essence, you are asking your reader to complete a series—then you can use points of ellipsis.

GOOD: Red slipped out in the middle of the night, watched the grass until he saw movement, focused on this night's prey,...; a few hours later, he jumped happily onto Lyn's pillow and licked himself contentedly.

GOOD: Ann and Ron went to San Gregorio, climbed a sand dune, found a quiet spot, spread out their blanket,...; a few hours later, they left the beach holding hands and beaming.[237]

THE PRINCIPLES FOR LUCID WRITING here are that you should use points of ellipsis to indicate missing material in quoted passages and in series, that..., and that you should use them with a correctly placed period when appropriate.

Copyright © Paul Fusco/Magnum Photos

237. Based on a suggestion by Peter Gordon.

87 *Last*

Y OU CAN RUN INTO problems when you use *last,* because the word is ambiguous. For that reason, you should use more precise terms.

You should use *preceding, previous, foregoing,* or a similar word, if that is what you mean.

> UGLY: The argument given in the last paragraph has been used for centuries.

> GOOD: We shall now prove the theorem that we gave in the previous chapter.

> SPLENDID: We shall now present the details of the algorithm that we introduced in Section 11.2.

When you write *the last paragraph,* your reader does not know whether you mean the preceding paragraph or the final paragraph (which might be several pages ahead), so you should divulge that information. When you write *the last chapter,* you fog your writing in the same way; writing *the previous chapter* helps matters but still requires your reader to note which chapter she is reading, so that she will know which the previous one is. Using explicit cross-references, with numbers, is always your most informative option. For cases other than cross-references, use the most specific phrase that you can find.

> UGLY: During the last performance, the leading man tripped and fell off the stage.

GOOD: During the previous evening's performance, the replacement actor[238] walked off the stage, tearing off her costume in disgust.

You should use *final, concluding, closing, terminal, endmost,* or a similar word, if that is what you mean.

UGLY: The algorithm ensures that all the nodes are visited, but it does not necessarily visit the last node last.

GOOD: The concluding presentation was the one that you just heard; it was not the one that you heard at the end of the evening.[239]

UGLY: Richard scolded, "You should have taken the last right turn."

GOOD: Betsy responded, "Do you mean the final turn before the bridge, or the turn that we just passed?"

THE PRINCIPLE FOR LUCID WRITING here is that *last* usually is the last word that you should choose, because it is ambiguous; you should instead choose a word that conveys the specific meaning that you intend.

---

238. *Actress* is a gender-specific word; use *actor* for both genders.

239. The last presentation was the one that you last heard; it was not the one that you heard last. Ah, yes.

88 *Focus On*

Y OU SHOULD AVOID overusing *focus on*. There is nothing intrinsically wrong with the term, but it gets so much use that it has become thin and flimsy. When you do use *focus*, be certain not to compound your sins by coupling it with *upon*.

UGLY!: This paper focuses upon the telecommuting option in the business world.

UGLY: Our discussion focuses on decision making under extreme time pressure.

GOOD: We shall concentrate on the positive aspects of marriage in this dissertation.

UGLY!: The user is encouraged to focus upon the menu options by the highlighted blinking text.

UGLY: The user is forced to focus on the text that she has selected for editing.

GOOD: The user is forced by a loud alarm bell to pay attention to the warnings.

GOOD: During user testing, the camera is positioned such that the lens is focused on the user's face.

UGLY!: The model focuses upon the acid-rain component of air pollution.

UGLY: The model focuses thought on the key variables.

GOOD: The model emphasizes the idiotically complex relationships among variables.

SPLENDID: The model makes alarmingly clear that political variables often are weighted more heavily than are functional variables.

UGLY: Some people think that Max is far too focused on his work.

GOOD: Several people think that Lyn is obsessed with Max.

THE PRINCIPLE FOR LUCID WRITING here is that you should avoid the worn-out *focus on*, and should instead ~~focus on~~ turn your attention to more interesting alternatives.

89 Troublesome Plurals

THERE ARE numerous plurals that you do not form by simply adding s to the singular.

You should use Latin and Greek plural endings in formal scientific writing. Generally, whenever you are writing for an academic audience, whatever your topic or intent, I recommend that you use Latin and Greek plural endings. Whatever decision you make for a given manuscript, cleave to it consistently.

Singular	Plural	Singular	Plural
alga	algae	ellipsis	ellipses
alumna	alumna	erratum	errata
alumnus	alumni	formula	formulae
antenna	antennae	fulcrum	fulcra
appendix	appendices	index	indices[240]
automaton	automata	interregnum	interregna
axis	axes	matrix	matrices
bacillus	bacilli	medium	media
bacterium	bacteria	memorandum	memoranda
cactus	cacti	minutia	minutiae
cortex	cortices	narcissum	narcissi
curriculum	curricula	nimbus	nimbi
datum	data	nova	novae

---

240. Do not use *indices* for the plural of any index other than the sort used in mathematics; for an index in which you may look up items, for example, use *indexes*.

Singular	Plural	Singular	Plural
nucleus	nuclei	syllabus	syllabi
papyrus	papyri	symposium	symposia
radius	radii	thesaurus	thesauri
radix	radices	trauma	traumata
schema	schemata	ultimatum	ultimata
stigma	stigmata	vertex	vertices
stratum	strata	vita	vitae

Similarly, you should form the plural of words ending in *is* by converting the *i* to an *e*.

Singular	Plural	Singular	Plural
analysis	analyses	mitosis	mitoses
antithesis	antitheses	parenthesis	parentheses
crisis	crises	synopsis	synopses
diaeresis	diaereses	synthesis	syntheses
exegesis	exegeses	thesis	theses

When you form the plural of a word ending in o, you should determine whether to use s or es by looking up the word. If it is not in the following list, use the first spelling given in the most recent edition of *Webster's Collegiate Dictionary.*[241]

Singular	Plural	Singular	Plural
bongo	bongos	dodo	dodoes
bravado	bravadoes	domino	dominoes
bravo	bravos	echo	echoes
casino	casinos	ego	egos
crescendo	crescendos	embargo	embargoes
desperado	desperadoes	fandango	fandangos

---

241. I recommend that you use *Webster's Collegiate Dictionary,* Springfield, MA: Merriam-Webster; that is the dictionary of choice for most publishers in the United States.

Singular	Plural	Singular	Plural
fiasco	fiascoes	ratio	ratios
flamingo	flamingos	salvo	salvos
ghetto	ghettos	so-and-so	so-and-sos
halo	halos	solo	solos
impresario	impresarios	tuxedo	tuxedos
logo	logos	veto	vetoes
manifesto	manifestos	virago	viragoes
obligato	obligatos	volcano	volcanoes
palmetto	palmettos	yo-yo	yo-yos
peccadillo	peccadillos	zero	zeros

You should also be careful in forming the plural of a term comprising more than one word (even if the term has been closed up and is now set as one word); you are particularly likely to get caught in forming the plural of foreign terms (especially French for the compound terms, and all languages for simple words).[242]

Singular	Plural
aide-de-camp	aides-de-camp
beau	beaux
bête noire	bêtes noires
cause célèbre	causes célèbres
chargé d'affaires	chargés d'affaires
comrade-in-arms	comrades-in-arms
court martial	courts martial
eau de cologne	eaux de cologne
fait accompli	faits accomplis
falling out	fallings out

---

242. An interesting plural is that for the singular *fish*. *Fish* is the plural for many fish of one species; *fishes* is the plural for many fishes of different species. Thus, *All the fishes in the deep blue sea* is correct. *There are other fish in the ocean* is also correct, provided that you agree that only fish of the same species are of interest.

Singular	*Plural*
governor general	governors general
grant-in-aid	grants-in-aid
jack-of-all-trades	jacks-of-all-trades
master-at-arms	masters-at-arms
member of congress	members of congress
mother-in-law	mothers-in-law
nom de plume	noms de plume
notary public	notaries public
objet d'art	objets d'art
passerby	passersby
petit four	petits fours
portmanteau	portmanteaux
runner up	runners up
sackful	sacksful
shovelful	shovelsful
spoonful	spoonsful
tableau	tableaux

THE PRINCIPLES FOR LUCID WRITING here are that you should use Latin and Greek plural endings unless you have a good excuse not to do so,[243] and that you should use care in forming plurals from terms that end in o, that comprise more than one word, or that are foreign.

243. A good excuse is that certain publishers explicitly forbid certain Latin plurals, for example. The same publishers also usually forbid Latin abbreviations (*e.g.*, *i.e.*, *etc.*, *ibid.*, *op. cit.*, and so on). Apparently, they believe that the reading public is not sufficiently literate to handle such terms or endings.

363

90 Around

AROUND HAS its own proper uses; it does not mean *about* or *approximately* or *circa*.

BAD: The program uses around 4 MB of RAM.

GOOD: The disk has approximately 4 MB available.

BAD: Max wanted to hold the meeting around 2 o'clock in the morning.

GOOD: Brian suggested that perhaps meeting at about 2 in the afternoon would be more conducive to mutual comprehension.

SPLENDID: Suresh thought they should hold the meeting around the table, but he kept his mouth shut.

BAD: Lyn's first terminal was an antique Heath kit, from around 1950.

GOOD: Punch cards were in use years ago, circa the era of the flower children.

BAD: Sometimes it seems that there are around 7000 skunks living in Max and Lyn's bedroom ceiling.

GOOD: Lyn thinks that the plasterboard will probably give way if roughly three more skunks join the fray.

GOOD: Lyn thinks it advisable not to walk around the house in heavy boots, because she does not want to startle the skunks.

THE PRINCIPLE FOR LUCID WRITING here is that there are sufficient serviceable words that mean *approximately, more or less, about, roughly, generally,* and *in round numbers;* you should not drag *around* into service for this purpose, but should rather let it stand for its own perfectly respectable meaning. Draw a circle around this principle to be sure that you understand the idea.

91 Nose

I̲ᴛ ɪꜱ critically important that you sniff out your audience before you begin writing. A good writer develops *nose,* which is her ability to identify the audience sufficiently well that she can pitch her prose accurately.

 You should *assess the backgrounds of your readers*. For example, if you are writing about the use of computers in a medical application, does your audience comprise physicians, or computer scientists, or both? If you are writing a proposal, will your readers be from your own discipline, conversant with technical terms, or will they be from diverse backgrounds, several of them possibly being oblivious to work in your field? If you can identify a homogenous audience, your task will be simpler. If, however, your readers are heterogeneous, then you must not ignore that state of affairs. As you write, you must explain concepts that will not be familiar to various segments of your audience. If necessary, you can insert sections or chapters that you label as optional, explaining that people who already have this expertise or are familiar with this information should skip ahead. For short conceptual explanations or definitions, you can use footnotes or end notes.

You should *judge your readers' levels of expertise*. For example, if you are writing a textbook about a given kind of VLSI circuit, you must decide whether you expect your reader already to know the basics of, say, transistor physics, or whether you will provide a tutorial. In a book, you should state in your preface what is the prerequisite knowledge for understanding your writing. You can give the same information in a précis of a talk.

You should determine *the degree to which formality is required*. In a doctoral dissertation, addressing your audience as *you guys* probably will not sit well. Less obviously, slightly casual phraseology may not be appropriate. If, in contrast, you are writing an electronic-mail message to your colleagues to invite them to play on the company volleyball team, formal wording may be a turnoff.[244]

You should be aware of *the political and religious climate*. Your blockbuster introductory joke poking fun at the Church will win you no popularity contests when you give a talk in the Bible Belt. Certain forums are appropriate for political riposte; others are decidedly not. As numerous stories of disastrous comments made by holders of public office show, there are circumstances under which it is wise to watch your mouth.[245]

You should *know enough about the demographics of your audience* that you can avoid inadvertently insulting or infuriating a large proportion (or even a small proportion) of your readers or listeners. For example, if your audience comprises a large portion of people who are out of work, talking about bleeding hearts who worry about the joblessness problem, which you know does not exist, may earn you a few well-deserved rotten tomatoes. If many of your readers are students, *sophomoric* may or may not be the adjective you want to use. If you are hoping to sell your book to a major corporation, you may not want to make wisecracks about that corporations' products during your presentation at corporate headquarters.

---

244. I use the term advisedly, as an example of casual phraseology.

245. See previous note.

THE PRINCIPLE FOR LUCID WRITING here is know thine audience.

92 Literal and Virtual

Many writers use *literal* or *literally*, and *virtual* or *virtually*, to mean the opposite of what these words literally mean.

You should *not* use *literally* as though it is a *magnifier* (as are *extremely, substantially, hugely, vastly,* and so on). *Literally* means *not metaphorically.*

BAD: There was a literal explosion of science at the new research laboratory.

Had there been a literal explosion — a chemistry experiment gone awry, perhaps — there might have been fatalities. We presume that the writer's intended meaning was metaphorical, so literally is precisely the least appropriate word to use in this context.

BAD: Lyn was literally green with envy when she thought about the fun Steve and Judy were having in Geneva.

Had Lyn been literally green, Lyn would have been rushed off to the nearest tertiary-referral center to be displayed on grand rounds.

BAD: Brian was literally surfing in cyberspace.

Brian died of oxygen deprivation.

BAD: It was literally as cold as a witch's teat.

Here, our objection is simple: We can only assume that the writer does not have the requisite experience to make such a claim.

BAD: The faculty meeting was literally a barrel of laughs.

Can you form the appropriate mental image?

BAD: A worm can literally bring the Internet to its knees.

But can it cause the Internet to roll over and play dead?

BAD: Lyn's computer literally threw a tantrum.

For that matter, it is not clear how anyone can throw a tantrum, or at whom the tantrums get thrown.

BAD: BB was literally quaking in her boots when Marina visited to give the cats their vaccinations.

BB is not a puss'n'boots.

BAD: Max literally came unglued.

Let us hope that Lyn was there to glue him back together again.

The correct use of *literally,* as I said, is to underline that you do *not* mean metaphorically.

GOOD: When Max heard the news, his jaw literally dropped.

I intend you to understand that Max stood there with his mouth open.

GOOD: When Richard decided to take up sky diving, Betsy was literally left holding the baby.

When Richard returned to earth, he diapered Sophia.

GOOD: Lyn was literally dancing with frustration.

A jitterbug of discomfort was on display.

 You should use *literal* to mean *accurate,* or *word for word,* as well as to mean *not metaphorical.*

GOOD: I would like to know the literal translation of *mors certi, amor incerta est.*

GOOD: Marina's holiday greeting card included a picture of a literal horse's hindquarters.[246]

 You should use *virtual* to mean *in effect,* or *in essence*—to denote an object that serves the same purpose or has the same function as, but for whatever reason is not the same as, another. *Virtual* is the antonym of actual, so you should never use *virtual* to mean *actual*— or *literal.*

BAD: The crowd in *The Shadow* is in a virtual panic.[247]

BAD: During a nasty argument about summer plans, Lyn was so frustrated that she virtually sat down on the trail and refused to budge.

GOOD: The system provides a virtual address space.

GOOD: Swapping out to virtual memory can free up RAM.

GOOD: *Virtual reality* used to be called *artificial reality,* but the latter phrase is an oxymoron.

SPLENDID: Peter donned the goggles and literally got lost in virtual space.

You can also use *virtually* to mean *almost entirely* (but not entirely), or *nearly.*

GOOD: A good text formatter takes virtually all the work out of compiling a table of contents.

---

246. The horse in question, Ruby, was properly attired in a sign reading *Marketing,* Curly was labeled *RAM Drive,* Swix sported *Payables,* and Spud was billed as *Collections.* In the picture used, Ruby had been turned around.

247. Had the crowd been almost in a panic, the sentence—from a newspaper headline—would be correct. The crowd, however, was in a full-fledged panic.

GOOD: When Lyn is working hard, she munches and knoshes and grazes during virtually every waking minute.

GOOD: Max was so exhausted that he was virtually on the verge of collapsing into the linguini.

GOOD: Lyn was so delighted with the spindle candlesticks that she was virtually flying around the room, bestowing kisses on Max every time she passed him.

THE PRINCIPLES FOR LUCID WRITING here are that you should use *literal* when you mean *not metaphorical* (when you mean *metaphorical,* you should say so), and should use *virtual* to mean *not actual* or *nearly.* You should take this principle literally, at virtually all times.

93 *Semicolon*

T<small>HE</small> S<small>EMICOLON</small> connects two sentences that are closely related to each other.

You should use a semicolon when what follows constitutes a complete sentence.

When what follows is a fragment (a *fragment* is a string of words that does not constitute a sentence), you must use a comma or an em dash. Note, however, that you can use an em dash in various cases where a semicolon also would be correct.

When what follows is a complete sentence, you always have a choice between a period and a semicolon.[248] You should use the semicolon when the two sentences constitute a single thought, or are otherwise *highly related*. The semicolon *tightly couples* the two sentences.

> BAD: Although this monitor reputedly can handle oodles of colors, it provides only black-and-white desktop patterns; checks or vertical lines, for example.
>
> *The portion after the semicolon is a fragment, rather than a sentence.*
>
> GOOD: Floppy disks can be a risky medium, because you can easily destroy data inadvertently—by parking a disk on a magnet, for example.

---

248. In certain cases, you also have the option of a colon.

374

GOOD: This interface is not particularly friendly—none of the command names mean anything in English.

GOOD: This equation is difficult to parse; that is, I cannot figure out what it is supposed to do.

GOOD: This machine is difficult to use; for example, it crashes whenever you change windows.

BAD: Holly wanted to live on a farm with plenty of chickens; and to have a stellar career as well.

The portion after the semicolon is a fragment, rather than a sentence.

GOOD: Misha enjoyed hot night life—yet he did not mind chickens quite as much as he pretended that he did.[249]

GOOD: DeeDee wanted to have her own car; she liked chickens just fine, but she did not think that they were a good transportation medium.

GOOD: Holly and Misha cooked yet another humongous meal—and refused to let anyone help them to wash the dishes.

GOOD: Lyn always hung out in the kitchen while Misha was trying to cook; she pestered him incessantly, and kept trying to lick the spoons.

GOOD: Max's head was throbbing. Lyn's heart was sinking.

If the condition from which Max suffered was not causally related to the sensation that Lyn experienced, then the period is correct.

---

249. He says that he likes them with garlic sauce, for example.

SPLENDID: Max's heart was throbbing; Lyn's head was swimming.

The sentences are highly related, so the semicolon is more appropriate than a period would be.

 Note that you should use a semicolon before complete-sentence clauses that begin with *for example* and *that is;* it is a common error to use a comma in this situation, rather than a semicolon. If the clause is a fragment, use an em dash, rather than a comma.

BAD: Misha was wildly busy these days, that is, he was sleeping even less than usual, and worrying even more than usual.

GOOD: Misha was preparing a wedding feast; that is, he was cooking 493 scrumptious dishes, each from his own special recipe.

BAD: We need to handle negated condition elements, that is, not just the joining of positive conditions.

GOOD: The left memory contains with each token a negation count; that is, it holds a count of the number of tokens in the right memory that match the left token.

BAD: Suresh found several unusual bugs, for example, the program displayed an upside-down diagram each time that he entered a value.

GOOD: Brian and Adrienne's wedding was well attended; for example, Brian's 4967 cousins all showed up.

BAD: Len spent much of his time making up rules, for example, no spending without money in hand.

GOOD: Brian supposed it was a good idea to have somebody looking after the details—for example, sending out invoices.

BAD: Sheila was losing weight, that is, she was getting even slimmer, and everyone wondered whether she would blow away some day.

GOOD: Brian was trying to fatten up Sheila—that is, he kept bringing her chocolate-chip cookies—but the technique did not seem to be effective.

BAD: Boris realized that true programmers, for example, people who knew at least one programming language, were a valuable resource.

GOOD: Devon was being hassled by every machine in the office—for example, the copier and the telephone network—but she managed to laugh about it all.

 You should use a semicolon to separate portions of an in-sentence list when at least one portion contains a comma. Note that you should use commas, rather than semicolons, for cases in which no portion itself contains a comma.

BAD: The model contains a *decision node,* which is a choice that you have to make, an *outcome node,* which indicates what will happen, and various *variables,* the values of which influence the outcome.

Two of the commas (the second and the fourth) should be semicolons.

BAD: When you build a model for a complex public-policy problem, you need to consider numerous variables and their influences on one another; to handle uncertainty; and to include political concerns explicitly.

The two semicolons should be commas.

GOOD: When Schmöe does consulting, he has to do knowledge acquisition, which can be tedious or demanding or most often both; model building, in which he makes all the assumptions explicit; and persuading, in which he attempts to convince people not to be such bloody idiots.

BAD: Lyn decided that she would do her own page makeup because she wanted to work directly with an expert, and to develop the design interactively, because she wanted to see the page layout, so that she could check that it pleased her, and because she wanted to be able to add graphics to the text.

Two of the commas should be semicolons.

BAD: Peter decided to give Lyn her head because he had faith that she would produce a high-quality book; because he always tried to be kind to her; and because he knew that trying to rein her in was a hopeless task.

The semicolons should be commas.

GOOD: Peter then worried whether it had been wise to permit Lyn this concession, given that Lyn had not done page layout for a book previously, although she had been working in publishing for many years; given that the publisher had a major investment in the book, having already begun the production process; and given that Lyn could be immensely pigheaded when she was convinced that she was right.

BAD: Homer decided not to complete his doctorate because he was bored with the program; he wanted to get back to being a doctor; and commuting from San Jose was a drag (in addition, he soon moved to Portland to take an even more inviting job).

The semicolons should be commas.

BAD: Greg left the program also, but for different reasons: he already had his doctorate, having earned one by dint of hard work, he wanted to pursue his own specialty, and he liked the snow and the fall leaves on the east coast (in addition, he was well suited to his new position).

Two of the commas should be semicolons.

GOOD: Maria came back to the program to earn her doctorate, but she also had to deal with partying all the time, at her house and other people's; with running around town with Geoff's daughter, who was in her first decade; and with listening sympathetically to all her friends' troubles (in addition, she had to work like a madwoman).

THE PRINCIPLE FOR LUCID WRITING here is that you should use a semicolon to set off a portion that is itself a complete sentence; furthermore, you should use a semicolon to separate in-sentence list entries when at least one entry contains a comma.

94 Code

You should determine in advance what conventions you will use for setting computer code, and for the names of various computer-related entities, be they procedures, commands, objects, rules, or menu selections. Using typographical distinction to indicate what is code (or the name of an entity) and what is English will greatly enhance the readability of any document that mixes computer languages with English.

When you begin to write, you will find it helpful to create and maintain a specifications document that lists and explains the conventions that you choose. Use the specifications as you write. Then, insert them at the beginning or end of your primary document, so that your reader also will know what conventions you are following. If you are working on a multiauthored document, you will find the specifications invaluable.

You can follow common convention and use a monospace typeface for code, to differentiate code from text. Monospace typefaces assign to each letter the same amount of horizontal space; they look like the type produced by old typewriters. `Courier` and `Typewriter`,[250] for example, are monospace typefaces. If you use a monospace typeface for code, use that face to set single words of code within a text line, to set code words within figures, and to set bodies of displayed code.

---

250. The code in this book is set in `Monotype Typewriter`.

If your base text is set in a serif typeface, you may also be able to use a markedly different sans serif typeface for your code.[251]

UGLY: Use Procedure LoadSymptoms when you desperately need an illness.

GOOD: Use `procedure LoadDiseases` when you are desperately tired of it all.

UGLY: The /File Xtract command allows you to save in a worksheet file a portion of your current worksheet.

GOOD: The `/System` command allows you to use the operating-system commands without quitting 1-2-3.

UGLY: Describe what the functions of begin and end are.

GOOD: Do not forget to begin with **begin** and to end with **end**.

---

251. *Serifs* are the small flourishes at the upper and lower ends of the stroke of a letter. Examples of sans (without) serif typefaces are

Gill Sans
Helvetica
Avant Garde

Examples of serif typefaces are

Garamond
Times Roman
Palatino

People have extraordinarily strong opinions about serif versus sans serif type—about whether or when you should use either, about which is easier to read, and so on—but everyone disagrees.

Also note, somewhat tangentially, that characters from different typefaces that are set in the same point size may appear to be of disparate sizes, in both height and width. In the types shown, for example, Garamond is much smaller than is Avant Garde.

UGLY: You can use man in this situation.

GOOD: Be aware that **grep** might be too powerful to use in this situation.

GOOD: When she was learning the system, Lyn became unconscionably excited at the mere thought of trying to grep her man.

You should determine a *consistent style* for comments in code. For example, you should decide whether to capitalize the first letter, and whether to use end punctuation. You may wish to have two styles; for example, you could capitalize the first letter and use a period for comments that comprise one or more sentences, and could use all lowercase letters and no period for comments that are fragments.

GOOD:
```
{ InitDoprior(Qcase.Age, Qcase.Sex);
   is expensive and is not needed. }
```

GOOD:
```
{creates hypothesis tree until finding
   all h's with Rscore<Rtotal*RminF, }
{ or until maxH hypotheses have been
generated }
```

GOOD:
```
end;{ procedure DoCase}
```

GOOD:
```
procedure LoadDiseases;
         { Loads in all existing not-yet-
            loaded diseases from KB, gets
            priors and initializes stuff. }
```

GOOD:
```
{Get f's associated diseases in links }

DumpChildren(QMRfindingID[f], level2,
diseasesForF);
{ for each disease d linked with f }
   for i := 1 to diseasesForF.Num do
```

```
GOOD: baby b                 //Construct a baby
      b.year_born            //Specify when it was
                               born
      b.percent-loaded //Specify how full
                               it is
                         //Display age using
                               a person member
                               function:
      cout << "The baby is" <<b/age ( )
      << "years old. " << endl;
GOOD: (startup
        (strategy mea) ; Set the conflict-
                           resolution strategy
        (make …)        ; Initialize working
                           memory with make
        (make …)        ; Provide commands
        (@wombats.wm)) ; Execute file of
                           make commands
```

Note that you should use only a few (at most) parallel styles; do not mix styles indiscriminately. That is, do not set both sentences and fragments, both lead capital and lead lowercase letters, both end puncutation and no end puctuation, and so on, without apparent logic. (Note that unapparent—opaque—logic is not helpful.)

```
BAD: //Get f's associated diseases
     //Gets f's associated diseases
     //f's associated diseases
     //diseases associated with f are found
     //Use to get f's associated diseases
     //Shows how to get f's associated
       diseases
     //a get command, for f's associated
       diseases
     //diseases associated with f
```

You should set large blocks of code as separate *numbered figures* or *programs* or *boxes*, rather than simply inserting them into the text. This approach makes page layout easier than with unnumbered blocks, and also makes it easy to cross-reference your code by number. Use double numbers (4.3, 8.5, 14.3, and so on) within a book chapter or proposal section, if your level 1 heads use double numbers. Use single numbers (7, 5, 19, and so on) in a journal article or whenever your level 1 heads use single numbers.

> GOOD: Box 7.3 contains the code for the new fathers' baby-sorting algorithm.

> GOOD: Figure 6.4 shows the code for analyzing the stock-market figures.

> GOOD: See Program 94.1 for an example of the subject of this example.

Once you have set up your code in a numbered figure, program, or box, call out the code by number, as shown; do *not* write

> BAD: The spouse-sorting algorithm follows:

> BAD: The new code for analyzing the annual percentages is as follows:

> BAD: See the following Program for an example:

> BAD: The above code will do your laundry in its spare time.

> BAD: For a solution to the dirty-dishes problem, see below.

You should give each figure or program or box a *caption* that identifies the code, and that provides any other useful information about the code.

> GOOD: Figure 4.5. Code for disentangling organizational snarlups. Note that crossed hierarchical divisions and parallel movement create massive confusion, whereas longitudinal flow leads to stabbed backs.

GOOD: Program 18.4 Code segment showing intelligent use of comment lines to document assumptions in the model. In a complex model, explicit assumptions allow the model to evolve over time, as new data become available. Hidden assumptions make the model inflexible, sentencing it to a short useful lifetime.

The principles for lucid writing here are that you should set code in a `monospace`—or **otherwise differentiated**—typeface, should set comment lines consistently, and should set large chunks of code as numbered figures, programs, or boxes.

Courtesy Richard Adamo

Courtesy Dona DeP. Oppenheim

```
procedure RefineSearch (RminStart, Rfactor,
  RminStop: RealType; maxH: LongInt);

  { A breadth-first search algorithm using
    stepwise refinement: }
  { It repeats a search to end with RminFraction,
    from RminStart, decreasing by Rfactor, }
  { until RminFraction< RminStop or until the
    total number of hypotheses > maxHyps}

      var
        n: LongInt;
    begin
      InitStats(o);
      n := 1;
      RminFraction := RminStart;
      repeat
        Search(RminFraction, maxH);
        RecordStats(n);
        if QSPavailable then   { Record error as
                                 SD relative to
                                 QS results. }
          RecordError(SdError(diff, QSPdiff^));
        RminFraction := RminFraction * Rfactor;
        n := n + 1;
      until (RminFraction < RminStop) or
        (NumHalloc >= maxH);
      PrintStats(outfile)
    end; { RefineSearch }
```

Program 94.1 A portion of the RefineSearch procedure used as an example of one way to set large chunks of code.

95 Comparatives

WHEN YOU USE a term that implies a comparison—*bigger, faster, quieter, funnier, more enticing, less annoying,* and so on—you should specify against what the comparison is being made. Otherwise, you leave out critical information; often, your reader cannot tell what is being compared, even if you can.

You should not leave the compared-against object to your readers' imaginations; you should *identify* it clearly.

> BAD: Betsy wanted a larger confidence interval.
>
> GOOD: Richard was hoping for a higher correlation coefficient than the one that they had obtained in the initial study.

> BAD: Unix was weaker in interprocess communication because its only facility for that activity was the pipe.
>
> GOOD: Now, 4.3BSD provides a richer set of facilities than that available previously.
>
> SPLENDID: That is, 4.3BSD uses a set of interprocess-communication facilities considerably more comprehensive than that available previously.[252]

---

252. A [*richer* or *more comprehensive*] *set...than that available previously* is an awkward construction; it is more precise to say *a set* [*richer* or *more comprehensive*] *than that available previously.*

BAD: Forming the product of two fuzzy sets is more interesting.

GOOD: Defining the intersection, union, and complement of fuzzy sets is more difficult than is identifying a fuzzy feeling.

BAD: A data-compression utility lets you store more data.

GOOD: A spell program helps you to correct more typos than you would catch if you were to rely on your own proofreading.

BAD: The primary benefit of this approach is that using the latter[253] makes calculation of large multivariate models simpler.[254]

GOOD: The primary drawback of this search method is that using the method requires more storage space than does employing other methods.

---

253. *It* set here would refer to the *benefit,* rather than to the *approach*—an error common in formal writing. If you find *the latter* awkward, you can repeat *approach,* you can use a different term (for example, *... is that use of this technique makes...*), or you can recast to avoid the problem (for example, *This approach has the primary benefit that its use makes...*).

254. As an example of the problem similar to that detailed in the previous note, a reviewer suggested, as a splendid example, *The main benefit of using computers is that it simplifies the text-editing problem.* In fact, that sentence would be classified as bad, and the construction is one of the most common errors that people make in using pronouns. The *it* in the sentence refers to the *benefit,* rather than to the *using.* Compare that construction with those that I have used in this example set.

You should specify the comparison either by using a *verb* or by using a *noun* (perhaps plus a verb).

BAD: BB's temper is shorter than Red's.

We do not know the answer to "shorter than Red's what?"

GOOD: Lyn's hair is longer than Max's is.

We have added a verb, which implies the noun hair.

GOOD: Lyn's hair is longer than Max's beard.

We have added a noun; note the difference!

BAD: Holly has a wider grin than Misha.

GOOD: Holly has a deeper understanding of chickens than Misha does.

GOOD: Holly's hat is bigger than Misha's briefcase.

BAD: Max's models are always more elegant than the general populace.

GOOD: Max's thinking is often more insightful than is that of the general public.

GOOD: Max's intuition is generally less accurate than is Lyn's gut feeling.

BAD: Carver's one-derful pun is more painful than Peter's.

GOOD: Carver's theories are more outrageous and unsettling than are those in standard physics textbooks.

GOOD: Carver's enthusiasm is more infectious than Max's cold is.

You should use the construction *as ... as* for positive assertions, and *so ... as* for negative assertions.[255] In either case, you should also not omit the second verb.

UGLY: A resolution graph created with a deletion strategy is not as big as one created with an unconstrained resolution rule.

UGLY: A resolution created with a decision strategy is so logical as a solution created with a decision strategist.

GOOD: Unit resolution is as easy to comprehend as is input resolution.

GOOD: Unit resolution is not, however, so easy to understand as is a New Year's resolution.

UGLY: Cosmo is not as tall as Reno.

UGLY: Reno is so friendly as Buttercup.

GOOD: Cosmo is not so heavy as Swix is.

GOOD: Spud is as ferocious as Swix is.

THE PRINCIPLE FOR LUCID WRITING here is that, when you use words that imply a comparison, you should specify precisely what you are comparing with what. Merely making your writing clearer is insufficient; you must make your writing clearer than it used to be, or clearer than your dearest colleague's, or clearer than mud.

---

255. I am grateful to Dona DeP. Oppenheim for suggesting this distinction.

96 *Tables*

TABLES CONVEY information that is represented most concisely in tabular form. Numbered tables are preferable, because you do not have to place them in the text column exactly where you mention them; furthermore, you can refer back to them unambiguously.

 You should set *tabular* tables; that is, your tables should have *columns* and *rows*. A list is decidedly not a table, and should not be mislabeled as one. Note also that a table is decidedly not a figure, and should not be mislabeled as one.

 You should *number* each table. Use double numbers (2.7, 14.5) if your level 1 heads (and thus your figures or examples or equations) have double numbers; otherwise, use single numbers (7, 16).

 You should *call out* every table in text; that is, you should refer to it by number at least once.

> GOOD: Table 6.7 shows the extraordinary demographics of the study population.
>
> GOOD: We collected numerous data on the feeding behavior of frantic finicky felines (Table 4.3).
>
> GOOD: Our results (Table 5) demonstrate clearly that, in the environment presented by our study design, the males are relationally disadvantaged, whereas the females are merely logically deranged.

You should write a *title* for each table. The title names the table, and tells your reader what information the table is designed to convey. Unlike a figure caption, a table title should not contain detailed information. Write the details in the accompanying text or, if necessary, in a table footnote. Your table titles should be tags—that is, they should be noun phrases, or names. Set the titles with the first letter capitalized, and end them with a period.[256]

> GOOD: Correlation between halitosis and broken dates; by age, gender, socioeconomic status, and ratings on the attractive-force scale.

> GOOD: Analysis of results of questionnaire study of mating behavior of suburban toms and queens.

> GOOD: Red's allergenic effects on former friends, by severity.

You should *label each column;* that is, each column should have a table head (as shown in Table 96.1).

You should use a *spanner*—a horizontal line—to indicate level 2 column heads.

> GOOD:

Hiccoughs per week			Age (years)
Choking	Laughing	Gurgling	

You should always indicate the *unit of measure,* when appropriate, in parentheses under the table column head.

> GOOD: Our table has three heads:

Elapsed time (minutes)	Distance (miles)	Awards won ($U.S. 1994)

256. This convention is counterintuitive, because a tag is not a sentence. Nonetheless, it is followed by most publishers. In contrast, when table titles are set with combined capital and lowercase letters, the period is omitted.

Table 96.1 Example table with various components.

Table column head [head of household[b]]	Head[a] above spanner [members of household]		
	Subhead [adult]	Subhead [child]	Subhead [child]
Side head			
entry[c]	entry	entry	entry
entry	entry	entry	entry
entry	entry	entry	entry
Side head			
B	U	G	S
C[d]	A	T	S
[Cathouse]			
[Max]	[Lyn]	[Red]	[BB]
[Doghouse]			
[Sachiko]	[Peter]	[Hobart]	[Pokey]
[Busy house]			
[Pat]	[Marina]	[Reno]	[Cosmo]

[a]A head set above a spanner subsumes the subheads set under the spanner.
[b]For tax purposes only.
[c]Entries may be words or numbers; numbers should be aligned on the decimal point (even if the decimal point is not set).
[d]CATS is an alternative classification system to BUGS. (Would you have preferred to read *CATS in Writing?)*

Legend: Head: table column head; Side head: table side head; Entry: table-body entry; B: bad; U: ugly; G: good; S: splendid; C: cosmic; A: acceptable; T: terrible; S: stupid.

Source: Adapted from Dupré L, *Not All Examples Are Amusing,* Woodside, CA: Homespun Press, 1994. Used with permission.

You should choose and stick to a *style* for table column heads. You will need a style for level 2 column heads, as well as one for level 1 heads, if you use spanners. I recommend that you capitalize only the first letter, although you can use combined capital and lowercase (that is, with most words capitalized), or all lowercase letters, or even all capital letters (but only if you have a powerful reason for using them). You should be consistent within and across tables.

You should insert a *rule*—a line—between the column heads and the main body of table text. You should also insert a rule at the bottom of the table text (see Table 96.1).

In the *body* of the table, you should not set vertical rules to divide the columns from one another, and you should not set horizontal rules to divide the rows from one another (see Table 96.1). Such flourishes are matters of design, and the publisher may choose a design that includes them; however, you should stick to the basic format of two and only two horizontal rules in the table body (plus one after the table title, if you wish; spanners do not count as horizontal rules). This format is not only simple, but also easy to comprehend. (If you are creating your own design, you can set excess rules if you wish, but I recommend that you not do so.)

In the table body, you should choose a style for your *entries;* I recommend that you use all lowercase letters, but you can also set only the first letter as a capital, or you can set only the first letter as a capital for the first column and set all other column entries in all lowercase letters. Whichever style you choose, be consistent, within and across tables.

In the table body, you should *align* columns of numbers on the decimal point, rather than centering the entries. You should also set zeroes before decimal points (0.6, rather than .6).

In the table body, you should choose whether to use *periods* at the ends of complete-sentence entries, and should stick to your choice within and across tables.

 In the table body, you should use *side heads* to group row entries. Side heads are, in essence, level 1 heads within the first column of your table.

 You should use *table footnotes* to explain and annotate the material in your figure. Use small superscripted letters to identify your footnotes, unless your publisher has set a different style. The footnotes should be set under the second rule, above the legends and credit lines.

GOOD: [d]Data have been adjusted slightly to meet the stringent publication-related tenure requirements at the first author's institution.

 You should use *legends* to explain any abbreviations or acronyms in your table, whether or not the latter are defined in the accompanying text. Choose a style for your legends, and stick to it. A simple one is

GOOD: RAM: random-access memory. RAW: random-access woman. RAC: random-access cat.

You should set legends under footnotes, at the bottom of the table.

If you prefer, you can simply set footnotes that contain expansions of any acronyms or other abbreviations in your table.

GOOD: [a]BAA: blind access alley.

[b]BAH: blonde actress hiccough.

 You should use *credit (source) lines* for tables that you have derived from a source other than your own head. You should get permission to reproduce the table from the copyright holder, which is usually—but not always—the publisher. The permission usually will specify how you should set the credit line. If it does not, you can use one of these styles:

> GOOD: *Source:* Adapted with permission from Dupré L. A longitudinal study of nail biting among editors. *Journal of Editorial Nightmares,* 48(2), page 67.

> GOOD: *Source:* Data from Gordon P. *Authorship and Hardship: A Guide to Debunking Your Authors.* Cambridge, MA: In Vain Press, 1982; used with grudging permission.

> GOOD: *Source:* Goldstein H. *How to Run Everything While Appearing Merely to Support Everyone.* Weekend Lake House, NH: New England Publications, 1994; used with permission.

You should set credit lines at the very[257] bottom of the table, under the second rule, and after the footnotes and legends. Alternatively, you can set the credit line at the top of the table, run into (that is, with no line break after) the table title.

THE PRINCIPLES FOR LUCID WRITING here are too numerous to rehash; you should simply remember that tables are highly constrained beasts, and you should take care to set them correctly and consistently.

---

257. This sentence demonstrates a reasonable use of the otherwise fuzzy *very.*

97 *Tense*

IF YOU DETERMINE before you write which tense you will use for different parts of your presentation, you will not find yourself switching midstream. In general, once you have chosen a tense for any given purpose, you should stick to it, and should not change to another without good cause.

Within a given discussion, you should maintain the same tense.

> BAD: Lyn was trying to reach Max. She dials the telephone, but the ring went on and on. She had wondered whether she will call him later. Or, perhaps she has just gone to bed.

> BAD: When you bring home the machine, first, open the box. Then, you see the display screen is broken. You will cry, but you will not despair. You have been thinking about the warranty. You realized that to repair the screen would cost only time.

> BAD: I would have tried again if I realized that you are hungry and want to eat dinner.

> GOOD: In general, you did not need to know the names of all the different tenses in which you could write; you needed only to be sufficiently awake to notice the one in which you found yourself, so that you could cleave to it.

When you write similar sections, you should use the same tense across sections. For example, if you are writing a book in which each chapter ends with a summary, you should always use the same tense in each of the summaries.

BAD: In this chapter [written in Chapter 4], we have discussed three-toed pollywogs.

In this chapter [in Chapter 5], we discussed newts.

In this chapter [in Chapter 6], we discuss tadpoles.

GOOD: In the following segments [written in the introduction to Chapter 4], we shall describe how Madeline learned figure skating.

In the following segments [in Chapter 5], we describe how Sophia learned to say "no" while shaking her head and performing forbidden activities.

BAD: In this chapter, we have described what happens when a man never takes a day off work. We examined in detail the health sequelae of such behavior, and we looked at whether they are reversible. We also analyze the reasons why a man might choose to work 7 days per week, despite the apparent irrationality of such a decision.

GOOD: In this chapter, we discussed the purpose served by the cat's purr. We described how a kitten's purr reassures the queen that the kitten is getting sufficient milk and other necessities, and how the queen's purr guides the kitten's sense of well-being in the world. We reviewed the research done to try to isolate and describe the mechanisms by which the purr is created. Finally, we gave explicit directions on the most appropriate ways to elicit the purr for further research.

When you describe research results, you should generally use the past tense to indicate that the results apply to only the population studied. If you want to imply that the results generalize, then you can use the present tense. Future tense clearly makes no sense.

> BAD: The analysis reported in the preceding section will show that Nick can differentiate shod from shoddy.[258]

> GOOD: The questionnaire showed that informaticists still had not reached consensus on how to differentiate *data, information,* and *knowledge.*

> GOOD: The results showed that 34 percent of maniacal CEOs reported happy marriages, 33 percent reported a divorce initiated within the past year, and 33 percent said they were too busy to participate in the study.

> GOOD: This study confirms that teenagers occasionally fight with their parents.

 When you inform your reader of what topics you intend to discuss in a later portion of your manuscript, you can use future tense. However, I find it simpler to use present tense in these cases. In any event, do not use the past tense, or you will confuse your reader.

> BAD: We discussed `do-while` loops in Chapter 10.

> GOOD: We shall discuss `do-unto-others` feedback loops in Chapter 9.

> SPLENDID: We describe the *do-it-now approach* in Chapter 4.

---

258. There are cases in which the future tense might be reasonable. For example, the import of the results might not be immediately apparent, leading to a construction such as "Our results will demonstrate that the hitherto accepted recommendations of the leading alchemist should be dismissed as folderol." Such a sentence, however, could be excused only if it appeared in introductory remarks. I detest the habit of confusing a *paper,* which reports research already carried out, with a *study.* Authors who speak as though the reporting of results itself constitutes an experiment are guilty of this sin.

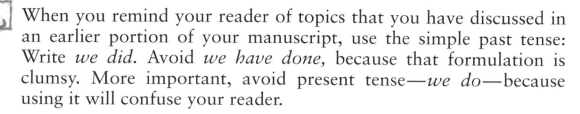

When you remind your reader of topics that you have discussed in an earlier portion of your manuscript, use the simple past tense: Write *we did.* Avoid *we have done,* because that formulation is clumsy. More important, avoid present tense—*we do*—because using it will confuse your reader.

> BAD: In Chapter 3, we present a lexicon for the subject.
>
> *When written in Chapter 5, this example is classified as bad; when written in Chapter 2, it is classified as good.*
>
> UGLY: We have analyzed the basis for the relationship, in Chapter 2.[259]
>
> GOOD: We already saw, in Section 4.7, that this problem is $\mathcal{NP}$-hard.
>
> GOOD: We described in Section 7.3 what a toothbrush had to do with the onset of Richard and Betsy's relationship.

THE PRINCIPLES FOR LUCID WRITING here are that you should pay attention to the tense in which you are writing, and should not change tenses without due cause (rather than without having had due cause).

---

259. The problem occurs only when you are referring to previous segments, sections, or chapters. There is no reason not to write, for example, *We have analyzed the basis in detail; now we will analyze the outcomes,* where you are saying that you have just done the analysis of the basis and are proceeding to the analysis of the outcomes.

98   *Abstracts*

YOU WILL NEED to write abstracts for journal articles and for most proposals.[260] An abstract tells your reader what the contents of your document are. Most important, it answers the question, "Why should I spend my time reading this document?"

You should keep your abstract *short, clear, and directly to the point.* Here, as everywhere else, you should avoid passive voice. Never tease your reader by promising to reveal interesting ideas later in your document; get the key ideas into the abstract. Avoid, insofar as practical, using technical language, and completely avoid using jargon, if you are able to distinguish these two classes within your discipline. Be aware of your audience, and write to its members. If your audience comprises specialists from numerous and disparate disciplines, explain your terms and concepts sufficiently that all readers can understand, without talking down to any group, and without blathering on about definitions when the terms are not required for an abstract.

Assume that your reader is tired, bored, and pressed for time. Give her a clear notion of what your document contains, and convince her that what you have to say is important.

---

260. Certain funding agencies do not include an abstract in their format for proposals; in such cases, you will need to use the first section (which may be Introduction or may be Specific Aims, or may be who knows what) to serve the function of an abstract.

You should provide in your abstract an *overall picture* of your topic. You should hit the *high points only,* without diving into details.

For example, you should not give detailed statistics; it is sufficient, say, to report that your results demonstrated that people who love chocolate are also highly sensual. In the body of your paper, you will presumably justify your conclusion. You should not introduce issues tangential or extraneous to your important conclusions. You should not justify, and usually should not even mention, the statistical measures that you used. In a proposal abstract, you should not speak about the qualifications of the members of the team that you will assemble to carry out the project. You should not give detailed time lines or budgets (you should not give budgets at all in an abstract unless you are explicitly asked to do so—the budget is the *other* proposal component that everyone reads).

BAD: The sample consisted of 190 women, of whom 14 had brown hair, 18 had blonde hair, 142 had red hair, and 16 had indeterminate hair.[261] To see whether there was a correlation between hair color and swimming ability, we first had to ensure that all observed hair colors were natural, or to determine what the original color of dyed hair was. We therefore set up an experiment to collect three hairs from each of the subjects. We utilized cloth caps, which we sewed by hand. We then also decided to investigate whether observable hair color correlated more or less, or to exactly the same degree, with swimming ability. <And so on.>

GOOD: When you set up an advertising campaign, it is important to identify the characteristics of the target market population. We engaged to determine for ProSwim

261. Always be careful when you give the breakdown of a population; frequently, the numbers in published research articles do not add up.

whether potential buyers would be more likely to identify with models whose hair was a particular color, and, if so, whether that would influence the decision to buy the product. We failed to demonstrate any correlations between hair color and marketing success, but we did determine that running ProSwim advertisements that showed muscular, heavily furred cats lounging by the pool greatly increased profits.

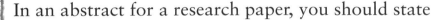 In an abstract for a research paper, you should state

- What the problem was
- What your hypothesis was
- How you tested it
- What the results were
- What your conclusions are

You describe how you tested your hypothesis by providing study design and measures; include this material only if these matters are simple—if they are not, skip them in the abstract.

GOOD: Decision making in medicine requires a method for handling uncertainty, because so many of the important variables have uncertain values. Influence diagrams provide one approach to representing uncertainty explicitly, so we decided to investigate whether we could use that representation to model the limited decision of whether to test for *Streptococcus* when a patient presents with a sore throat. We describe RedThroat, a decision-support tool that we have implemented in three college health-services clinics, where numerous patients who have sore throats are seen. RedThroat's recommendations compared favorably with those of six experts, and clinicians reported high satisfaction with the tool.

RedThroat demonstrates the feasibility of using influence diagrams to support physicians who are making treatment decisions in a limited medical domain.

In an abstract for a proposal, the contents are much the same: what the problem is, what your hypothesis is, how you can solve (or move your discipline closer to solving) the problem, what you propose to do, how you will measure your success, and perhaps what you will deliver and how quickly you will deliver it.

Of course, in a proposal for a business contract, your primary goal is to tell your reader why she needs you, or what you can do for her. Avoid making wild claims, however, without backing them up (wild claims that you can prove are dandy).

> UGLY: Your company needs a scheduling algorithm. Our company specializes in scheduling algorithms. We will write a great scheduling algorithm for you if you will award this contract to us.

The preceding example needs only a sentence or two of expansion to be reclassified as good. You could say, for example,

> GOOD: Dizzy Flights is attempting to shorten the turnaround time on its airplanes. The staff responsible for the task of turning around an airplane for the next flight would work more efficiently if they had a good scheduling algorithm to direct them. Our company specializes in scheduling algorithms. We have helped other companies with similar tasks, such as making up hotel rooms or turning around the space shuttle. We demonstrate in this proposal that we already have in place most of the techniques that would allow us to develop and deliver an efficient, custom-tailored algorithm for you quickly. Moreover, we can undertake the proposed work within a lean budget.

You should write your abstract as a *stand-alone chunk;* you should *not* consider it to be page 1 of your document. Thus, for example, you should not define in it abbreviations that you do not need to use in the abstract (you should define in the main text abbreviations that you use in the main text). If you do define and use abbreviations, you still should redefine them when you first use them in the main body of the document. You should not refer to other sections of the document in your abstract.

> BAD: As we demonstrate in Section 7, our work is so critical to the continued well-being of the citizens of this country that you should not hesitate to fund us, even though our overhead rate is 467 percent.

> GOOD: We demonstrate that *idiotfear* is a serious disorder that strikes 83 percent of males and 82 percent of females attempting to commit to a relationship for the first time, and that our six-step program can greatly reduce the severity of this disorder. Furthermore, we show that reduced severity correlates well with self-reports of improved happiness, increased self-esteem, and decreased nail biting.

You should *write your abstract last,* after you have finished writing the rest of your document. How can you summarize a document that does not yet exist?

There is another school of thought on this matter, and a valid one. The exercise of writing an abstract can force you to think through what are the key points of your document, and that is useful information to have in hand before you write anything else. If you are working with coauthors, an abstract to which you have all agreed can help to keep you all writing more or less the same document, rather than veering off down different conceptual alleyways. If you are working under an advisor or supervisor, discussing and refining the abstract can similarly ensure that you share a vision of the document's structure. If you choose to write your abstract first for these reasons, you should regard it as a working draft abstract. When you finish the document, you should revisit your draft abstract, and should rewrite it as necessary.

THE PRINCIPLE FOR LUCID WRITING here is that you should realize that your abstract is often the most critical part of your document; if you write it well, your reader may be seduced into reading the remainder of what you have to say. Keep it clear, simple, direct, and short.

99 *Neither Nor*

THE CONSTRUCTION *neither...nor* is troublesome for many writers. *Neither...nor* is the negative form of *either...or*.

You should use *nor*, rather than *or*, with *neither* (use *or* with *either*).

> BAD: Moisés had neither a default rule or a logical disjunction, because he was now dealing with a multitude of uncertainties.

> GOOD: Moisés could perform inference on neither logical propositions nor probabilities, because he had had no sleep during the previous week.

> BAD: Unfortunately, neither the time stamp or the global ordering is sufficient to determine that event *A* happened before event *B*.

> GOOD: Uninterestingly, neither a common clock nor a set of perfectly synchronized clocks is available in a distributed system.

> BAD: Lyn neither had eaten or slept since she started laying out her book.

> GOOD: Adrienne was neither anxious nor dubious about getting the project finished on time.

You should avoid using *nor* without the accompanying *neither*.

> BAD: Judy had not gone swimming that day—nor had she gone bicycling.

> GOOD: That day, Steve had neither cooked nor cleaned.

> BAD: You have not tried to increase the locality of reference to minimize seek latency, nor to improve the layout of data to make large transfers possible; no wonder that the file-system performance has degraded.

> GOOD: In my opinion, neither the global-policy routines nor the local-allocation routines were reasonable file-system–layout policies.

You should use care in the word order of a *neither ... nor* phrase.

> BAD: Neither has time sympathy nor malevolence.

> GOOD: Time has neither sympathy nor malevolence.

> GOOD: Neither time nor Richard has either sympathy or malevolence.

> GOOD: Neither sympathy nor malevolence is a property that we should ascribe to time.

> BAD: Neither Paul's internship nor her husband's lack of sleep disturbed Andrea.

> GOOD: Andrea was neither disturbed nor worried by her husband's career-training demands and resultant lack of sleep.

> GOOD: Neither her husband's severe sleep deprivation, nor the nature of his job, could ruffle Andrea's calm.

BAD: Neither is fluffy logic nor inane giggling forbidden in bedtime conversations.

GOOD: Lyn argued that neither fluffy logic nor pressing deadlines would excuse Max reading his e-mail rather than coming to bed.

GOOD: Lyn often neither is fluffy nor cares to be logical.

GOOD: Fluffy logic is not yet fully documented, and neither is Lyn.

THE PRINCIPLE FOR LUCID WRITING here is that you should use *neither* and *nor* together, and should take care in placing other words near them, or your sentence will be neither correct nor clear.

100 *Will Likely Be*

T HE PHRASE *will likely be* (or a different verb) is incorrect, although you will see it often. The correct constructions are either *probably will be* (or a different verb) or *is likely to be* (or a different infinitive).

BAD: Ted thinks that the human-genome project will likely be a good source of funds.

GOOD: Larry thinks that the laboratory's major project is likely to be funded.

GOOD: Darlene thinks that the students probably will be disappointed if their funding is cut without sufficient warning to allow them to think up alternate projects.

BAD: If you keep stomping on the keyboard, the output will likely be disturbing.

GOOD: If you put the floppy disk in the CD-ROM slot, the machine is likely to get confused.

GOOD: If you try to use a rat as an input device, your finger probably will be bitten.

SPLENDID: If you persist in trying to train the rat, she will probably eat the mousepad.

BAD: Lyn thought that the concert would likely be a bore.

GOOD: Lyn asked Moisés to play for her instead, because his music probably would be more interesting.

GOOD: Greg and Moisés agreed to play together, because they thought that the music was likely to lay the foundation for a good working relationship.

BAD: Red thought that stalking raccoons would likely be unnecessarily risky.

GOOD: BB thought that hiding in silk cocoons probably would be snug and comfy.

GOOD: Red and BB agreed that traveling in hot-air balloons was likely to be alarming.

THE PRINCIPLE FOR LUCID WRITING here is that you will probably now remember—and are likely to remember—to avoid the construction *will likely be*.

101 Importantly

MANY PEOPLE write *importantly* (an activity modifier, or an adverb) when they mean *important* (an object modifier, or an adjective). You should tune your ear to catch this error, because a portion of your audience will find the mistake intensely annoying.

 Importantly is an *adverb,* and you should use it as such, to modify a verb. It means to take a given action *in an important manner.*

> BAD: Importantly, the algorithm cannot handle situations with a large state space.
>
> GOOD: Sarah walked importantly in her cheetah bathing suit.

> BAD: Most importantly, note that smoking is permitted in only the specially installed dungeon.
>
> GOOD: Cleo managed to wiggle importantly as she showed off her new tricks to Helen.

Important is the word that you should use in all other cases.

> UGLY: It is important to note that[262] we used a sharply constrained domain in our trials.

---

262. The phrase *it is important to note that,* although technically correct, is awkward, clumsy, wordy, and generally inexcusable; that is why it has received the rating ugly.

GOOD: Most important,[263] remember to provide fresh food and water, and much petting, for your cats.

SPLENDID: Note that[264] we did not have time to write a coherent final report for this project.

BAD: Importantly, Max forgot to shut the door when he left the house, resulting in an influx of raccoons.

UGLY: It is important to note that, before Lyn joined Max in Woodside, none of the doors had locks on them.

GOOD: Lyn thought it important to install locks immediately, so she got out her power tools and went to work, even though Max was greatly amused by her paranoia.

SPLENDID: Lyn asked Max to note that the locks would be moot if the door were left wide open.

BAD: Importantly, for purposes of the analysis, we removed from the data set those points that represented outliers from the curve.

GOOD: It is important that we removed from the analysis values that were more than three standard deviations from normal.

SPLENDID: Note that we ignored unbelievable values.

---

263. Here, we get rid of the awkward phrase, and retain *important*.

264. For this example (that is, where we are looking for alternatives to *it is important to note that),* omitting the *important* and simply saying *note that* is the most powerful option.

SPLENDID: Certain of the examples in this book are exceptionally
silly, but they are importantly so.[265]

THE PRINCIPLE FOR LUCID WRITING here is
that it is important that you not use *impor-*
tantly when you mean *important*. You can
write importantly, which is arrogant, or you
can write important material.

Image based on a photograph copyright © Paul Fusco/Magnum Photos

Courtesy of Nicholas Iversen

265. Contributed at 2:00 A.M. by Max Henrion.

102 Since

THE WORD *since* properly denotes relationships of *time*. The correct term to use to indicate *causal* relationships is *because*. Use of *since* to mean *because* is so common that it is now on the verge of being correct. However, *since* is also overused in scientific writing; thus, you should consider resisting using *since* except to denote relationships of time. At the least, whenever you write *since*, ask yourself whether you are using it to denote a time relationship; if you are not, ask yourself whether you need to use it at all.

UGLY: Malcolm had to get home to babysit, since Gelareh needed to study, and that task is complicated when you are simultaneously trying to amuse a 6-month-old child.

GOOD: Malcolm woke up in wonderment, because Alyssa was sucking his nose.

GOOD: Neither Malcolm nor Gelareh has slept since Alyssa was born.

UGLY: Pipelining will not work in this case, since linkages are disallowed.

GOOD: You cannot use that software, because you do not have sufficient memory to run it.

GOOD: That trick does not work anymore, since we installed the version upgrade.

UGLY: The computer is useless, since it never had a mother-board.

GOOD: The CD player has been useless since Reno stuck his pacifier into the little slot at the front.

GOOD: The computer is useless, because Cosmo carefully put away his chewing gum in the disk drive.[266]

UGLY: Since she was awake, Lyn decided to get up and to work on her book, even though it was 3 in the morning.

GOOD: Lyn engaged in this odd behavior because there was no point in tossing around in bed and worrying, while Max groaned gently and tried to sleep.

GOOD: Max's sleep has been disrupted since Lyn started bouncing out of bed at all hours of the night.

UGLY: Max could not get into his car, since he had locked the door while his keys were still in the ignition.

GOOD: Max had been standing forlornly at the curb since he realized his predicament.

GOOD: Lyn was not at all worried, because she had an extra copy of Max's car key in her backpack.

---

266. This event occurred not long after Reno broke the needle off the turntable arm.

UGLY: Lyn is grumpy tonight, since she did not make much progress during the day.

GOOD: Lyn has been ecstatic lately, because Max has been bringing her flowers every night.

GOOD: Max has been idiotically happy lately, [ever] since he learned how to parachute.

SPLENDID: Since Red has stopped bringing her carcasses every night, because BB has been distracting him, Lyn has been ridiculously ecstatic.[267]

THE PRINCIPLE FOR LUCID WRITING here is that you should reserve *since* for time relationships, and should use *because* to indicate causal linkages, ~~since~~ because those are the correct uses of the words.

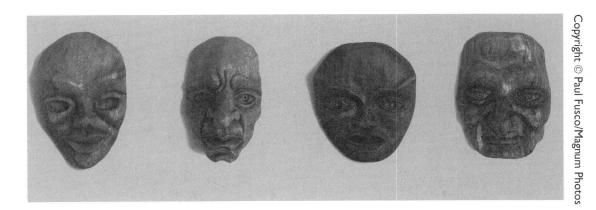

267. Contributed by Peter Gordon.

103 References

IT IS GENUINELY important that you adopt a consistent and complete style for references. References provide critical information for your readers, who may well wish to look up the citations that you give.

Often, the style is set by the publisher; in such cases, you should follow it compulsively. If you are submitting an article to a journal, for example, you should look at the authors' guidelines (or, if the guidelines do not specify a reference style, you should look at the references in the journal itself) to determine what style to use. If no style is set by the publisher, you can use the style given here, or you can use any style you wish, provided that it is complete and that you implement it consistently.[268]

You should avoid using *et al.* when you are choosing your own style, because it leaves out information. If someone (you, perhaps), later needs to convert your references to another form, she will have to find the missing author names. It is much easier to convert from complete references to those using *et al.*, if and when necessary. If you do decide to use *et al.*, however, use it consistently for more than two authors, for more than three authors, or for more than five authors. (Note that, if you are following a style that uses *et al.*, you

---

268. Certain people believe, for example, that including authors' full names is more important than is achieving simplicity. The guidelines suggested here, like those in the rest of the book, reflect my personal preferences.

418

must determine which rule is being applied.) Set *et al.* in roman type,[269] rather than in italic type, in the body of the reference.

> UGLY: Darwiche A, *et al.*
>
> GOOD: Dagum P, et al.
>
> SPLENDID: Pearl J, Goldszmidt M, Boutilier C.

 You should include the initials in the *authors' or editors' names*. The simplest style, and therefore the one that I suggest, is to give the last name first, followed by the initial, with no comma in between; to use no period after the initial; to use a comma after the initials; to omit the *and* before the final author; and to use a period after the initials of the final author.

> GOOD: Gordon PS, Dupré LA, Goldstein HM.

 You should set in italics, and in mixed capital and lowercase letters (called cap/lc in publishing), the *title* of the book or journal. Place a period at the end of the title.

> GOOD: Holloway, Q. *Heap Sort Applied to Garbage Collection.* New York: Sanitary Publications, 1987.
>
> GOOD: Codswallop TS. Textual deconstruction in *BUGS in Writing. Structuralist Critique,* 8(4):19–37, 1996.

You should give the *location of the publisher* as a city and two-letter state code, or as a city and country, followed by a colon, followed by the publisher's name. There are certain cities in the United States that are considered to be so well known as not to require the state code; examples are New York, Los Angeles, Boston, San Francisco, and Chicago. Again, if you are in doubt, include the state code; it is easy

---

269. That is, set *et al.* in roman type unless you are referring to the term, rather than to its denotation, as I am doing here; in such cases, italic type is correct.

to remove it later. Note that you must look in the book that you are citing to discover the publisher's location, because many publishers have separate houses in disparate locations. Use the *first* location listed on the title page.

GOOD: Englewood Cliffs, NJ: Prentice-Hall

GOOD: New York: Basic Books

GOOD: Reading, MA: Addison-Wesley

GOOD: Menlo Park, CA: Addison-Wesley

GOOD: Redwood City, CA: Addison-Wesley

GOOD: Wokingham, England: Addison-Wesley

You should use the short form of the *publisher's name*. In general, you should omit extra words such as *Inc.* or *and Sons*. If you are in doubt, spell it out; it is easy to shorten a long name later. Note that, if the publisher sets its name with an ampersand (&), you also should use the ampersand in your reference.

UGLY: Charles Scribner's & Sons, Inc.

GOOD: Scribner's

UGLY: Harcourt Brace & Company

GOOD: Simon & Schuster

UGLY: Harper and Row, Inc.

GOOD: Harper & Row

SPLENDID: HarperCollins

When you give a *reference to a book,* you should include the authors, the title, the publisher, the location of the publisher, and the year of publication. If relevant, you should also include information on the edition, the volume number, and the translator.

> GOOD: Henrion M, Dupré L. *How to Survive Against the Odds* (second edition). Translated by Barry A and deWitt K. Woodside, CA: Id Press, 1994.

> GOOD: Gordon P. *How to Sleep Through Meetings Without Anyone Noticing: The Sleep-Deprived Executives Guide* (volume 2). Cambridge, MA: Greener Grass Publications, 1981.

When you give a reference to a *journal article,* you should include the authors, the title, the journal name, the volume, the page number, and the year of publication. You can include the issue number if you wish, but if you include the issue number for *any* journal references, then you must include it for *all* of them. When you give the page number, you can give first page only, or you can give the range. If you give the range, be consistent in your choice of whether to repeat numbers (321–324 or 321–4). Use an en dash rather than a hyphen, for this (and any other) range.

> GOOD: Ahumada A, Doerner M. Recognizing that hospitality is an art. *Journal of Meaningful Friendships,* 14(3):45–62, 1994.

> GOOD: Mulligan J. You too can play volleyball. *Weekend Journal,* 3:467-8.

When you give a reference to a *chapter in a book,* you should include the authors, the chapter title, the editors of the book, the book title, the publisher, the publisher's location, and the year of publication (and edition, volume, translator, or any other applicable information). You can also include the page numbers of the chapter.

GOOD: Dupré L, Nims M. Questioning standard medical advice: Why two aspirins may not be the most sensible prescription. In Oppenheim G (ed), *The Etiology of Difficult Patients*. San Rafael, CA: Noodle Age Publications, 1993.

When you give a reference to a doctoral *dissertation,* you should include the author, the title, the department, the university, the location of the university, and the month and year of publication.

GOOD: Student A. *Feline and Insect Symbolism in* BUGS *in Writing.* Doctoral dissertation, Department of Lexical Studies, Humany University, Challow Altos, CA, January 1995. Also published in *SmartyPants Students,* 9(21):43–159, 1995.

When you give a reference to a *technical report,* you should include the author, the title, the technical-report number, the department, the organization, the location of the organization, and the year of publication. Note that, because a technical report is the equivalent of a journal article, rather than of a book, you should set the title accordingly; that is, following my style, in roman and lead cap.

GOOD: Lehman H. Assessment of stomach pain and heartburn in crying adults. Technical report #328965, GUI Laboratory, Let's Talk Consulting, Sunnyvale, CA, 1986.

When you give a reference to a paper published in the *proceedings* of a conference, you should include the author, the title, the editor of the proceedings, the title of the published proceedings, the page numbers, the location of the conference, and the month and year of the conference. Note that various proceedings are published as books by a conventional book publisher. In that case, you can reference them as books, giving the location and name of the publisher.

GOOD: Swerdlow C. Arrhythmias and outcomes. In Harrison
        D, Reed L (eds). *Proceedings of the Fourth Confer-
        ence on Arrogance and Fun in Cardiology,* pp 26–48.
        Beach by the Ocean, OR, May 1984.

 When you give a reference to a paper that has been *accepted* by a
journal but has not yet been *published,* you should include all the
information you would include for a published paper, but you
should *not* include the year of publication, even if you feel certain[270]
that you know when the paper will be published. Papers often do not
get published on schedule.

BAD: Goldszmidt M. Qualitative reasoning in quantitative
        crises. *Decision Theorist,* 1999.

BAD: Adamo R. Logical reasoning in emotional crises.
        Family Theorist, 1999 (in press).

GOOD: Dagum P. Qualitative reasoning applied to car diag-
        nosis: Putting years of research to practical use. *Deci-
        sion Theorist* (in press).

Note that books, too, often do not get published on schedule.
Always use *in press,* without a year, for any document that is in
production but is not yet published.

BAD: Dupré L. *A Litany of Prosaic Catastrophes.* Reading,
        MA: Addison-Wesley, 1999.[271]

BAD: Dupré L. *BUGS in Love.* Reading, MA: Addison-
        Wesley, 1999 (in press).

GOOD: Dupré L. *BUGS and Cats: A Guide to Debugging
        Your Relationship with Your Cat.* Reading, MA:
        Addison-Wesley (in press).

---

270. Note that feeling certain and being certain are different states.

271. Based on a suggestion by Carver Mead.

When you wish to give a reference to an *unpublished* paper, you *cannot* do so. A paper that has not been published should never be cited. Do not ever give a reference to a paper as *submitted*. Anyone can submit to a journal any twaddle that she chooses; only a paper that has been through the peer-review (or, in a non–peer-review forum, editorial-review) process and has been *accepted* can be cited in conjunction with the journal. Often, unpublished papers are made into technical reports, and you can refer to them in that way until such time as they are accepted. Otherwise, you can say only *personal communication,* just as you would if the results, data, anecdotes, or other text worth citing was communicated to you orally, by electronic mail, or by any form of communication other than through a published report. In general, you should set references for personal communications as footnotes to the body of the text, rather than giving a citation in text and setting them in the reference section, and thus making them look at first glance to have the authority of a true reference. If you are not using any footnotes, use square brackets.

> BAD: Rock J. Let's go now! *Travel Life* (submitted).
>
> UGLY: Rock J, personal communication, Mountain View, CA, 1993.
>
> *The two preceding examples would be set in the reference section, being cited in text as, say,* [Rock 1993].
>
> GOOD: J. Rock [personal communication] says "yes!"
>
> SPLENDID: Dr. Rock reported from Cern that the spins still do not balance out.[272]

When you give a reference to a user manual or other *documentation,* you should give the author (the company is the author if no authors are listed), the title, the company name and location, and the year of publication.

---

272. Personal communication, electronic-mail message, 1994.

GOOD: Brown D. *CEO User Manual*. Santa Clara, CA: The Startup Experts, 1992.

GOOD: Provan G, Pradhan M. *Diagnostic Assessment of Noisy Max and Leaky Max*. Final Report on National Science Foundation Grant #278543689-NOT. Los Altos, CA: Institute for Decision Systems Research, 1991.

GOOD: Institute for Decision Systems Research (IDSR). The Spaghetti Belief Network (QMR-DT). Interim report. Los Altos, CA: IDSR, 1994.

When you are faced with an *unusual* reference, such as one for an unusual medium or in an unusual format, just use your common sense. Be sure to include sufficient information that your reader can locate the source. Include the equivalents of author, title, publisher, location, and date of publication.

GOOD: Lumina Decision Systems (LDS). *Online Tutorial for Demos (Decision Modeling System)*. Los Altos, CA: LDS, 1994.

GOOD: Analyzum I. *BUGS in Writing: Fact or Fiction?* Course notes, Literary Forms 101. Tappan, New York: Local Junior College, 1996.

GOOD: Dupré L. Tasteful, affordable and appropriate office decor that meets rigorously tested feline-factors specifications. Internal memorandum. Woodside, CA: Writers' Madhouse, 1993.

GOOD: Arnold B. *Relationship of Bug Lists to Feature Lists: A Problem of Exponentiation*. Electronic mail to Software Sobbers BBoard, September 19, 1994.

GOOD: Anonymous. *Ode to a Departing Hangnail*. Poem graffito on the New York IRT downtown express, May 26, 1969.

GOOD: Adamo, R. *O Madam, I Am Adamo.* Leaflet posted in the men's restroom at Salivating Rumpled Idiots, an exclusive club dedicated to palindromic pastimes.

GOOD: *Windy Blustery Upbringing.* An anagram communication conveyed by revelation to L Dupré during a pilgrimage to the peak of Windy Hill, CA, undertaken in search of examples.

References are both demanding and boring, and sloppy writers often ignore them when submitting papers. Such writers may later discover that (1) journal editors do notice references, and may use such matters to decide a close call between peer reviewers; (2) reviewers always check the references to see whether they are cited, and become annoyed when their citations are missing, mangled, or incomplete; and (3) sorting out references after the fact is a massive headache. Finally, a word to the wise: Do not lift your references from other published sources. Rather, go to the original material. Numerous references are set incorrectly at a point in their citation history, and the errors are propagated as people use the erroneous lists to make up their own. It is exceedingly frustrating to make a trip to the library only to discover that a reference that you are chasing does not exist.

THE PRINCIPLE FOR LUCID WRITING here is that you should set your references with compulsive care. If your publisher suggests a style, use it; otherwise, use any you wish, but use it carefully and consistently.

104 Cannot Versus Can Not

THE WORD *cannot*, the opposite of *can*, is always set as one word. It is correct to set *can not* only when the negation applies to the activity, rather than to the ability to carry out the activity. In other words, you are saying

> X can do not A

rather than

> X cannot do A

where X is an agent, and A is an activity.

BAD: The current–voltage characteristics of this network can not be measured.

GOOD: The current I cannot be plotted as a function of V_1 and V_2.

GOOD: The current mirror can work, or, if our luck fails to hold, it can not work.

BAD: "It can not be!" cried Lyn, when she noticed the gentle swelling of BB's tummy.

GOOD: "Nonsense," Max responded brusquely. "Anything that can be also can not be."

GOOD: "BB," Lyn continued, ignoring Max's interpolation, "you cannot imagine how cute your kittens will be!"

BAD: Steve and Judy can not decide whether to come back to California from Geneva.

GOOD: Steve and Judy cannot earn much of a living in Geneva if they stay.

GOOD: Steve and Judy said to each other, "Either we cannot stay, or we can not stay by choice."

Either Steve and Judy are unable to stay in Geneva, or they are able to choose not to stay.

BAD: When Max realized that Lyn was planning to acquire 200 cats, he thought, "She can not be serious!"

GOOD: Lyn saw Max's expression and decided, "Although Max cannot stop me from pursuing that dream, I would do well to desist on my own."

GOOD: Max, immediately understanding what Lyn was thinking, said, "You can do whatever you wish. You can adopt a pride of lions, or you can not adopt excess felines."

THE PRINCIPLE FOR LUCID WRITING here is that you should set the negation of *can* as *cannot;* only if you intend to negate an activity, rather than an ability, should you use *can not.* If you cannot remember the principle, you have no choice; if you can not remember the principle (that is, if you can intentionally forget the principle), you do have a choice.

105 *Also*

T̲H̲E̲ ̲P̲L̲A̲C̲E̲M̲E̲N̲T̲ of *also* modifies the meaning of your sentence. Although no placement is technically incorrect, a given placement may convey a message different from that you intended to convey. Consider the following examples:

> GOOD: Max also loves Lyn.

> GOOD: Max loves Lyn also.

Clearly, the sentences have different meanings, as illustrated by these expanded versions:

> GOOD: Max is angry at Lyn; Max also loves Lyn.

> GOOD: Dona loves Lyn; Max also loves Lyn.

> GOOD: Max loves Sarah; Max loves Lyn also.

 You should place *also* before the verb when *also* modifies either the agent (for example, *Max)* or the activity (for example, *loves*).

Another way to think about this principle is that *he is one way and she also is that way. He also is* means *he too is <whatever>.*

> GOOD: Max and Lyn dashed to the restaurant because they were 20 minutes late; fortunately, Greg and Nicola also were late.

> GOOD: To certain people, marriage represents the key to bliss onearth;torelational-databasedesigners, MARRIAGE also represents a one-to-one bill-of-materials rela-

429

tionship that can be reclassified as an entity because of its associated relationships.

GOOD: Carver took Lyn out to breakfast at Late for the Train because he was longing for cheese blintzes; Lyn also was longing for them.

 You should place *also* after the verb (and, usually, after the recipient) when *also* modifies the recipient of the activity (for example, Lyn).

Another way to think about it is that *he is one way and he is also another way. He is also means he is, in addition, <whatever>.*

GOOD: Carver took Lyn out to dinner at Flea Street because he was in need of amusement; he was also hungry.

GOOD: Mappings between the business conceptual schema and the logical data model also may include naming differences, transformation operations on conceptual schema, interrelation of business rules, and actual conflicts.

GOOD: If you want to allow remarriages, then you must add date to the primary key; if you want to allow people to marry twice on the same date, then you must add another attribute also.

THE PRINCIPLE FOR LUCID WRITING here is that you should keep in mind that there is a major difference between *Lyn also loves Max* and *Lyn loves Max also.*

106 Nonwords

I‍T IS NO LONGER correct to hyphenate words beginning with *non*. There is a general trend over time in English to close up with hyphens words that were formerly two words, then to close completely words that were formerly hyphenated (baby sitter to baby-sitter to babysitter; bed room to bed-room to bedroom; play school to play-school to playschool). Thus, although hyphenating *non*words was acceptable in the past, it is now outdated.

 You should not use the hyphen at all in most nonwords.

> BAD: Non-monotonic reasoning can be useful for this particular non-trivial problem.
>
> GOOD: Spencer could not imagine a noncaffeinated life.
>
> SPLENDID: Lyn has been happily sans caffeine for 1 year.

 Note that you should also not use the hyphen in nonwords in which the second word begins with an *n*.

> BAD: Use non-numerical analysis when you cannot obtain the exact values.
>
> GOOD: Several nonnuclear sources of power are available.

You should use an en dash when you are modifying a term that itself contains more than one word.

> BAD: That is definitely a nonpolar-bear lifeform.

BAD: He has a decidedly non-money-market approach to investment.

The mark after non *is a hyphen; it should be an en dash.*

GOOD: The furry mass is also a non–grizzly-bear creature.

Note that a nongrizzly bear is a bear of a type other than *Ursus horribilis*, whereas a non–grizzly bear[273] is an object that may be *any* entity other than a grizzly bear.

You should use an *en dash* when you are modifying a word that begins with a capital letter.

BAD: A non-Unix operating system might allow a more intuitive interface.

The mark after non *is a hyphen; it should be an en dash.*

BAD: His speech is non English.

GOOD: Use non–Monte Carlo sampling methods.

THE PRINCIPLE FOR LUCID WRITING here is that you should not hyphenate nonwords unless the second term consists of multiple words or begins with a capital letter. For example, a non–Dupré system might use a hyphen in *nonwords*.

---

273. There is no hyphen here between *grizzly* and *bear* because *grizzly bear* is a noun here; in the preceding example, *non–grizzly-bear* is an adjective, and thus takes the hyphen.

107 Missing That

MANY PEOPLE leave out *that* from sentences in which it belongs. The result is sloppy prose.

UGLY: Devon believes Max is sleeping late every morning.

GOOD: Devon believes Max when he says that he is working at home in the morning.

This example is intended only to show you what is wrong with the preceding example.

GOOD: Max believes that Brian is a superb programmer.

UGLY: Red (a massive chunk of feline masculinity) thinks rats are crunchy and tasty, and should be caught whenever possible.

GOOD: BB (whose delicate chocolate-point body is a mere feline wisp) thinks that rats are hairy and scary, and that they should be avoided at all costs.

UGLY: Soren has not yet tried the Chinese medicine Lyn recommended for his stuffy nose.

GOOD: Soren used to recycle the old plastic milk bottles that Lyn collected when he did her the favor of changing the oil in her aging Volkswagen (these days, she has to do it herself).

UGLY: The program deletes the first number it encounters.

GOOD: This program now matches the second variable that it instantiates.

UGLY: Judy had trouble opening the beer bottle Steve handed to her, because she had torn a ligament in her thumb; she gave up quickly, because she knew she could injure herself again if she continued.

GOOD: Steve enjoyed opening the bottle that Judy gave back to him in disgust; he promptly poured it into Judy's bath, because he had heard that bathing in beer produces euphoria.

UGLY: Red thought it important to wake Lyn up on time every morning, so that he could get his treat.

GOOD: BB thought that it was important to sleep on Lyn's feet all night, so that she could insist on being petted every time that Lyn turned over.

UGLY: The code on lines 45 through 47 ensures a process seeking an exclusive lock will not deadlock against a lock that it already holds.

GOOD: The processes that are waiting at lines 40 and 50 are awakened when other processes release the locks.

UGLY: Lyn and Max hated the squawking, shrieking neighbors' radio they could hear every morning.

GOOD: Lyn and Max loved the squawking, shrieking blue herons that nested in the redwoods above their house every year.

THE PRINCIPLE FOR LUCID WRITING here is <u>that</u>, although it is good practice to leave out unnecessary words, it is not good practice to leave out words <u>that</u> belong in your sentence.

108 All Of

USING THE stilted and longer term *all of* is silly when *all* alone will suffice. Use *all of* only when *all* alone makes no sense.

UGLY: Marina could not afford to spend all of her time driving back and forth over the hill.

GOOD: Pat could not afford to spend all his money on the new mile-wide television set.

GOOD: Reno told Buttercup and Ruby and Hershey and Caltrans that most or all of them were silly.

The most [of] *requires the* all of *here.*

UGLY: All of the processors receive the same input.

GOOD: Not all types of lateral inhibition can lead to self-organization.

GOOD: Today, only one-half of the circuits were delivered, but all of them will be here by tomorrow.

The all of *is necessary to match the* one-half of.

UGLY: Lois complained that Soren was spending all of his time under the car again.

GOOD: Soren pointed out that all the cars were running well all the time.

GOOD: Not all of Soren and Lois's hobbies are related to cars—but most of them are.

The all of *is necessary to match the* most of; *you will almost always need to write* all of them.

THE PRINCIPLE FOR LUCID WRITING here is that you should omit the *of* after *all* unless doing so leaves your sentence stranded without all its meaning intact.

109 *Utilize*

NEVER *utilize* any object without considering whether simply *using* it will suffice. *Utilize* implies making use of a commodity that would have been wasted or otherwise unused.

GOOD: Dona utilized old newspapers for the papier-mâché.

GOOD: David and Claire utilize almost all their garbage, either by composting it or by recycling it.

UGLY: To enter data into the program, the physician utilizes a light-pen to make selections from multiple-choice displays.

GOOD: To start the program, the nurse uses a mouse device to double-click on the icon.

GOOD: To run this program, we can utilize the terminals that have been sitting in the closet for years.

UGLY: Pat utilizes a posh telephone system to do triage on client calls in his consulting business.

GOOD: Marina used snazzy pink and blue hair coloring in her personal self-enhancement program.

GOOD: Cosmo and Reno figured out how to utilize Marina's worn-out stockings to make a toy, over which Swix and Spud then spent the afternoon growling.

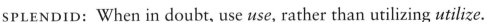

UGLY: Michael utilized his hard-earned money to attend an important conference.

GOOD: Shellee used her cash to buy gorgeous trash on a shopping trip with Lyn.

GOOD: Michael and Shellee utilized their spare room to make comfortable the friendly mutt who had followed Michael home in the snow.

UGLY: The doctor insisted that her patient utilize the latest physiotherapy equipment.

GOOD: The patient decided to use a couple of old bricks stuffed into a fanny pack.

GOOD: The hospital maintained a history of bed-utilization statistics.

SPLENDID: When in doubt, use *use,* rather than utilizing *utilize.*

THE PRINCIPLE FOR LUCID WRITING here is that you should *utilize* only those articles that would otherwise have lain stagnant, and should *use* everything else. Use this principle to avoid utilizing jargon.

110 Dissertations

A DISSERTATION is a *document* that you write as part of the fulfillment of requirements for a degree (it can also be simply a discourse on a subject). A *thesis* is an *assertion* that you have presumably validated or proved, and on which you report in a dissertation.

Your research advisor should give you guidance on writing your dissertation. Academic centers and even departments within one institution differ widely in requirements and preferences. I list here just a few points that you may want to consider when you undertake writing your dissertation; for that matter, you probably should consider them when you write any large document.

 You should *write and agree on with your advisor a detailed outline* before you set pen to paper or finger to keyboard. In most cases, this advice will seem ridiculously trivial and obvious; all to the good. Just be aware that there is no point in your writing reams of prose, only to discover that your advisor had in mind a substantially different approach to your topic.

You should *set with your advisor a schedule* for you to deliver pieces and for her to respond with comments. Make every attempt to be realistic when you set deadlines, and to adhere to your schedule. If your schedule slips, however, immediately draft and agree to a new one. Do not keep setting back a due date by small increments, and failing repeatedly to deliver, because then you and your advisor will get into the habit of thinking that it is normal to miss deadlines. Rather, if you miss a deadline, reset it by a sufficiently large increment that you will make it next time.

You should allow time in the schedule for any editing or proofreading that you intend to have done. You should also determine when you will deliver what to other members of your committee, and when they will respond.

 You should *adopt a formal, careful author's voice.* Casual, informal slang or jargon is not at home in a dissertation. I do not mean that you should write dry, boring prose. On the contrary, you should ideally write amusing, fascinating, exciting prose that is impeccable in style, syntax, and semantics. By all means include jokes, but only subtle, erudite ones.

You should *number your chapters and heads,* using double numbers (such as 4.6, or 8.2) for level 1 heads, figures, tables, and other numbered elements. Use as many levels of heads as is logical; each head should cover more than a paragraph of material, but can cover as little as two or three paragraphs if the subject matter is notably different from that of the preceding subsection. Heads keep your reader oriented. *Write small* (but more than one-sentence) *to medium paragraphs*—do not run on for pages without a break in your thought.

You should think about and *include figures, displayed lists, tables, boxes,* and any other elements that will break up the text visually and conceptually. There is something daunting about page after page of full-margin text: It *looks* boring.

 You should *determine early what design elements* (types of text, from a visual point of view, such as displayed equations, tables, boxes, equations, heads, lists, and so on) *you will use,* and should develop *design specifications* for each.[274] Most text editors allow you to define *environments,* which are design specifications that you implement automatically by choosing the environment. Do your best to implement a simple, graceful design, but do not fall into the trap of procrastinating about writing by tinkering with the design elements and wasting hours fiddling with the presentation.

You should *write informative heads.* Insofar as possible, you should tell your reader in the head what she can expect to find in the accompanying prose. However, you should not write overly long heads; a good rule of thumb is that most of your heads should not wrap— they should cover just one line.

You should *write at least one or two sentences of introduction* after a head on a given level and *before a subhead.* That is, you should never have two heads (for example, a 1 head and a 2 head, or a chapter title and a 1 head) following each other without intervening text. Missing introductions are both visual and conceptual errors.

If you want to add class to your dissertation, you should *write a good index.* Indexing is a difficult task, because you must think about what concepts are subsumed by what other *umbrella* concepts. Sometimes, the umbrella concept will not even appear in your document, but will be the word that the reader looks up to find the subsumed concept. (For example, the reader may look for the umbrella concept *beer,* whereas the words appearing in the text may be *ale* and *stout.*) For this reason, you should not rely on automated indexing. Automated indexing creates plenty of other problems as well—I have seen indexes that match almost exactly the heads of a

---

274. You should first find out the extent to which your formatting and design are dictated by your academic institution.

document[275] and are virtually useless. One technique for writing an index is to maintain a card file as you go along, jotting down key terms that you wish to index, one per card. When you are ready to create the index (after you have finished writing), you sort the cards into related-concept piles, choose umbrella terms, and write from the piles. I realize that this old-fashioned, manual method will not appeal to certain machine-dependent people, but you can use the idea to create a computer-based system of your own.

You should *write an abstract* that sums up all the important points in your dissertation, without giving details or justifications. Anyone who reads it should know, for example, what you did, what the rationale was for what you did, what the results were, and what conclusions you drew.

You should *state at the outset what your hypothesis was and how you tested it.* Your hypothesis might be twofold in that you might have a *general hypothesis,* such as that automated bibliographic search can help people to stay abreast of developments in their field, and a *limited hypothesis,* such as that a given technique will permit faster searching on bibliographic databases. You should state both explicitly.

You should *say at the outset what your measures of success are.* You are asserting a hypothesis (or several of them), so you need to say how the work you have done has, after all, tested those hypotheses. (You may argue that you and other scientists are not in a position to test the general hypothesis directly; you may have to rely on common sense, references, or future work to demonstrate it. You should, however, be in a position to validate your limited hypothesis.)

---

275. The indexes in this book match identically the heads, but I hope are not useless.

If you are implementing a system or algorithm or function, for example, you should explain the validation study that you used, and should justify the measures that you applied. Obviously, validating a laboratory system is quite different from testing a hypothesis about economic systems or cognitive function, but in each case you should state with your hypothesis what the measure of success is.

 You should *indicate in what order you will discuss the various segments of your exposition,* and you should explain why you have chosen that order. Use chapter numbers, so that your reader knows where to find the information she seeks (you should also include a table of contents, of course). Reveal the structure of your document to your reader, so that, at any given point, she will know why she is reading what you have to say. Periodically remind her of the structure as you move from chapter to chapter.

Within each chapter, you should similarly begin by saying what you will discuss; use section numbers, and say what is in which section. State the logic of the order that you have chosen. You can also begin each chapter with an abstract of the chapter, which describes succinctly what the chapter covers.

You should never tease your reader about what you will reveal later in your dissertation. That style is appropriate for a mystery story, or for a magazine article on an epidemiologic study. In a dissertation, there should be no guessing games, and no readers still wondering about what you are babbling as they begin Chapter 4.

You should *write in the first person,* and should *avoid passive voice.* Be extremely clear about what work you did and what work was performed by other people, including by members of your team in a research laboratory. A dissertation is a highly personal document; in it, you are reporting primarily your own work. Be sure to take credit (or blame) where it is due.

If you have a long *background section,* detailing work by other investigators, you should *think carefully about in what tense you will write it.* In most cases, it is easiest and most appropriate to use the simple past.

 You should *put material in the appropriate section.* Do not get ahead of yourself, and start talking about the results of a clinical study before you have described the population from which you sampled. Do not tell your reader what statistical tests you ran while you are reporting the data collected. Do not draw conclusions in the analysis section. Progress logically, purposefully, and in an organized fashion, putting in each section the material promised by the heads.

 You should *not say that you are in the midst of further work.* A dissertation should report a block of work already performed. In your conclusions, when you tie your work into a larger picture of work in your field, you can suggest future research that would carry on where you have left off, or would investigate other paths that your work has uncovered. Do not, however, say that you are undertaking that future research.

You should *set your references compulsively* as you go along, and should not leave them until the last minute when you will be far too frantic to deal with them. In many dissertations, the reference section is huge. You should also, generally, *use a citation style that gives the names of the authors and dates of publication,* rather than just a number. It is especially important in a dissertation that your reader be able to tell, without page flapping, whom you are citing, as well as the date of publication of the cited work.

Be sure to get your references from the original sources; that is, do not simply copy them from other reference lists, thus propagating the errors of other authors.

You should *consider having your work edited by a professional copy editor.* Naturally, after you have absorbed the contents of this book, you will be a superb writer who almost never makes any errors. However, no one can edit her own work completely, so you should have at least one person read for style, syntax, and semantics, in addition to the various people who will read for content.

Whether or not you work with an editor, you may wish to hire a *proofreader,* who will read for typos and spelling errors.

You should *keep this book next to your bed or favorite armchair,* or wherever you might be tempted to look at it frequently. Once you are writing, you will find that the principles contained in it take on alarming relevance.

You should *be prepared for thesis psychosis,* a disorder that strikes most people at least once while they are writing their dissertations. Thesis psychosis is a type of loss of nerve; it may take the form of writer's block (untreated, it can last for years), or it may manifest itself in various forms of bizarre behavior.

When you write your dissertation and graduate, you move from being a student to being an expert scientist in the community. That transition can be terrifying. In addition, when you write your dissertation, you may be exposing yourself for the first time; you are publicizing your ideas, and your expertise, to be read and critiqued by other experts. Furthermore, if you are like most degree candidates, your dissertation is a much bigger and more complex writing project than any you have ever attempted. The combination of these factors alone is sufficient to cause most people to lose about a year to thesis psychosis, on the road to earning their degrees. If you know that the condition is normal—that, even if you are raving mad, so are most people who are doing what you are doing—then you are well equipped to ride out your bouts of depression or mania.

Concentrate on writing in small increments, and do your best not to worry about exposing your ideas. Tell yourself that nobody ever reads dissertations anyway, that the dissertation requirement is just an archaic form of torture, a rite of passage. Tell yourself that, after you have finished up and have earned your degree, you can edit your dissertation sensibly, and have it published as a book. Tell yourself whatever makes you feel competent, and get back to work.

THE PRINCIPLE FOR LUCID WRITING here is that you should approach your dissertation as you would any other writing project: with this book in your hand, with as many colleagues and advisors of various ilks as you can muster, and with a number of friends on tap to provide support.

111 Issue

FAR TOO MANY writers use *issue* as a catchall term. I recommend that you use more specific and more meaningful terms instead.

UGLY: The staff decided to sort out the kitchen-detail issues.

GOOD: Everyone agreed that the kitchen was a mess, and that a roster for kitchen cleanup would repair frayed tempers and soothe agitated souls.

UGLY: Health-care issues are at issue in this year's discourse.

GOOD: Problems in health-care delivery and the lack of universal health-care insurance have caught the nation's attention this year.

UGLY: We will address funding issues in Section 6.

GOOD: In Section 6, we will analyze our financial needs, present our budget, and describe alternative sources of funding.

UGLY: Semantics issues are separate from syntax issues.

GOOD: When you write, you must pay attention to both content and form; however, the structure of a sentence often determines the meaning.

448

UGLY: Max had privacy issues about his diary, and Lyn had back issues of her favorite magazine.

GOOD: Max and Lyn discovered, as have couples throughout history, that relationships elicit strong feelings, and that a companion may range from infuriating, to discombobulating, to disconcerting, to comfortable, to exhilarating, to ravishing.

THE PRINCIPLE FOR LUCID WRITING here is that you should avoid writing about issues, and should instead describe your subject clearly and in sufficient detail. This principle is not an issue.

112 Terms for Human–Computer Interaction

WHETHER YOU are writing a textbook, a research report, or a manual, you will need to style *consistently* the terms that you use to describe a user's interaction with a computer. For example, you will need to determine and follow a style for

- Names of keys
- Names of commands
- Names of buttons on the display
- Names of menus
- Menu items
- Names of functions
- Names of variables
- Names of objects
- Names of nodes or arcs in a diagram
- Names of windows
- Strings that the user is to type
- References to actions, such as selecting, clicking, double-clicking, dragging, or highlighting

- Code
- Comment lines in code
- Placeholders for words to be filled in, such as names of files or passwords

The primary principle here is that you should think about how you will set these terms *before* you begin to write your document. If you put off the decisions, you will find that retrofitting a manual, for example, is a major headache. Instead, begin by making a list of all items in your document that require a style decision. Make and note each decision, and then follow the style compulsively. As you write, you will find that these design specifications evolve; just maintain the specification document, and keep it at your elbow as you write.

There are no absolute rules covering styling of such terms. There are, however, several principles that you should follow.

You should use *typographical distinctions* intelligently. For example, you can set code in a monospace typeface, such as `Courier` or `Typewriter MT`[276] (or, if your base text is set in a serif typeface, you may be able to differentiate code by setting it in a MARKEDLY DIFFERENT or sans serif typeface). You can set commands in `monospace`, **boldface,** or *italic* type (or in a combination thereof). You can set strings that the user is to enter in **boldface** or *italics,* or in a typeface markedly different from that of your base text. You can also use underlining to differentiate terms, although underlining looks messy in many typefaces.

> GOOD: Enter **yes** to indicate that the program is to continue.
>
> GOOD: Max typed `I lust after your huge-screen color monitor.`

---

276. In this book, code is set in `Typewriter MT`.

GOOD: Lyn typed *I lust after your mouse.*

GOOD: Use `chmod` in this case.

GOOD: Press **f6**.

GOOD: Open the **Options** window.

 You should use *styling distinctions* intelligently. For example, you can capitalize the names of keyboard keys. You can set placeholders by using angle brackets <> to enclose a description of what the user is to fill in (for example, `<hostname>`). You can style menu entries precisely as they are shown on the screen. You can set menu names with a Lead cap or in Mixed Capital and Lowercase Letters. You can set Boolean operator names (such as AND and OR) in SMALL CAPITAL LETTERS.[277] Note that you can combine styling differentiation with typographical differentiation.

GOOD: I would prefer that you AND the expressions that have been ORed.

GOOD: You can use the following form: `while <Boolean expression>`.

GOOD: Enter <password>, then press Return.

GOOD: Open the `Diagnoses` pull-down menu, and select `Carpal-tunnel syndrome`.

GOOD: Double-click on the Calculate button.

GOOD: The LOVE node is connected by bidirectional arcs to the PEACE and HOPE nodes.

GOOD: Use the **PageDown** key to navigate the document.

GOOD: Use the Option + Arrow key to move an object one point at a time.

GOOD: Select the image using the *magic wand* tool.

---

277. The classifications BAD, UGLY, GOOD, and SPLENDID in this book are set in small capital letters.

You should *not distinguish terms that do not require special treatment.* The least helpful approach is to clutter up your page with boxes, 12 different typefaces, underlining, boldface and italic type, and so on; the result will be to confuse, rather than to clarify.

BAD: First, type

| I want to go Home. |

Then, PRESS

(**Enter**)

or (*Return*)

THEN type in another <u>line</u>:

UGLY: ***Start up* BRAINDEAD**™ by *Clicking* on the <u>Do It</u> **Icon.** Then, ***Answer*** "<u>YES</u>" in the *Dialog Box* that appears in the <u>Control</u>   *Window.* ***Type*** "*I think this example is a little weird,*" and ***Press*** <u>Enter</u>. ***Close*** the *Window* by *Clicking* in the *Close Box.* ***Press*** the <u>Cancel</u> **Button** in the *Dialog Box* that appears in the corner of your **Screen.**

Note that, unlike the first example, the second example is classified as ugly: The terms are styled consistently!

THE PRINCIPLE FOR LUCID WRITING here is that you should determine at the outset of your writing project how you will style terms used to describe human–computer interaction. You should **use the** *arsenal* of **methods** FOR `differentiating` <u>type</u>, but you should use it wisely, so that the result is clarification of your meaning.

113 So Called

T̲HE TERM̲ *so called* can be troublesome; in addition, it is generally unnecessary. Wise writers avoid using it.

 You should generally simply leave out *so called;* you will convey the same meaning without it. Note, by the way, that *so called* is hyphenated (only) when it serves as an adjective.

> UGLY: The so-called gold standard in this case is determined by common practice.

> GOOD: The gold standard in this case is determined by a set of measurements locked in a secure vault.

> UGLY: In the so-called spanning-tree algorithm, all the bridges regularly exchange special frames.

> GOOD: The spanning tree is used in source rooting.

> UGLY: So-called interprocess communication can be difficult to set up.

> GOOD: Interpersonal communication is the most difficult process to set up.

If you think that simply writing the word provides insufficient indication that you are, in effect, naming an entity, you can use *italic letters* or "quotation marks." (If you are using italics to differentiate a word from its denotation, use italics in this case; if you are using quotation marks for that purpose, use quotation marks in this case.)

UGLY: The first two years of marriage constitute the so-called adjustment phase.

GOOD: The second through the nineteenth years of marriage constitute the "working-it-out phase."

GOOD: The twentieth through the fiftieth years of marriage constitute the *peaceful-coexistence phase.*

UGLY: Brian could not tolerate the so-called happy face.

GOOD: Bob was not worried about *losing face.*

GOOD: Tom was concerned about "business face."

If you sometimes feel compelled to use *so called,* do not introduce redundancy by also using italics or quotation marks.

BAD: This so-called "resort" gives you towels so thin and small that they could not keep a cat warm.

BAD: The so-called *special maharaja dinner* comprised so many dishes that Max and Lyn were quite ill after they gobbled down the feast.

GOOD: The so-called computer consisted of three highly competent elderly clerks and three abaci.

BAD: Len made up a so-called "cashflow forecast analysis predictive no-surprise spreadsheet system."

BAD: Devon did not think much of the so-called *telephone redirecting and messaging system.*

GOOD: Max enjoyed working on the so-called space-mating
project at NASA.

THE PRINCIPLE FOR LUCID WRITING here is
that you should avoid using *so called*. If you do
use it, do not also use quotation marks or italics
to delineate the so-called term under discussion.

114 Note That Versus Notice That

IF YOU WISH to call special attention to a sentence, use *note that* for strong emphasis, or *notice that* for weaker emphasis; avoid the awkward phrase *it is important to note [or notice] that.*

Note that is the strongest and most efficient of the four phrases, and will convey your point without adding excess verbiage. Thus, if you want your reader to mark your words, choose *note that*.

> UGLY: It is important to note that our sample contained only two data points, so the results may not generalize to the entire population.

> UGLY: Notice that the previous discussion does not apply to nerds who own more than seven machines.

> GOOD: Note that we have truncated the code subtly for this example; were you to run it as shown, you would get spaghetti for results.

> UGLY: It is important to note that Max had been severely provoked before he threw his smelly old socks at Lyn.

> UGLY: It is important to notice that sure signs that BB is feeling unfriendly are laid-back ears, bared teeth, and loud hisses.

> UGLY: Notice that, whenever Max has had a tough day, BB acts as though he is her implacable enemy: She threatens to scratch his eyes out if he comes near her (unless he offers to scratch her tummy).

> GOOD: Note that, when Max is relaxed and happy, BB climbs on his lap, rolls on her back, and wriggles invitingly (which always causes Max to scratch her tummy).

In many cases, even *note that* is redundant; the strongest option often is to let the content of your sentence speak for itself. Presumably, if you are doing your job as a writer, your reader is noting everything you say, so you should not have to flag a sentence to get her attention.[278]

> GOOD: In interpreting our report, note that seven of the 10 subjects in this study died before all the data had been collected; thus, the results may be slightly unreliable.

> SPLENDID: Seven of the 10 personnel managers in the meeting asphyxiated before someone took an ax to the specially designed nonopening window; after years of careful study, we have concluded that the recently installed air-circulation system may be suspect.

> GOOD: Note that Patrick had spent hours making careful suggestions, only to be told that he was reading an outdated document. You should always double-check to be sure that a certain idiot has not given you misleading information.

278: I use *note that* frequently in this book, not because I do not credit you with paying full attention, but because I have chosen it as a device to let you know that I am introducing points subsequent to the main one under discussion. (The main points are set off with the device of the *icon,* well known in techy land.)

SPLENDID: Patrick stayed up all night trying to figure out how fly sputum was related to the topic under discussion. You should never assume that a writer has her act together sufficiently that each word that she writes has any business taking up the space in which it sits.

 You should use *notice that* when you are writing about the act of noticing or are inviting your reader to notice a point in passing. *Notice that* is weaker than *note that;* it implies more dawning realization than sudden focus.

GOOD: Over the course of an uncomfortable year, Misha noticed that, since he moved to Oregon, he was not hiking as much as he used to do.

GOOD: In the middle of a rainy morning, Holly noticed that the goats had been gnawing on the corners of the picnic table.

GOOD: As you have been reading the examples in this book, have you noticed that Holly and Misha used to live in Palo Alto, California, but now (to Lyn's sorrow) live in Portland, Oregon?

THE PRINCIPLE FOR LUCID WRITING here is that, when you wish to flag a sentence for your readers' attention, you should first ask yourself whether the content is not sufficient to call attention to itself without your aid; if you determine that the sentence needs a flashing light, use the simplest choice: *note that*. In addition, you should notice that *notice that* has its uses too.

115 Affect Versus Effect

MANY PEOPLE have difficulty using *affect* and *effect* correctly; the two words have usefully different meanings.

- The noun *effect* denotes the result of a process, event, or activity. An effect is coupled with a cause.
- The verb *effect* denotes bringing an object into existence, or bringing about a state of affairs.
- The noun *affect* denotes an emotional state.
- The verb *affect* denotes producing an influence on, or producing an effect. It also means having a fondness for, or pretending.

Here are examples of correct uses:

GOOD: The most pronounced effect of Lyn's vacation in Canada was a noticeable disappearance of Lyn's sense of humor.

Here, effect *is used as a noun.*

GOOD: Max effected a change in Lyn's outlook by meeting her at the gate and holding her extremely tight while blocking the egress of all the other passengers.

Here, effect *(effected) is used as a verb.*

GOOD: Max having thus arranged Lyn's homecoming, Lyn's affect changed to one of buoyant delight and glee.

Affect *is used as a noun.*

GOOD: Max and Lyn thus demonstrated how companions can affect each other's moods.

Max and Lyn wield considerable power over each other's happiness. Affect *is used as a verb.*

GOOD: Lyn is unable to affect indifference to Max.

Lyn is unable to pretend that Max has no effect on her. Affect *is used as a verb.*

GOOD: Lyn affects silk dresses.

Lyn sports dresses that float around her as she moves; this predilection for silk might qualify as an affectation. Affect *is used as a verb.*

You should be extremely careful when you use either *affect* or *effect*: Be sure that you are writing what you intend to write. Let us look at another set of examples.

BAD: The affect of AVL rotation can be used to keep a tree in height balance.

GOOD: The effect of pointer manipulation is to rotate the subtree whose root is the pivot node.

Here, effect *is used as a noun.*

BAD: Steve affected his pseudosabbatical only after months of negotiation.

GOOD: Judy effected a career change by quitting her job in Palo Alto and prancing off to Geneva without any idea of what she would do next.

Here, effect *(effected) is used as a verb.*

BAD: Lyn carves wooden masks whose faces conspicuously reflect each person's effect.

GOOD: Lyn was always puzzled when she tried to envision a flat affect sufficiently precisely to carve a mask of a person in such a state.

Here, affect *is used as a noun.*

BAD: Betsy's social life was effected by Sophia's disinterest in bottle feeding.[279]

GOOD: Richard's ears were affected by Sophia's penchant for beating biscuits to a pulp with a rubber hammer.

Here, affect *is used as a verb.*

BAD: Lyn's torn meniscus, which resulted in her doctor proscribing her daily 13-mile run on pavement, effected Lyn's ability to reach her destination quickly.

GOOD: Lyn's fractured vertebra, which resulted in her doctor prescribing bed rest, affected Lyn's ability to reach her toes quickly.

Here, affect *is used as a verb.*

BAD: Max effected disdain for mushy love and romance.

If we intended to assert that Max caused other people to feel such disdain, the sentence would be correct.

---

279. If the sentence was intended to report that Betsy's entire social life was brought into being by this particular preference of Sophia, it would be classified as good. We assume, however, that the intention is to report that Betsy's social calendar was moderately constrained during the first 6 months of Sophia's life.

GOOD: Red affected disdain for mushy cottage cheese and romaine lettuce.

Here, affect *is used as a verb.*

BAD: Joe effected classy horn-rimmed spectacles and subtle cologne as insurance against mediocrity.

GOOD: Joe affected a shaved head as an antidote for tedium and as a guard against hot-headed behavior.

Here, affect *is used as a verb.*

SPLENDID: Lyn's weight lifting effected numerous calluses, giving her hands a calloused[280] texture that affected the people with whom she shook hands.

SPLENDID: Diane's desire to appear intelligent affected her speech; affectation was the effect.[281]

THE PRINCIPLE FOR LUCID WRITING here is that you can affect your reader adversely if you do not effect correct usage by using words that have a good effect on her ear and thus positively influence her affect.

---

280. Note the difference between callus (a pad of thickened skin) and callous (thickened, hardened, or having calluses, whether physically or emotionally).

281. Contributed by Peter Gordon.

116 Indices Versus Indexes

ALTHOUGH BOTH *indices* and *indexes* comprise more than one index, the type of index denoted is different.

You should write *indices* to refer to a number or expression that indicates a position or location in mathematics.

> BAD: In the expression A = {a_1, a_2, a_3}, 1, 2, and 3 are indexes.

> GOOD: You can also use letter indices, as in b_i and b_j.

> BAD: The indexes 1 and 5 indicate that c_{15} lies in cell 1 of column 5.

> GOOD: The indices 5 and 2 indicate that B_{52} is located in the fifth row of the second column.

You should write *indexes* to refer to the plural of all other types of index. The alphabetical lists (those at the back of this book, for example) in which you look up entries are indexes, as are indicators and pointers.

> BAD: The indices for the book series were not well conceived, so it was impossible to look up relevant terms.

> GOOD: An indexer is a person who has expertise in writing book indexes.

464

BAD: A man's swagger may be one of several indices of his machismo.

GOOD: The conditions of a cat's teeth, eyes, and fur are good indexes to the cat's health.

THE PRINCIPLE FOR LUCID WRITING here is that you should distinguish between the indexes in, for example, a book, and the indices in, for example, mathematical expressions.

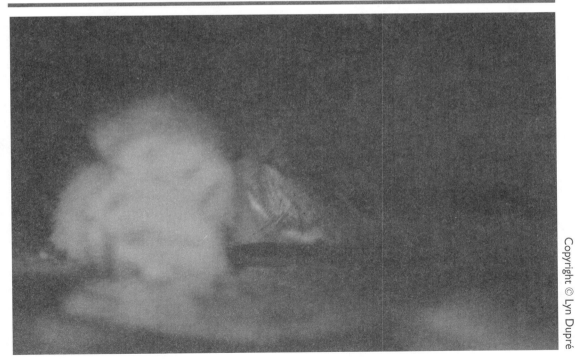

117 Solidus

THE SOLIDUS is the forward slash (/). Used in English terms,[282] it means *and or*. Most publishers in the United States forbid any (even correct) use of the solidus with English words, perhaps because people so often use the solidus incorrectly. For that reason, I recommend that you avoid setting the solidus in English terms in your formal writing.

You should generally avoid using the solidus. Many writers use the solidus erroneously to indicate an equal-weighted pair. The correct punctuation mark in this case is an en dash.

> UGLY: The input/output device drivers must be custom tailored.
>
> GOOD: Our new system's input–output processes are accomplished by numerous peripheral devices.

> UGLY: The read/write head crashed into the disk.
>
> GOOD: The read–write head is a critical component of a standard computer system.

---

282. In contrast to being used, for example, in mathematical expressions.

466

UGLY: Our network uses a client/server configuration.

GOOD: Avoid using a client–server setup at home.

UGLY: The overall structure for this pointer-chain represen-tation is an example of parent/child organization.

GOOD: The parent–child structure is based on values for the supplier grocery stores.

UGLY: Lyn requested that John use a red/maroon back-ground for the cover art.

GOOD: Lyn asked John whether he had used brown–orange or burnt orange for the background of the central graphic on the front cover.

UGLY: The doctor/patient relationship often inhibits, rather than encourages, communication.

GOOD: The provider–client relationship can be enhanced if the physician remembers that she is offering a service to a person who retains control and decision-making authority.

UGLY: Richard has developed a fine-tuned distinction for describing his current mental/emotional status: the great/OK dichotomy.

GOOD: Lyn's constant superb–lousy toggling was wearing down Richard's patience.

GOOD: Carver reexamined continuous and discrete theories.

You should generally avoid using *and/or,* for several reasons. First, the term is redundant: It means *and and or or.* Second, most publishers do not allow its use, presumably because of that redundancy. Third, you can usually write simply *and* without losing any meaning; if it is important to indicate that only one option may be operative, you can assume an *inclusive or* and simply write *or.* Fourth, there are other (less problematic) ways to express the same thought. You may find it difficult to break the *and/or* habit, but why use a confusing, redundant term when you can avoid doing so?

UGLY: Sophia likes to be held and/or to eat.

GOOD: Madeline likes to play and to eat.

UGLY: You can use a mouse and/or a keyboard for input.

GOOD: If you cannot make a selection, then the mouse or the keyboard is not functioning properly.

SPLENDID: Max or Lyn will cook dinner, and one or both of them will wash the dishes.

THE PRINCIPLE FOR LUCID WRITING here is that the solidus (forward slash, /) means *and or;* you should not use it for equal-weighted pairs or in the redundant term *and/or.*

118 Equations

decorative horizontal rule

DISPLAYED EQUATIONS are equations that are set after and before a line break; they are *broken out* from the text line. Your primary challenge in setting equations (be they displayed or in text) is likely to be finding and learning to use a text editor that can handle them. Here, we shall look at a few basic ideas for setting displayed equations.

Regarding the complex rules governing setting of mathematical expressions, Ronald Barry[283] gives this advice:

> The symbology of mathematical expressions has an immense scope, and adequate coverage might easily require an entire book. At the turn of the century, a candidate for a doctoral degree in mathematics was required to demonstrate at least a rudimentary knowledge of the 20 or so distinct areas of mathematics extant at that time. Today, there probably are about 3000 different areas of study in mathematics, and the field continues to expand, with no sign of slowing down; it is impossible for most people to be acquainted with more than a few of these areas, and even professional mathematicians are unlikely to be conversant with more than a few hundred of them. The practice nowadays, for any branch of mathematics that might conceivably be construed as being outside the mainstream, is for the author of a work on that subject to provide, at the beginning or end of the work, a glossary of terms used throughout

283. Personal communication, Palo Alto, CA, 1994.

that work. I suspect that this approach will continue to be the optimum available technique for any reader who is not intimately familiar with the subject matter, and perhaps even for those readers who do possess such familiarity.

I heartily endorse Barry's advice: Provide a glossary or list of conventions used in any document for which the reader may find such a guide helpful.

You should set variables in italic type, and constants in roman type.

GOOD: $\displaystyle\sum_{0 \le k < n} \lfloor \sqrt{k} \rfloor .$

GOOD: $f_\nu(x) = (0 \le x < v) .$

GOOD: $\log_{10}\eta + C' + \dfrac{B'}{T} .$

GOOD: $F_C + {}'\mathrm{F}'_R = 0 .$

You should set mathematical expressions in figures in the same way as you set them in text. Do not, for example, set your variables in roman type in the figure label (set them in italic type, as in the text); do not set an expression in capital letters, or mixed capital and lowercase letters, in the figure label, and set it in lowercase letters in text. Figure 118.1 is an example of this mistake (the X and Y should be lowercase italic letters *(x* and *y)*, rather than capitalized roman letters); we classify it as bad.

You should be consistent in whether you use end punctuation in displayed equations. I recommend that you do use it, because it gives your reader more information. You should set a period, comma, semicolon, or no punctuation at the end of a displayed equation, depending on where that equation falls within a sentence.

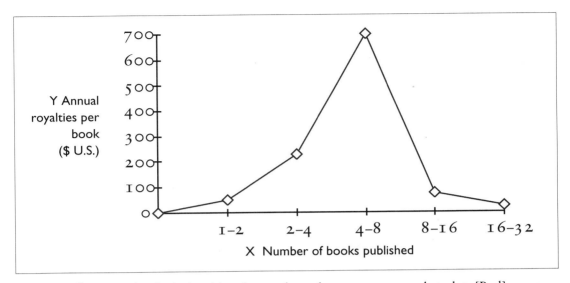

Figure **118.1** Graph of relationship of annual royalty amount to market glut. [Bad]

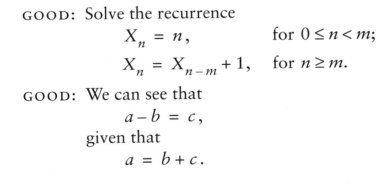

GOOD: Solve the recurrence

$$X_n = n, \qquad \text{for } 0 \le n < m;$$

$$X_n = X_{n-m} + 1, \quad \text{for } n \ge m.$$

GOOD: We can see that

$$a - b = c,$$

given that

$$a = b + c.$$

You should set a colon before a displayed equation *only* when the words preceding the display call for a colon. At minimum, they should constitute a full sentence. It is a common mistake to set a colon before a display, when the colon makes no sense as a punctuation mark. When no colon is called for, you should set whatever punctuation is correct—a comma, semicolon, period, question mark, or exclamation mark each might be a candidate.

GOOD: The equation is as follows:

$$c = \frac{d - m}{2}.$$

GOOD: We revert to the old equation—

$$g_n = g_{n-1} + g_{n-2}$$

—in this difficult case.

GOOD: Let us try to prove directly that

$$\sum_{d/m} \vartheta(d)\, n^{m/d} = 0 \pmod{m},$$

without using the clue that this equation is related to necklaces.[284]

Abbreviations are not variables; you should set abbreviations in roman type.

GOOD: $V_{\min} = k\,(\min(V_1, V_2) - V_b).$

GOOD: The carrier density at the source end of the channel is denoted by N_s; that at the drain end is denoted by N_d.

GOOD: One way to use the notion of a standard regression coefficient in a measure of uncertainty importance is in $U_{\mathrm{SRC}}(x_j, y).$

You should generally indent displayed equations from the left margin, by a consistent distance, unless the design for your document specifies otherwise. When you have several equations in a display, you should align them on the equal-to sign (or equivalence, or greater than, and so on).

---

284. This example was stolen, with bits and pieces from other examples in this segment, from Graham RL, Knuth DE, Patashnik O, *Concrete Mathematics*. Reading, MA: Addison-Wesley, 1989.

GOOD: Consider the following relations:

$$a + b = c,$$
$$b \geq c,$$
$$f - g \neq h,$$
$$d \approx e + f.$$

You should number those equations to which you will want to refer in text. Use single or double numbers, depending on the numbering of your level 1 heads. Set the number right justified, in parentheses. There is no need to number other equations.

GOOD: You are undoubtedly familiar with the following equivalence:

$$a^2 + b^2 = c^2. \tag{5.4}$$

GOOD: For sets A and B,

$$A \cup B \neq A \supseteq B. \tag{18}$$

When you *refer to numbered equations,* you should capitalize the word *equation,* and should not set parentheses around the number. It can be confusing to your reader to use words other than *equation* (such as *equivalence)* in these references, so I recommend that you avoid doing so when practical. Also, do not refer by number to an equation that you are about to show — call it *the following equation,* or *the next equation.*

UGLY: We can use Equivalence 6.7 and Assertion 6.9 to obtain Equation 6.10.

GOOD: Equation 4.3 is the general case, valid for $n > 1$.

GOOD: Using Equation 6 and our common sense, we obtain the following:

$$\sqrt{a - b/c} + \overline{)d} \neq \phi \pm \varphi. \tag{9}$$

THE PRINCIPLE FOR LUCID WRITING here is that, after you have mastered a text editor that can handle equations, you should review the few simple ideas presented in this segment.

119 *Half*

WHEN YOU REFER to one of two parts that constitute a whole, remember that those parts must be equal to deserve the title *one-half*.

You should write *one-half of*, rather than *half*, in most cases. When you write about *the other half*, it is acceptable to use either form. It is also acceptable to use *half* in well-known terms or phrases. Finally, you should use *half* to denote the result of halving.

> UGLY: To Max's dismay, half the expansion boards were defective.
>
> GOOD: To Max's delight, one-half of the expansion slots were unused.
>
> GOOD: To Max's relief, the store let him pay half the price down, and the other half in time payments.

> UGLY: Greg spent half his weekend looking for a reasonable deal on a used car.
>
> GOOD: Nicola spent one-half of her day working, and one-half gardening.
>
> GOOD: Nicola spent half her evening sewing, and the other half making jewelry.

BAD: Steve and Judy had only half a pillow between them on the camping trip.

GOOD: Steve and Judy had between them only one-half of a problem; they solved it without fuss.

GOOD: Steve and Judy did their tax returns together so efficiently that they cut in half the time usually required.

GOOD: Given just half a chance, Judy was able to carve out a new career.

GOOD: Steve's faith in his other half was unshakable.

 You should set the numeric fraction (for example, $\frac{1}{2}$),[285] rather than spelling out a fraction (for example, one-half), in situations where you would use a numeral for whole numbers (such as with units of measure). You can also, if you prefer, write 0.5.

---

285. There are four ways to set a fraction: case, solidus (or shilling), built up, and piece. You should generally set case fractions in technical material, if you have the capability to do so. In nontechnical material, you can set your fractions solidus or piece.

- A *case* fraction is set with the numerator directly over the denominator; the entire fraction is the height of one text line:

$$\tfrac{a}{b}.$$

- A *solidus* fraction is set with a solidus, or slash:

$$a/b.$$

- A *built-up* fraction is set with the numerator directly over the denominator, with each element being at least the height of one text line:

$$\frac{a}{b}.$$

- A *piece* fraction is set with a solidus, but the numerator and the denominator are on different horizontal and vertical lines:

$$a\!/_b$$

or

$$^a\!/_b.$$

BAD: Max promised to go hiking for half a day.

Because a day is a unit of measure (of time), you should set numerals with it; thus, you should set the fraction ½²⁸⁶ or 0.5.

BAD: Lyn's temperature went up by half a degree every time she saw a mistake in her manuscript.

Degrees also are units of measure.

BAD: The horse had only half a furlong to go.

Furlongs too are units of measure.

BAD: Lyn could not fathom how far down half a fathom was.

Fathoms too are (surprise!) units of measure.

GOOD: Lyn promised to be ready for bed in ½ hour.

GOOD: To BB, a shrew that stands 0.5 inch tall is a monstrous and frightening beast.

GOOD: The entrance to the walkway in the ceiling above Lyn and Max's bed was just ½ centimeter too small for Red to squeeze through.

GOOD: Lyn ordered 0.5 cubic feet of feathers for her nest.

GOOD: Sophia is now more than ½ year old.

You should write *moiety,* rather than *one-half,* to refer to a portion of a whole that has been divided into two unequal parts.

BAD: Cosmo tried to grab the bigger half of the cookie.

GOOD: Marina gave Swix the greater moiety of the can of dog food, because Swix weighed more than Spud and was also dominant.

---

286. I have set the fractions in different ways to give you examples; see previous note.

BAD: Reno quietly fed Swix the smaller half of his broken candy bar.

GOOD: Pat gave the lesser moiety of his waking time to his family, but he provided quality to make up for lack of quantity.

GOOD: Lyn and Max were never sure which of them was the other's better moiety.

THE PRINCIPLE FOR LUCID WRITING here is that you should use *one-half* to refer to one of two equal parts, and should use *moiety* to refer to one of two unequal parts that together constitute a whole. The greater moiety of this book comprises more than one-half of the segments.

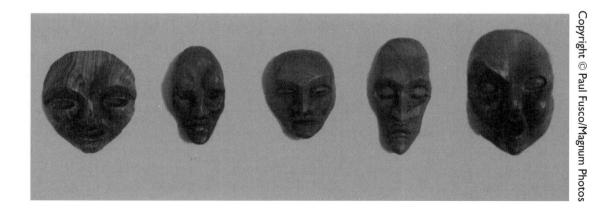

120 Media

THE WORD *media* denotes more than one *medium* of the sort that is a member of a broadcast system, or that stores or transmits information (the human variety takes the plural *mediums*). Never use *media* as a singular, as that error is anathema to many readers.

BAD: The news media has turned out in force to cover the latest scandal.

GOOD: The media have been preoccupied lately with their image.

BAD: Tape is a good backup media.

GOOD: A floppy disk is a convenient storage medium.

BAD: The media is the message; it strikes a happy media.[287]

GOOD: The medium can be massaged to present what you want presented.

BAD: Max uses various mediums to give his talks.

GOOD: Lyn is learning to program multimedia applications.

---

287. Contributed by Richard Adamo.

SPLENDID: Max and Lyn consulted several mediums to determine which media reported medium-rare as the most delicious way to cook a steak; they also discovered which news medium had tried to use petri dishes and old socks as the media in which to grow the meat for the barbecue, and an ice locker as the storage medium for the leftovers.

THE PRINCIPLES FOR LUCID WRITING here are that you should use *media* to denote more than one medium, unless the medium is a human being, and that you should never use *media* as a singular. Either use numerous writing media, or use just one medium.

121 Not Versus Rather Than

IN SENTENCES such as the following, the correct term is *rather than*, rather than *not:*

> BAD: Max wanted to build decision models all day, not to participate in video conferences, meetings, and conference calls; however, he did what was required.

> GOOD: Lyn wanted to write all day, rather than to fuss with supervising schedules, making telephone calls, and sending faxes; therefore, she spent her time editing photographs.

> BAD: The logic example demonstrates *modus ponens*, not *modus tolens*.

> GOOD: You should use proof by refutation, rather than resolution theorem proving.

> BAD: Use the vocabulary of nodes and links, not that of frames and slots.

> GOOD: The `Siamese` class is a direct subclass, rather than a direct superclass, of the `Cat` class.

BAD: Red wanted to capture bats, not to enrapture them.

GOOD: Lyn preferred to thank Red for his generous gifts, rather than to chastise him for murdering every creature in sight.

BAD: Richard was drinking far too much coffee, not sipping herbal tea and fresh juice.

GOOD: Lyn was taking calming Chinese herbs, rather than shooting herself in frustration.

THE PRINCIPLE FOR LUCID WRITING here is that you should use *rather than,* rather than *not.*

122 Visual Aids for Presentations

W HEN YOU give a presentation, you will usually prepare slides to accompany your talk. (I shall use *slide* here loosely, to mean any visual aid—a transparency, for example, or a poster, a videotape, or a presentation on a CRT screen or LCD display.) Certain people prepare their slides first, and then talk from the slides. Other people conceive the talk, and then make up their slides.[288]

There are two schools of thought[289] on the role that your slides play in your presentation. One school says that the speaker should be the focus of attention, and that therefore the slides should merely make a few points here and there, but should not cover comprehensively the content of the talk. I agree that the speaker should be the focus, provided that the speaker is worth the attention, but I do not agree with the deduction regarding the slides. The other school of thought, to which I adhere, is that the slides provide the anchor for your audience, which has a tendency to drift downstream or onto the banks. Thus, I recommend that you make your slides comprehensive, and—more important—use them to keep your audience oriented.

I believe that few people are able to listen to most lectures without a break in their attention; excellent speakers hold audience attention

---

288. There are numerous techniques for public speaking, and numerous courses and books that will teach you those techniques. If you give presentations often, you would do well to avail yourself of these resources.

289. There are probably 185 perspectives, or more.

most of the time, and average speakers hold audience attention fitfully. Your slides, however, can bring up the stragglers.

In addition, certain people absorb ideas more easily via visual representations; other people find aural information easier to grasp. By using comprehensive, clear slides, and also speaking well, you will reach both factions of your audience.

Because your slides constitute a shorthand version of your talk, they are *highly concentrated information carriers*. All the principles for disambiguating and otherwise clarifying your writing apply in spades to slides. Take a great deal of care developing your slides.

I leave out here the audience-softening joke slides and other grab-and-hold tricks that you may want to use.

If you have available *software for preparing presentations*, you may find that using it is extraordinarily helpful. Resist the temptation, however, to use all the bells and whistles suddenly available to you, as otherwise you will create a din above which your meaning cannot be heard.

Use color sparingly (or at least gently). If you are not versed in how different colors and combinations of colors attract and pull the eye, you may want to read up on the subject; failing that alternative, look at your slides from a distance and ponder what happens in terms of understandability when you switch colors. Also *use color consistently*: If an object is green in one slide, do not color it red in the next. Do not distinguish by color creatures that have no right to be distinguished—if, for purposes of your talk, they form a homogeneous group, then color them to so indicate.

Use highlighting, animation, blinking, and other options thoughtfully. In general, use only one or two visual furbelows per slide, and use only those that have an excuse for being there.

If you or your organization has a *logo*, you should place the logo in the same corner on every slide. An example of a logo follows:[290]

If you have no logo, put your name or other appropriate identification (for example, the name of your laboratory) on the slides. The following slide[291] has a logo in the lower-right corner. The subsequent slides have various forms of identification.

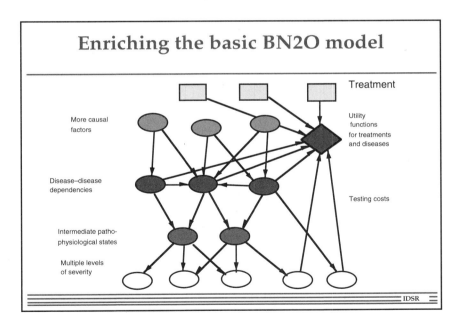

---

290. Logo courtesy of Max Henrion, Lumina Decision Systems, Inc., Los Altos, California.

291. Slide courtesy of Max Henrion, Institute for Decision-Systems Research, Los Altos, California.

Your *opening slide* should say who you are, at what organization you work (if that information is relevant or existent), and what the title of your talk is. The title, whether humorous or deathly serious, should tell anyone reading it what the content of the talk is.

Ten Tips for Preparing Slides

Lyn Dupré
Corporate Helping Hand, Inc.

CHH

 Your *second slide* can be the general question or problem that you will address, if that information is not covered by the title.

How can you keep your audience awake?

Exciting Talks, Inc.

Why should your audience care?

Exciting Talks, Inc.

How can *you* stay awake while you talk?

Exciting Talks, Inc.

 Your (second or) *third slide* should outline the structure of your talk. It should indicate what you will talk about in what sequence; when you show it, you should say why you have adopted that sequence—why the sequence is a logical and helpful way to teach the subject.

OVERVIEW OF OPERATING-SYSTEMS CONCEPTS

1. MULTICS
2. OS/MVS
3. VMS
4. Unix
5. MS-DOS
6. OS/2
7. Mach

Avi Knows

All your *subsequent slides* should be tied to the outline slide. One excellent technique is to keep the outline in a small box in the corner of each slide, with a highlighter indicating where in the outline you are.

1. Selecting a Breed
  1.1 Cat Shows
  1.2 Literature
2. Preparing Your Home
  2.1 Bedding
  2.2 Food and Water
  2.3 Litter Box
3. Selecting an Individual
  3.1 Gender
  3.2 Personality
4. Bringing Home Your New Kitten

3.2 Personality

- Watch how the kitten interacts with her littermates.
- Play with the kitten and pay attention to how she interacts with you.
- Read the kitten's tail language.
- Time the interval between initial petting and loud purring.

Dupré Felines

If you do not use the outline box, then, at minimum, the titles of the subsequent slides should match exactly the titles given in the outline, and you should switch back to the outline slide as you move from segment to segment, reminding your audience of where you have been, where you are, and where you are going. Assume that, at any given time, someone in the audience has just woken up from a day-dream and is trying to figure out about what you are talking; help her out!

 Your *conclusion slide* should hark back to the title or question–problem slide. Show that you have answered the question or solved the problem, and say (briefly) how you have done so. Again, assume that your listeners will have forgotten the question, even though you have reminded them several times.

How I Have Kept You Awake Today

1. Vigorous delivery style
2. Intelligent use of visual aids
3. Careful audience selection
4. Friendly interface; excellent rapport
5. Good explanation of topic

Exciting Talks, Inc.

 You should include a *final slide* that indicates how your work ties into a relevant bigger picture, and what future research on the subject might be, if your conclusion slide does not provide this information. If you are giving a business presentation, you can tie your work to overall goals such as big profits. If you are sharing personal experiences, you can say what (or when, or why, or how, or where) other people might be in similar situations.

10. Decision Analysis (DA) for the Masses

I have demonstrated today why making the general public conversant with the principles of DA would encourage rational approaches to public policy.

To put DA in the hands of everyone, we need tools that

• Are inexpensive and easy to use

• Guide users not versed in the application of statistical measures

• Guide users not familiar with the principles of DA

• Provide summary advice and rationales on demand

Decision Makers Anonymous

You should include a *severely limited* amount of information per slide. There is no point at all in displaying an elaborate slide with great blocks of text—or intricate diagrams—that no one can read. The problem is that people will try to read it anyway, and will stop listening to what you are saying. Limit each slide to a small, digestible chunk of information. Use simple, easy-to-follow graphics and comprehensible symbols. For text, use large type that can be seen and read easily by someone sitting in the back row, even if that someone's vision is not like that of a hawk. Thus, the following three slides are classified as ugly.[292]

292. Text for the third slide courtesy of Max Henrion, Lumina Decision Systems, Inc., Los Altos, California.

Tables of Contents for *BUGS in Writing*

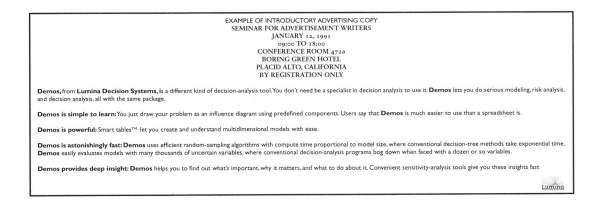

You should use simple graphics liberally. A few boxes with arrows between them can show system flow, for example, much more concisely than can a word description. Be sure that the labels on your graphics are clear (be sure that there are labels!), and that they are tied unambiguously to components of the picture. (In the following graphic, the words CPU, Main Memory, Disk Controller, and Ethernet Controller are labels.)

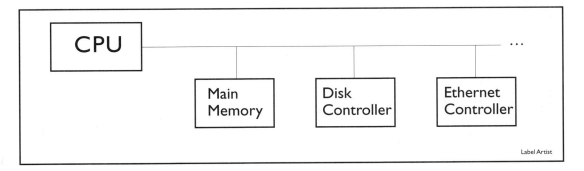

You should use absolutely consistent styling and spelling of words in both text and graphics, and should proofread every slide numerous times. Check also that your slides are loaded in the correct orientation. As the following example shows, slides are a highly distilled form of writing, and any booboos will be disturbingly noticeable.

3.4 Any Booboos will be distirbingly Noticeable

Embarrassing Moments, Inc.

You should use parallel heads and structure. By *parallel structure*, I mean that your slides should be obviously consistent in layout and in use of design elements. Do not set lists in seven different ways; set them in one way. Do not give *questions* and *answers* on one slide, and *problems* and *solutions* on another that has similar content. Do not set the title of one slide in lowercase letters, the title of another in all capital letters, and the title of a third in both capital and lowercase letters. Do not set one title as a question, the next as an assertion, and the next as an imperative. Do not number a few slides, and then forget to number the remainder. Be sure slides follow logically.

---

1. Can software help?

Hard Sell, BSA

---

2. Certain packages can indeed help

Hard Sell, BSA

---

3. You should buy software!

Hard Sell, BSA

---

4. Thou shalt not steal thy neighbors' software.

Hard Sell, BSA

Paid Announcement by Buy Software Associates.

Hard Sell, BSA

Red decided to spend the winter of '95 curled up in front of the second-floor heater, to the relief of the small creatures with whom he shared the house.

Non Sequitur

You should explain complex ideas as simply as you can. This principle applies to all expository writing, and your ability to carry it out is usually a function of how well you understand your subject matter. Keep asking yourself, as you prepare your talk,

- How can I make this concept simple?
- How can I explain a new idea in terms of the old paradigms?
- With what familiar terms can I describe this notion? Which parts of my presentation are sufficiently new that they require a new name?
- Are the different components of my presentations — be they study groups, hardware modules, cows, or newt populations—sufficiently delineated? Are they named such that my audience will know which is which?
- What analogies can I use to help people grasp this notion?
- How can I tie this idea to everyday situations with which people are familiar?
- What examples can I use to clarify meaning?

Use examples liberally, and be sure to make them both simple and *explicit*. In the following examples, there is nothing technically wrong with the first version; the second version, however, is more explicit and thus should be easier for the audience to follow.

UGLY: "We consider how to recombine objects, where certain objects are incompatible with others in the absence of a third. The recombination rules are that only two objects at a time can move from one set to another, and that A must be one of those two objects."

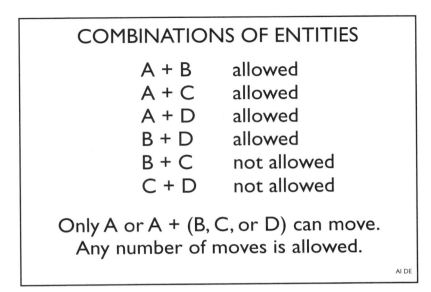

COMBINATIONS OF ENTITIES

A + B	allowed
A + C	allowed
A + D	allowed
B + D	allowed
B + C	not allowed
C + D	not allowed

Only A or A + (B, C, or D) can move.
Any number of moves is allowed.

AI DE

GOOD: "Lyn has to transport three entities from the downstairs bedroom to the upstairs deck: Red, a mouse, and a wheel of cheese. How can Lyn get all the entities onto the deck without allowing any of them to be eaten?"

DECK-TRANSPORT DETAIL

1. Lyn can carry only two objects at a time.
2. Without Lyn's supervision, the mouse will eat the cheese.
3. Without Lyn's supervision, Red will eat the mouse.
4. Red is not partial to cheese.
5. Lyn is in good shape, and does not mind running up and down the stairs.

HOME AI

UGLY: "The patient presents with a symptom, and the doctor must decide whether to order diagnostic tests, and whether to prescribe treatment."[293]

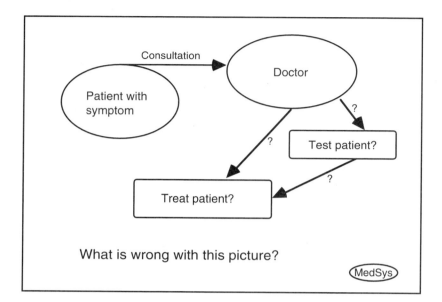

GOOD: "The client presents with a history of fainting, and the physician must help the client to decide whether to undergo arrhythmia mapping (an invasive procedure that incurs risk, cost, and discomfort to the client), and whether to take antiarrhythmic drugs (which have unpleasant side effects and also incur cost to the client)."[294]

---

293. Note that, to Dr. Henrion's disgust, the slide does not contain an influence diagram.

294. The purpose of this example is to emphasize that explicit examples are more interesting and informative than are vague examples. The other content changes are irrelevant to the point under discussion. In its irrelevant contents, the example makes no claim to being realistic; it does, however, present a point of view that may spark discussion.

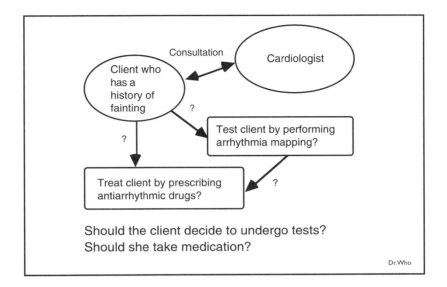

If you have a function or equation or other symbolic language in your slide, point to it (or highlight it), and express it in natural language as you talk.

GOOD: "The probability of A given that we know that B has occurred is equal to zero."

$$p\,(A|B) = 0$$

GOOD: "If a particle is subject to some external force, then, during the time that it is free, it will accelerate in keeping with Newton's law: Force is equal to mass times acceleration."

Newton's Law

$$f = ma$$

Figs, Inc.

GOOD: "Woman laughing plus woman singing may cause man to feel relieved."

$$W_L + W_S \longrightarrow M_R$$

Highly Likely

 You should generally not give your audience handouts at the beginning of, or during, your talk, unless you prefer to see only the tops of people's heads. You should particularly not give your audience the outline and slides of your talk, because people will be happily browsing along three slides ahead of you, and will miss what you have to say. The exception occurs when you want people to take notes on your talk; in that case, giving your listeners an outline that has sufficient blank space for them to write their notes may be a good option. I urge you to consider, however, that the most interesting speakers are sufficiently easy to understand and to remember that their audience usually desists from note taking early in the talk.

It is often helpful to provide the outline, slides, or other supporting material at the *end* of your talk. You can then suggest that people *not* take notes as you speak, pledging to supply the notes yourself.

Of course, if you have specific material that you want people to examine and that you cannot project or otherwise display, then you will have to resort to handouts. Do not, however, have the audience pass around objects during your talk. Again, imagine having the woman sitting behind you tap you on the shoulder and hand to you an object that you are supposed to examine and then, by further shoulder tapping, pass on; clearly, such distractions are not conducive to your paying attention to a lecture.

As you speak about a list of items, such as the outline of heads for your talk, you should avoid displaying the entire list at the start. Rather, block out the entries, and reveal them one at a time as you get to them. This technique keeps your audience with you, instead of flailing into the woods ahead of you without a guide. You can use a piece of cardboard, for example, to block a transparency (but pay careful attention to it at all times, as otherwise you will cause alarming visual distractions). Another technique is to use a piece of transparent, colored (yellow is good) plastic to highlight the item that you are discussing.

GOOD: "Lyn has to balance carefully two steaming, full mugs of ginseng tea while trotting down an extremely narrow, steep staircase; in addition, she must neatly avoid the two cats who are wreathing around her ankles."

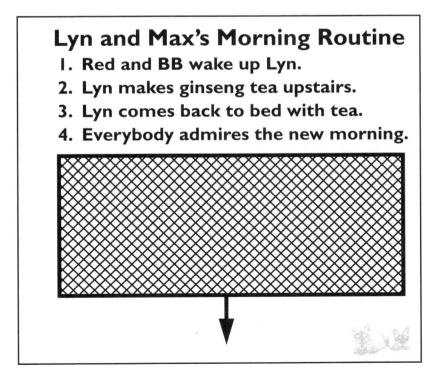

When doing so is practical, you should alter your slide as you speak, such as by highlighting, pointing to, sticking colored pins in, or otherwise drawing attention to different parts of the display.

GOOD: "I shall now discuss the processing step." [Click.]

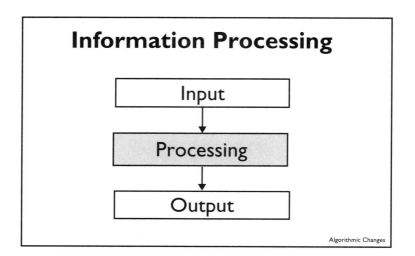

You should speak about each item in each slide; if you have nothing to say about it, it should not be there. Do not show a list with six entries, but mention only two of them. Do not show a diagram with four components, but speak about only three. By the same token, do not show a diagram with four components, but speak about five of them.

BAD: "So, there is one important lesson for you to learn."

1. Squeeze toothpaste from bottom.
2. Smell flowers from above ground.
3. Keep your feet under you when your head is in the clouds.

Sappy Advice

BAD: "There are three considerations in modem communications: the modem speed, data-bits setting, and stop-bit setting."

Modem Communications

1. Line speed
2. Data-bits setting
3. Parity-checking setting
4. Stop-bit setting

Papa Bell

 You should use the same terms in your talk as those you use in your slide. Do not label an object in your slide as a *magazine,* and then talk about a *journal.* Do not write about a *diskette,* and then talk about a *floppy disk.* Do not diagram a *gallery-visiting algorithm,* and then talk about *traveling sales representatives.*

BAD: "Let's look now at the odds–ratio version of Bayes' rule. By using this rule, we calculate, from the prior on A, the posterior probability on A given B."

Odds–Likelihood Formulation of Bayes' Theorem

$$O\,(Q|P) = \lambda O(Q)$$

We can thus compute the posterior odds on Q (given P) from the prior odds on Q (that is, from the odds on Q before we observed that P was true).

Odd but Unlikely

When you point to a component or entry on an overhead-projection transparency or computer screen, you should point to the display at which your audience is looking; do not hunch up over the projector and point to items on it. First, you should be seeing what your audience is seeing; your enormous forefinger jabbing at entries in a list may look quite different from what you imagine. Second, you should be talking to your audience, rather than to your projector.

 When you point to an item, you should always point to the side of— rather than across—your body. That is, if you are standing on the right side of the display, from the audience's perspective, then you should use your right hand to point; if you are standing on the left, use your left hand. If you point across your body, you will automatically turn your back on your audience, and will start mumbling into the screen.

 If you use a pointer, you should be careful what you do with it. Do not play with it, wave it around, or otherwise distract people with it. Do not open it and close it against your body, stick it in your pocket, or otherwise risk the gaffe of gaffing yourself.

You should use your hands thoughtfully. Intelligent use of gesture is a superb way to engage your audience; flapping about like a duck that walked into a beehive is not. Do not ball your hands into fists in your pockets; if you want to use your pockets at all, use only one at a time, and never fiddle with the contents. If gesturing is not coming easily because you are tense, gesture with one hand at a time. In general, you should neither hold your hands close to your body (signaling tension) nor throw them wide (like a politician).

You should not hide the front of your body when you speak, tempting as that option may be if you are a shy speaker. Step out from behind the lectern. Uncross your arms. Put down your papers.

You should stand to one side of your display, so that you do not block it, forcing your audience to wriggle and squirm as people try to read your slides. If you wish, you can turn off the display when you want your audience to focus on you, and turn it back on again when you want people to switch attention back to it.

You should make eye contact with your listeners. Do not mumble along staring down at your notes; avoid sounding and looking like you are reading a paper (even if that is what you are doing). Find a few people in each area of the audience—front, back, right, and left sides—who look sufficiently interested and friendly that you can look at them directly without flinching. Shift your gaze as you make different points.

You should try to have fun! A speaker who looks haggard, frightened, bored, disgusted, or otherwise distressed is unlikely to capture the imagination of her audience, and is likely to make everyone else feel uncomfortable as well. Convey your enthusiasm for your topic, and give your audience a reason to care about what you have to say. If you prepare a talk that you find interesting, chances are that your audience will find you interesting.

THE PRINCIPLE FOR LUCID WRITING here is that you should apply all the principles in this book to your slides, taking even more than your usual care, **becuase errors will be expecially obvious.**

123 Plural Abbreviations

W HEN YOU WRITE about the *plural* of an object that you have denoted by an abbreviation, you should form the plural correctly.

 You should (usually) simply append an s to an abbreviation to form the *plural*; you should not use an apostrophe.

> BAD: All the CPU's and printers in the office went berserk simultaneously.
>
> GOOD: The CRTs seemed to be unharmed and the LCDs were winking merrily.

> BAD: If you live in New York, you know how often the IRT's run.
>
> GOOD: You have only one VAX machine? I expected you to have at least several VAXEN.

You should use apostrophe s to indicate a *possessive* formed from a singular abbreviation.

> BAD: DARPAs (now and formerly ARPAs) old system caused the problem.
>
> GOOD: Correct usage would dictate that you write "you know how often the IRT's trains run."
>
> GOOD: The fans at the rock concert reported seeing a UFO's lights.

You should (usually) use s apostrophe to form the *plural possessive* of an abbreviation. (If you would pronounce the plural possessive as ending with *esses,* then you should use s apostrophe s.)

> BAD: The fans rioted because they were afraid of the three UFO's contents.
>
> BAD: The IRSs' staff members are not always perceived as angels by the tax-paying public.
>
> GOOD: The UIDs' forms were difficult to comprehend.
>
> GOOD: The BONEs's inventor was a desperate woman.

THE PRINCIPLE FOR LUCID WRITING here is that you should form the plural of an abbreviation by adding simply s (might there be several BUGSs?); you should form the possessive of a singular abbreviation by adding apostrophe s (Lyn is BUGS's inventor); and you should (usually) form the plural possessive by adding s apostrophe (but will she be INSECTs' or BUGSs's inventor?).

124 Style Sheets and Spell Checkers

CAREFUL AUTHORS and editors use a *style sheet* to maintain consistency in the way that they handle all the various terms for which they must make a style decision. A style sheet comprises sections that describe styling decisions on such matters as those in the following list. Note that I give here only representative examples; the list is by no means comprehensive.

- *Figures:* How are they numbered (single or double numbers)? What kind of captions (tags or sentences) do they have? How are the labels styled (cap/lc, lead capital, or all capital letters)? What convention do you use for calling them out (do you use *see* for repeated callouts?)?

- *Tables:* How are they numbered? How are the titles styled? How are the column heads and side heads styled? Are the entries left justified or centered? How many rules (horizontal lines) do you set? What type do you use for the notes? How do you punctuate legends?

- *References:* Do you use periods with the authors' initials? Where do you put the date? Do you spell out the month for a conference? Do you abbreviate *department* or *university?* Do you give inclusive page numbers? Is there a comma or a period at the end of a title? Are journal titles set in quotation marks? How are titles styled?

- *Punctuation:* Do you use the series comma? Do you capitalize a full sentence after a colon?[295] Is there a comma after an introductory clause? How do you set displayed lists?

- *Abbreviations:* What abbreviations do you use? How are abbreviations defined? Do you set in boldface type the definition of an abbreviation? Do you permit Latin abbreviations in parenthetical remarks? Do you define common abbreviations for units of measure?

In addition, the style sheet contains an alphabetical listing of all words or terms about which you have made a style decision. For example, if you decide to capitalize the names of functions in your document, that is a style decision. Or, if you decide to follow the style that I suggest, and to close up *non*words (that is, not to hyphenate them), that too is a style decision. Another (correct) style decision would be to omit the hyphen in two-word adjectives that follow the noun that they modify—to write *I want to make the program cost effective,* rather than *...program cost-effective.*

I recommend that you develop your own style sheet; you will probably want to maintain it online, so that you can search it easily. Think of your style sheet as boilerplate material for the functional specification of your document. That is, when you design a document, you must think first about what it will do (such as sell a product, convey a research finding, or obtain funding for you). You must decide for whom you are writing it, and what you want to say in it. You must determine what the structure of your document will be. Finally, you must choose how the elements of your document—sentences, figures, tables, references, and so on—will be styled. That part of the specification is your style sheet.

---

295. That is, do you capitalize the first letter after a colon if what follows the colon constitutes a sentence? I recommend that you do.

Your style sheet should contain numerous entries for all the terms about which you make a decision over the course of your authorship lifetime. Nearly every principle in this book would feel at home in your style sheet. In addition, your discipline or company, for example, may use many terms not covered by any dictionary. You and your colleagues will find it immensely helpful to maintain a shared style sheet, so that all manuscripts produced by your group are consistent with one another. When you work on large, multiauthored documents, a style sheet is indispensable.

You should make frequent and intelligent use of your *spell checker*.[296] Obviously, running the spell checker will catch many typos and spelling errors. In addition, you can use your spell checker to help you implement your style sheet. When you spell check your document, simply add to the spelling dictionary all terms that you have chosen to style in a certain way. Using your spell checker certainly will not eradicate all styling inconsistencies, but it will at least reduce their number.

THE PRINCIPLES FOR LUCID WRITING here are that you should develop and maintain a style sheet, and that you should use your spell checker to help you to implement that style sheet.

---

296. You may also want to use a *smell checker*, defined by Richard Adamo as someone who sniffs your manuscript to see which parts stink; such people often are known as copy editors.

125 Maybe Versus May Be

T HE ADVERB *maybe*, set as one word, means *perhaps*; it is short-hand for *it may be that*.[297] You should write the verb *may be* when you mean that an entity may or may not have a characteristic, or that a state of affairs may or may not occur.

BAD: May be Red seems extraordinarily fat.

GOOD: Maybe Red's appearance is deceptive, and he is solid muscle, right down to the tip of his tail.

GOOD: It may be that catfood with a high percentage of fat is not the most helpful diet for a cat who can no longer squeeze through a (large) cat door.

GOOD: The bald stripes on Red's back may be the result of his trying to use the cat door regardless of its obviously meager dimensions.

BAD: Max thought that may be Lyn's intents and purposes were nefarious.

GOOD: Max realized that maybe Lyn's intense purposes were to his liking.

GOOD: Intense intents may be most effective in getting the job done.

---

297. The noun *maybe* —*the wedding tomorrow is a maybe*—also is one word.

BAD: May be there was a good reason why *copy-on-write* was not included in 4.2BSD Unix.

GOOD: Maybe there is another way to create processes without exorbitant overhead.

GOOD: When large processes fork, copying the entire user address space may be expensive.

BAD: Gabe thought, "May be this bright shiny object is good to eat."

GOOD: John suggested to Gabe that maybe not all objects on the ground were edible.

GOOD: John may be happier now than he was before Gabe was born; he certainly is busier!

BAD: Joe was thinking that may be he should not have volunteered to undertake the design of Lyn's book.

GOOD: Lyn was thinking that maybe they would get the book out the door without any further significant catastrophes, disk crashes, or other unnatural disasters.

GOOD: You may be wondering whether Lyn was writing this book 3 minutes before she sent the files to the shop for filming.

Be careful not to introduce redundancy when you use either *maybe* or *may be*.[298]

---

298. Weasel wording in research papers often contains such redundancies, as the authors backpedal from commitment of any kind.

UGLY: Joe may perhaps decide to do his laboratory work before the semester begins.

UGLY: Lyn found maybe approximately 200 spiders in her shower stall this morning.

UGLY: Presumably, you may perhaps be able to see that, seemingly, this sentence apparently could maybe contain a redundant phrase.

THE PRINCIPLE FOR LUCID WRITING here is that you should set the term *maybe,* meaning *perhaps,* as one word. You should set the verb *may be* as two words. You may be still learning, but maybe you already know this principle.

126 Figures

Y̶OU SHOULD USE figures wisely as an adjunct to your writing.

A figure is a graphic representation of information; it might be a photograph (halftone), a drawing (line art), or a shaded or screened picture. It might be a screen dump or a computer-generated graph.

It would not be appropriate in this book to discourse at length on the design of good figures. You should be aware, however, that there is a wealth of information on this topic, and that it is worth the time to make yourself conversant with that material. A good start would be to read Tufte's book:

GOOD: Tufte ER. *The Visual Display of Quantitative Information.* Cheshire, CT: Graphics Press, 1983.[299]

A list of written entries is not a figure—it is simply a list. If, for whatever reason, you wish to highlight or number a list, or any other block of plain text, you can enclose the text in a box; do not, however, call it a figure.[300] Similarly, you should label and number tabular material as a table, rather than as a figure. You can handle a section of computer code as a figure, or you can label and number it as a program (Program 4.5, Program 6), even if it is only a fragment

---

299. Tufte's book is good; the citation for it is set correctly, so it too is good.

300. Figures can consist of simply text if, for example, they are reproductions of advertisements; in other words, there are exceptions. The problem arises when you try to use slides or transparencies, which often properly comprise only text, as figures.

of a program. In general, you should set code in a `monospace typeface`, or in another typeface obviously distinguished from that of your base text (for example, in a **sans serif typeface** if the base text is in serif typeface).[301]

Figure 126.1 should be an intext list or a box, rather than a figure. We thus classify it as bad.

- Sleeping in Lyn's lap while Lyn works
- Hiding under the bed while Lyn runs errands
- Clawing her way up Max's jacket sleeves to nest in the top of his closet
- Sleeping on top of Lyn's computer monitor while Lyn works
- Hiding under the bed when Lyn has visitors
- Grooming Red on the couch next to Lyn, while Lyn reads British mystery stories
- Rolling on her back to flirt with Max
- Sleeping on Lyn's pillow when Lyn is asleep in bed
- Having four adorable kittens

Figure 126.1   Several of BB's favorite activities. [Bad]

301. I recommend monospace (typefaces in which each letter takes up the same amount of space, like letters printed by an old-fashioned typewriter) because its use for this purpose is currently common, and readers will understand what you are doing. There are other ways to distinguish code from base text, and many people have strong opinions about which ways are preferable. I do not like to set code in a type size different from that of base text, for example, whereas other people find down sizing acceptable. The primary point is that you should choose a convention that will clearly distinguish code from base text and that also will be legible, and you should use your convention consistently.

Figure 126.2 should be a table, rather than a figure; it too is classified as bad.

Red	BB	Comments
Floppy; apparently boneless	Stiff; highly angular	Difference in mellowness
Purrs when spoken to	Runs when spoken to	BB purrs for Lyn easily
Flame point	Chocolate point	Both are gorgeous
Weighs 21 pounds	Weighs 7 pounds	Both are well fed

Figure 126.2   Notable differences between Red and BB. [Bad]

Figures 126.3 and 126.4 are set correctly as figures; we classify them as good. Note that, in Figure 126.3, the legend (the key explaining what is signified by the differentiators, such as dots, lines, colors, shading, and so on) is contained within the figure; it also could be set in the caption.

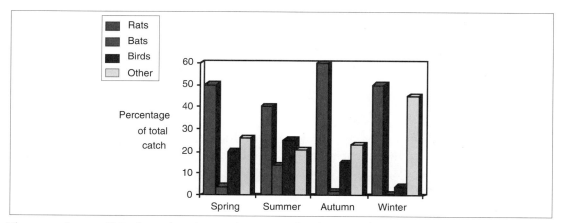

Figure 126.3   Bar chart showing various prey that keep Red amused (by percentage of total catch) during the four seasons. *Source:* Dupré L. What's waiting on the rug in the morning. *Fluff and Fancy,* 4(4): 444. [Good]

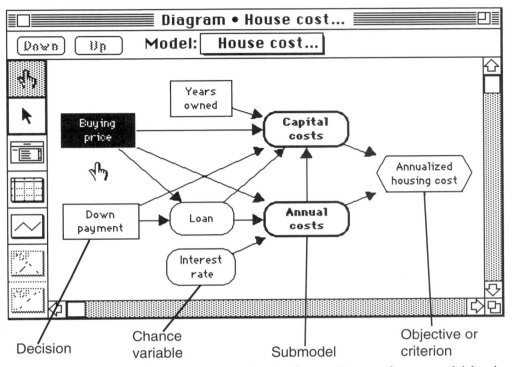

Figure 126.4   Demos screen showing a top-level influence diagram from a model for the decision of whether to buy a house. Rectangles represent decision nodes, rounded-corner rectangles represent chance variables, thick-line outlined rounded-corner rectangles are submodels, and the diamond node is the objective or criterion node for the decision. The arrows show relations, or influences, between nodes. Courtesy of Max Henrion, Lumina Decision Systems, Inc., Los Altos, California. [Good]

You should number all figures. In an article, proposal, or other short document, you can use single numbers (Figure 2, Figure 6), or you can use double numbers by section (Figure 4.1 being the first figure in Section 4, and so on). In a book chapter, you should use double numbers (Figure 5.3 being the third figure in Chapter 5, and so on). The primary reason for numbering figures is to make layout easy. An unnumbered figure must be set in the text column precisely where the callout for it appears, whereas a numbered figure may be set several pages away if necessary.[302] In addition, you can refer precisely to a numbered figure later in the text.

> BAD: The following figure shows Peter, Lyn, Sarah, and Max, shrieking on Splash Mountain at Disneyland.

> GOOD: Figure 12.4 shows that, in 1993, SIGGRAPH was held in geographic proximity to Disneyland.

 You should call out all numbered figures in text.[303] To *call out* a figure is to refer to the figure by number.

> GOOD: Obelisk's robust cash-flow forecasts for 1994 are shown in Figure 3.2.

You should use figures whenever, and only when, *a graphical representation will communicate information more clearly and effectively than will a word description.* A figure that is difficult to understand or confusing will serve only to reduce communication with your reader, whereas your goal should be to enhance communication. A good figure is simple and easy to grasp. For these reasons, you should not have a long description in the text that conveys the same information as does a figure. If you present a chart showing the percentage of entrepreneurs who have nervous breakdowns, classified by years since incorporation of their companies, do not add a

---

302. Obviously, if you are not submitting your manuscript to a publisher (or if you are, at any rate, doing your own page makeup), then you may not be concerned about this difficulty. However, many new authors do not apprehend the layout problems that can arise once a book or article is in production.

303. An excellent technique is to search on words such as *Figure, Table, Program,* and *Box* when you finish your document, to ensure that you have called out each element in the proper order. Also, most text formatters will let you use automatic numbering for all numbered elements in your document. I strongly recommend that you use that option if it is available; machines are considerably more accurate counters than are humans.

redundant paragraph presenting the same data. *The figure caption plus the figure should, in most cases, stand alone.*

Figure 126.5 is not a useful figure; we classify it as ugly.

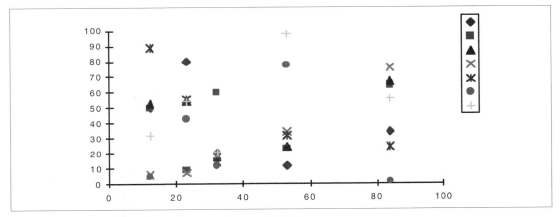

Figure 126.5   Data from the experiment. [Ugly]

Figure 126.6 is equally useless, primary because it has no legend, and the caption explains little. It receives the classification of ugly.

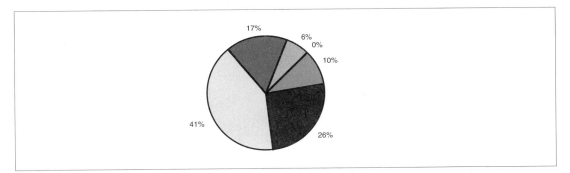

Figure 126.6   Suggested breakdown of revenue sources. [Ugly]

Figure 126.7 may at first look reasonable, but closer inspection reveals a presumed mismatch with the caption. Therefore, we classify this figure as ugly.

Figure 126.7   Scatterplot showing relationship between system reliability and computational effort. A nonlinear relationship can be observed, despite the two outliers. Photograph courtesy of Misha Pavel. [Ugly]

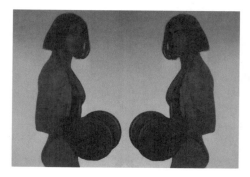

This figure has neither a callout nor a number (at least it has a caption!). A reader might well wonder what it is doing here. [Bad] Image based on a photograph copyright © Paul Fusco/Magnum Photos.

Figure 126.8 is a reasonably clear figure;[304] we classify it as good.

---

304. It is not, however, a clearly reasonable figure.

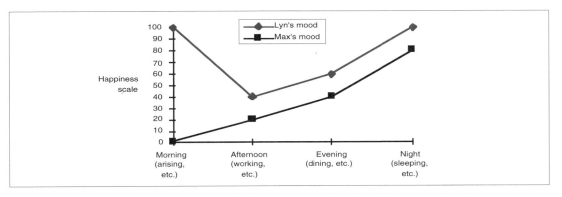

Figure 126.8   Max and Lyn's moods, by period of the day. The vertical axis represents a scale from O (no happiness) to I OO (full potential happiness). We could discern in the data set no causal influence of the two variables on each other. [Good]

You should label your figures clearly and consistently. A figure label is a word or phrase that indicates what an object in a graphic represents. You might be labeling the axes of a graph, or the nodes of a network, or the components of a VLSI circuit. You might be labeling a house and a happy family, or a flowchart for a system. Figure 126.9 contains numerous labels.

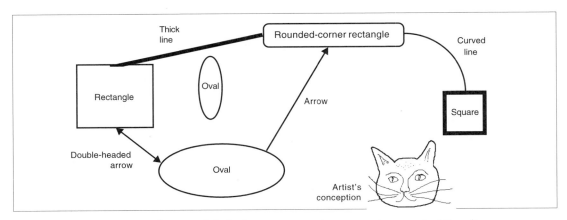

Figure 126.9   Figure showing labels for common shapes and artwork. [Good]

You should *style* all labels in all figures consistently, in terms of capital versus lowercase letters. That is, you should set all labels in mixed capital and lowercase letters, with only the first letter capitalized, or in all lowercase letters. Do not set the labels for one figure in all capital letters and set those for the next figure all lowercase letters, and do not mix styles within one figure.

The labels in Figure 126.10 are inconsistent, both internally and as compared to the caption. In addition, the information in the caption does not match that in the graph. Finally, the unit of measure being used in the vertical axis is not specified. With no hesitation, we classify it as bad.

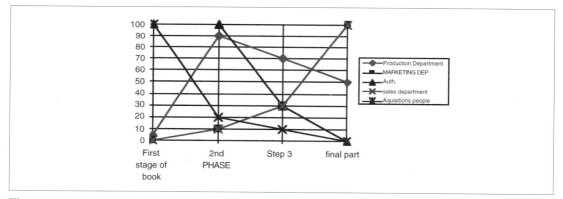

Figure 126.10   Rate of activity of various entities involved in the creation of a book, by phase. In the first phase, the author and sponsoring editor are working at full tilt. In the second phase, the division of sales and marketing begins to develop a strategy, while the author continues to write and the sponsoring editor supervises. The production supervisor comes on board in phase 3, and the author and editor theoretically have less work to do. In phase 4, manufacturing takes over producing the bound book. By phase 5, sales representatives and marketing specialists are doing the bulk of the work. [Bad]

Occasionally, a figure may be complex enough to require several levels of labels, just like levels of heads in a text outline. For example, if your figures primarily show systems in which there are objects taking up cells, you may wish to label the systems with level 1 labels, and the cells with level 2 labels. In that case, choose a logical style for each level, and stick to it. Figure 126.11 has two levels of labels ("compilers," for example, is level 2, whereas "APPLICATIONS PROGRAMS" is level 1).

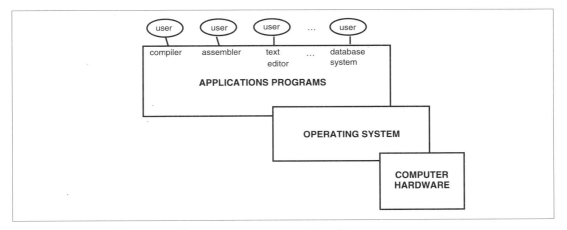

Figure 126.11   Key parts of a computer system. [Good]

You should style and spell all labels as they are set in the text (with the exception of capital versus lowercase letters). That is, if you refer to an object as a *layer cake* in the text, do not label it as a *torte* in your figure. Also do not label it as a "layer cake" (do not add or delete quotation marks; match the way that the term is set in the text), or as a *layer-cake* (do not use spelling or hyphenation different from that set in text). Be sure to set all variables in italics, just as they are set in the text, and to set other mathematical symbols in the same way (do not use a times sign in the text and an asterisk in the figure). If you are distinguishing certain terms in the text by using a different typeface—for example, if you are using a monospace face for code, or for names of variables—then be sure to use that typeface for such terms in your figure labels as well. The default face probably will be different from the text default, as most designs call for separate type-faces for figure labels and base text.

Figure 126.12 shows mismatch between labels and caption text (the source note also is suspect); we classify it as bad.

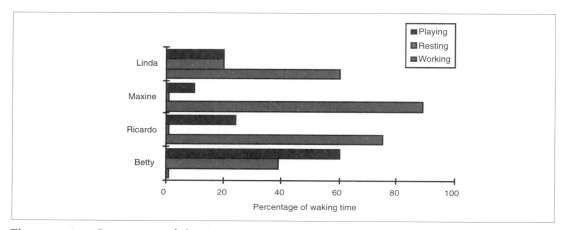

Figure 126.12 Percentage of daytime spent on various activities for Lyn, Max, Richard, and Betsy. *Relaxation* includes daydreaming and spacing out. *Effort* includes only those activities undertaken for financial remuneration. *Fun* is the default category. The researchers have been criticized for failure to recognize that certain hard work is performed gratis. *Source:* Adapted from Oppenheim G & Hoffman, G. *Past-Life Regression Therapy,* Tappan, New York: Confide Press, 1990. [Bad]

Figure 126.12 BB as a kitten. Can you spot the obvious problem with this figure? [Bad] Photograph copyright © Lyn Dupré.

You should not set the title of your figure within the figure itself.[305] The title belongs in the caption.[306] Figure 126.13 contains a title; we classify it as ugly.

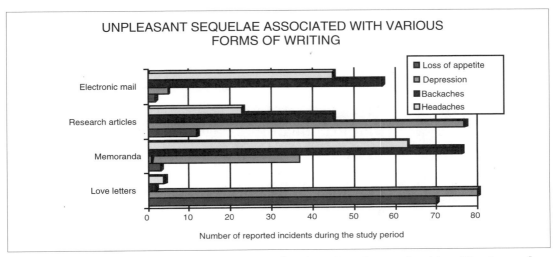

Figure 126.13   Unpleasant sequelae associated with various forms of writing. The data reflect subjects' self-reports of correlations between incidents of the four unpleasant effects (loss of appetite, depression, backache, and headache) and the type of writing that the subjects had done during the study period. [Ugly]

Note that you may find it helpful to design your figures before you begin writing. Just as certain people can talk from any set of slides, certain people can write from any set of figures. Experiment with designing your figures before you write your text; see whether the technique works for you.

305. In slides, transparencies, or other visual aids for presentations, you should have a title. Thus, be careful if you convert a slide to a figure: Remove the title!

306. In traditional book publishing, the *art manuscript* and the *text manuscript* are separate entities. The art manuscript contains the figures; the text manuscript includes, in addition to the regular text, figure captions, tables, boxes—in other words, everything else. The figure title should not be part of the art manuscript.

If that technique does not work for you, consider writing your prose and drawing your figures as you get to them, rather than after you have finished writing. You may find that the process of drawing the figures changes your thinking about the text presentation. In addition, you may find it difficult to reconstruct what you wanted to show when you wrote a section 6 months ago.

THE PRINCIPLE FOR LUCID WRITING here is that graphical representation is a powerful and compact form of communication that you should use wisely. You should choose figures that are clear, simple, and easy to understand, and that are labeled carefully.

127 Gender-Specific Words

ORDS THAT assign a gender that is not relevant to a role will annoy a portion of your readers, and will seriously insult another portion. There is no reason to indicate the gender of people whose job or activity has nothing to do with their gender. Instead, use alternatives that are not gender specific.

When possible, use a word that is genderfree. When no such alternative is available, construct a gender-neutral word by, for example, substituting *person* for *man,* or coin a plausible word. Even if the word is awkward, it shows your reader that you are sensitive to gender bias and to other gender-related issues; a large portion of your readers will applaud your sensitivity.

> UGLY: Richard suggested, "The problem with manned space flights is the human factor."
>
> GOOD: Betsy devised a schedule to ensure that the playschool would be staffed at all times.

> UGLY: The car salesman complained to his manager that he was being used to sell man-made objects.
>
> GOOD: Lyn worked out an alternative algorithm to solve the traveling-salesperson problem, in which the salesperson must visit numerous towns to sell artifacts.
>
> GOOD: The sales representative for Avon forgot his sales kit, which contained various manufactured items.

UGLY: Dona missed the ferry from Quadra Island to Campbell River because a "men working" sign demanded that all cars detour to the other end of the island.

GOOD: Every summer, as soon as tourist season starts, "crew working" signs go up all over Quadra, because the roads keep subsiding.

UGLY: Len suggested that they discuss manpower requirements for the consulting job, now that Bob had done an analysis of the man-hours required.

GOOD: Max replied that he would prefer to discuss staffing, based on person-hours.

UGLY: The stewardesses called a strike when they were asked to pick up after the passengers; they objected to serving as garbagemen. Firemen stood by nervously.

GOOD: The sanitation workers refused to cross the flight attendant's picket line; the stench soon became unbearable, so the firefighters burned the garbage.

UGLY: As chairman of the company, Max decided to cancel the launch party because he was too busy to attend.

GOOD: Theodora drove home at far above the speed limit after the board tabled her nomination for chair.

UGLY: The policemen closed the highway to accommodate the congressman's campaign entourage.

GOOD: The congressional representatives who favored privacy and those who favored security collided on the information superhighway; police officers were called in to clean up the mess.

UGLY: The new motherboard perturbed the man–machine interface.

GOOD: The team developed a parentboard that allowed them to try an innovative human–computer interaction.

UGLY: Since Max's maid left town, the spiders had taken over, and the mailman refused to come to the door.

GOOD: Lyn suggested that they either advertise the spiders on the World Wide Web or hire a new housekeeper, so that the mail carrier would resume service.

UGLY: Misha's friends were always trying to introduce Misha to his fellow countrymen, when such people visited from the mother country.

GOOD: Moisés visited his homeland with his compatriots.

UGLY: The waitress tripped, and spilled hot coffee on Brian's laptop.

GOOD: The waiter then dried Brian's laptop vigorously in the hope of preserving her tip.

GOOD: Brian was unimpressed by the waitperson's clean-up techniques.

GOOD: Brian and Adrienne hired 10 waitrons for their wedding reception.

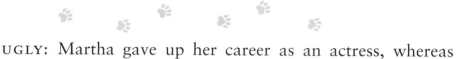

UGLY: Martha gave up her career as an actress, whereas Anthony maintained his commitment to being an actor.

GOOD: Whenever Anthony played opposite the new actor in the troupe, she tried to upstage him.

UGLY: The landlord wanted to evict the rats from Max and Lyn's ceiling, but Lyn was unwilling to harm any living creature, and she was unable to locate an experienced pied piper.

GOOD: The property owner worked with Lyn and Max on various alternatives, such as smoke, a trail of cheese, or nightly banging of cymbals. They selected, as the method most likely to be effective in ridding the house of creatures, continual broadcast of closed-loop demos from computer trade shows.

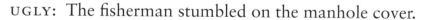

UGLY: The fisherman stumbled on the manhole cover.

GOOD: The fish catcher[307] fell down the hole in the street because the subterranean-access lid was missing.

---

307. *The New York Times* has accepted *fisher;* a 1994 headline read "With both sexes enjoying the sport, 'fisher' is losing a suffix." [Lawson C. Fly fishing's lure draws women to the river, *The New York Times,* August 8, 1994, p. A1.]

UGLY: The businessmen all came to the conference wearing their most conservative gray suits.

GOOD: The businesspeople and the managers attended the exotic erotic ball wearing unexpected costumes.

UGLY: The question of whether women should be clergymen has excited much debate.

GOOD: Even individual clergy—particularly those who are women—hold different opinions.

UGLY: The speaker at the feminist meeting closed by saying, "I'd like to thank you guys for attending my lecture"; several of the ladies threw rotten tomatoes at him.

GOOD: "I wish to remind our speaker," said the meeting's leader, "that there is only one guy present tonight among all the people here."

UGLY: Mankind has always waged war, a testament to man's inhumanity to man.

GOOD: Lyn wondered whether humanity's inhumanity was evident to other *Homo sapiens*.

Note that there are respectable gender-specific terms that denote roles that *are* related to gender.

GOOD: The mother nursed her daughter while the cock crowed and the peacock spread his fan.

GOOD: The queen kittened while the bitch whelped.

GOOD: Max and Lyn decided to walk around, rather than through, the field in which stood a massively imposing and ferocious-looking bull.

THE PRINCIPLE FOR LUCID WRITING here is that, to mistress the art of writing, you should avoid using words that are gender specific when the roles that they denote are not gender related.

128 Continuous Versus Continual

THE WORDS *continuous* and *continual* have different meanings. *Continuous* means *occurring without interruption; continual* means *recurring steadily.*

UGLY: The neighbors' continuous boisterous, drunken late-night parties often kept everyone awake.

GOOD: The continuous cacophony of the heron chicks allowed Max and Lyn to do away with their stereo system throughout spring.

UGLY: The simplest resistive network is a one-dimensional continual network.

GOOD: Continuous networks have characteristics different from those of discrete networks.

UGLY: A virtual-memory system continuously swaps data in and out of memory.

GOOD: The CPU, in a time-sharing system, is continually switching from process to process.

UGLY: Max decided that breathing continually was necessary for his continued well-being.

GOOD: Lyn was continually amazed by the difficulty of the publishing process.

UGLY: The continuous scrabbling noises in the bedroom ceiling startled Red whenever they commenced.

GOOD: The continual squawking, squeaking, and gnawing in the bedroom ceiling led BB to decide that, whatever else was up there, rats could not be ruled out.

SPLENDID: Lyn worried almost continuously that the continual gnawing and scrabbling, which continued to be broadcast throughout the bedroom every night, was leading to continuous degradation of the building's infrastructure, including the electrical wiring and the beams.

SPLENDID: If you drink carrot juice continually, you may notice a continuously deepening and disturbingly bright orange cast to your complexion.

THE PRINCIPLE FOR LUCID WRITING here is that, when you are writing, you should be on guard against errors continuously, to avoid making mistakes continually.

129 Fewer Versus Less

THE WORD *fewer* denotes a reduction in the number of a given collection of individual items; *less* denotes a reduction in the amount of a given stuff.

BAD: Steve thought that there should be less cars on the road than there were at rush hour on Highway 101.

GOOD: Judy thought that there should be fewer bicyclists per pack than there were on weekends on Arastradero Road.

GOOD: Steve and Judy agreed that there should be less traffic on the road on which they wanted to bicycle together.

BAD: I think that this manuscript would be dull as ditch-water if it had less clip-art inserts.

GOOD: Desktop-publishing systems allow you to move through the production process in fewer steps than are needed for traditional typesetting systems.

GOOD: Paul Brainerd could have been less successful, but he produced PageMaker™ and made a killing.

BAD: Reno wanted less toys than Cosmo did; Marina immediately checked Reno's temperature.

GOOD: Spud had fewer tricks in his repertoire than Swix had.

GOOD: Swix put up with less adult attention and more tail pulling after Cosmo was born.

GOOD: Cosmo ate less ice cream than Reno did; Cosmo had fewer stomach aches that week too.

THE PRINCIPLE FOR LUCID WRITING here is that you should distinguish between *fewer* and *less*; making fewer errors will allow you to feel less foolish.

130 Italic Type

I̲TALIC TYPE is the most common typographical distinction used in text. If you are familiar with its uses, you can write precisely by, for example, distinguishing terms themselves from denotations, and emphasizing phrases.

Whenever you set text in a typographical style different from that of your base text, remember to set the accompanying punctuation marks in the same style. Set an italic comma, period, quotation mark, parenthesis, question mark, exclamation mark, and so on next to an italic letter. The reason for this convention is that a roman period, for example, will not close up properly to an italic letter, but instead will leave an ugly space on the baseline. Therefore, most publishers adhere to this convention;[308] if you do not do so from the outset, you may find yourself backtracking tediously to reset zillions of punctuation marks.

The exception to the preceding convention occurs in mathematical expressions, where it is standard to set the delimiters (parentheses,

---

308. Certain people feel strongly that this convention represents a logical error. They argue that, if the text block associated with the punctuation mark is set in roman type, the appearance of a term in italic type should not influence the style used to set the punctuation mark. This matter is another about which people hold powerful opinions. You may choose to follow either convention, provided that your publisher is willing to give you the choice. As always, the critical point is to be consistent across your document. This book is set using the convention suggested in the text, to which Joseph Norman objected nonviolently.

angle brackets, caret marks, or whatever) in roman, regardless of the style of the alphanumeric characters between the delimiters.

You should use italics for *emphasis*. You should not use boldface type for this purpose, because boldface leaps off the page too energetically. Reserve boldface for key terms. Do not use boldface **and** italic type, in any event.[309]

> UGLY: When the visual effect of the pointer manipulation is to rotate the subtree whose root is the pivot node, the operation is referred to as **AVL rotation**.[310]

> GOOD: The insertion that unbalanced the tree structure occurred in the *left* subtree of the *left* child of the pivot node.

> UGLY: Steve cried, "**You** are disappointed that we can't go to the beach today? How do you think *I* feel? I **insist** that we build a sandcastle in our backyard, then."

> GOOD: Judy snapped, "I don't think that's a *compromise*; it's more of a *unilateral mandate*. I prefer that we stay in bed and contemplate *lotus blossoms*."

> GOOD: "You're right—your plan does have its appeal. Let's do that, then—and I *am* sorry," sighed Steve. "So am *I*," whispered Judy, as the couple started to cuddle in the most outrageous manner.

---

309. In general, you should use boldface and italic type only when you have a reason for each typographic style; for example, if you are using boldface to pick out key terms, and a key term is also a mathematical variable, then you would set that key term in italic boldface.

310. Of course, depending on context, *AVL rotation* might well be a key term; in that case, the example would be classified as good.

You should use italics (with no boldface) to indicate that you are speaking about the *word itself,* rather than about its denotation.[311]

UGLY: John's father Desmond, and soon many people, used **anathema** as a noun, to refer to what was left of a loaf after someone had removed only one-half of a slice of bread.

GOOD: Desmond once informed an overly inquisitive interviewer that the difference between a *flute* and a *lute* was eff off.

GOOD: *Space mating* is the interesting term used to describe certain of the research programs on docking in space.

GOOD: The name *ARPA* was changed to *DARPA,* and then back to *ARPA,* in response to the political climates.

GOOD: Many people fail to notice the plural in *National Institutes of Health*.

You should use italics on *foreign words* and phrases. Note that you should, however, set in roman type common foreign words. The test to determine whether a given foreign term is common is to look it up in the main body of *Webster's Collegiate Dictionary;*[312, 313] if it is

---

311. You can also use double quotation marks for this purpose; I prefer to use italic type.

312. Be careful: There are zillions of versions of *Webster's Dictionary,* but only the one published by *Merriam-Webster* (Springfield, MA) is used by almost all publishers in the United States. Because the (wonderful) unabridged version is updated less frequently, use the collegiate edition as your day-to-day reference. Tangentially, note that Webster's first *spelling and styling* is the *preferred* one, whereas the first *definition* is the *oldest* one. Subsequent definitions are given in order of common usage. Other dictionaries follow usage in the ordering of all the definitions, and some follow etymology in the order of spelling and styling. Whatever dictionary you use, be sure you know what the orderings mean.

313. If the phrase is listed in the "Foreign Words and Phrases" section, rather than in the main body, you set it in italics.

there, it is considered to be a word of the English language, whatever its derivation, and so should be set in roman type.

GOOD: Lyn still firmly believes that *amor vincit omnia,* provided that the *omnia* cooperate.

GOOD: The sign on Max's door read *Caveat Entrator.*

GOOD: Many couples spend years not only trying to get past their hang ups, but also determining *meum et tuum.*

GOOD: Blackford suggested the formation of an ad hoc committee to study the per capita incomes of physicians who serve on numerous committees.

GOOD: Doug reported [Owens et al., 1994] that his guideline work was so rough these days that he was being absolutely forced to spend an entire week in Hawaii as a visiting professor.

GOOD: Lyn hoped that there would be no problems when the file was RIPed (i.e., they hoped that filmout would be performed without a hitch).

GOOD: Lyn was exceptionally lucky in having superbly competent and confident people (e.g., Max and Jan and Adrienne and Joe) on her book team.

GOOD: The study grant has been canceled without warning; *requiescat in pace.*

GOOD: In his `.plan` file, Moisés described his current plan as "*Dolce far niente.*"

GOOD: "Ah Red," exclaimed Lyn as she saw what Red had wrought in Max's study, "*à bon chat, bon rat.*"

You should set titles of books and movies in italic type.

UGLY: Lyn insisted on seeing Streetcar Named Desire again and again, until Misha finally sat down with her and went through the film frame by breathtaking frame.

> GOOD: Lyn failed to appreciate "The Beverly Hillbillies."

> GOOD: Max, having had a deprived childhood in London, failed to appreciate fully both *Gone with the Wind* and *The Wizard of Oz*.

> GOOD: Max, who values privacy highly, did not always appreciate fully the anecdotes in *BUGS in Writing*.

You should set *variables* in italic type. Note that, when you set the plural, you should merely add an s—you should not set an apostrophe unless you wish to indicate ownership.

> GOOD: We denote by m_{Monday} Lyn's mood on a given Monday.

> GOOD: Red sat down on the $(n-1)$th step.

> GOOD: Calculate all the a_is.

> GOOD: You can calculate x's value if you know y's.

You should not use italics to indicate that you mean the shape of a letter, rather than the letter itself. Rather, set such letters in sans serif type.[314] If your base text is set sans serif (and you are doing your own typesetting), set the letter in boldface type, and in a different sans serif typeface.

> UGLY: Most people have moderately S-shaped spines.

> GOOD: When Jackie sits on Dona's windowsill, he is perfectly A-shaped.

> GOOD: Lyn and Max have an **L**-shaped dining room.

---

314. *Serifs* are the small flourishes at the upper and lower ends of the stroke of a letter. Examples of sans (without) serif typefaces are Gill Sans, Helvetica, Avant Garde; examples of serif typefaces are Garamond, Times Roman, and Cochin.

People have extraordinarily strong opinions about serif versus sans serif type—about whether or when you should use either, about which is easier to read, and so on—but everyone disagrees.

THE PRINCIPLE FOR LUCID WRITING here is that you should set in italic type words that you wish to emphasize, words under discussion, foreign words, and variables.

131 Truncated Words

WHEN YOU truncate words without reason, you transmute respectable words into annoying jargon or overly casual phrases that will strike many of your readers as unpleasant. Respect the language by using the full form of its words.

You should write *laboratory,* rather than *lab.*

> UGLY: Ted's lab has grown considerably in the past 10 years.
>
> GOOD: Carver keeps the size of his laboratory team fairly constant, because he believes that maintaining a small group encourages collaboration.

> UGLY: You should do all the lab work for your science courses on time.
>
> GOOD: AT&T and Lucent (once known as Bell Labs) attract a large number of highly qualified scientists.

You should write *memorandum,* rather than *memo.*

> UGLY: Some compulsive people file every memo they receive, even memos announcing children's birthday parties or luncheons.
>
> GOOD: Max prefers to toss each memorandum that he receives into the circular file, with all the other silly memoranda.

UGLY: Many people feel compelled to fill their memos with jargon; perhaps they think that jargon makes them appear to be experts, but it just makes them appear to be puffed-up silly twits.

GOOD: When you write memoranda, keep them brief and to the point.

You should write *mathematics* or *mathematical,* rather than *math.*

UGLY: If you wish to write about your work in math, you need the wherewithal to set math equations.

GOOD: Even though Lyn once majored in mathematics, she could not understand any of the mathematical expressions in the book that she found.

UGLY: Backward math induction goes from n to $n - 1$.

GOOD: We use mathematical induction to prove that a statement about integer n is true for all $n \geq n0$.

You should write *advertisement,* rather than *ad.*

UGLY: The space for the ad for the condo cost $500.

GOOD: The advertisement for the condominium covered a full page.

UGLY: When she first arrived at her new job, Karen placed an ad in the singles column.

GOOD: When they had time, Max and Lyn designed the new advertisement for Demos.

You should write *network,* rather than *net.*

> UGLY: The company's wide-area net took far too many years to implement.

> GOOD: The arcs in a belief network represent dependencies.

> UGLY: The Hopfield net is similar to the matrix associative memory.

> GOOD: Carver has moved on from his highly respected work on neural networks.

You should write *telephone,* rather than *phone,* for the noun (denotes an object), adjective (modifies an object), and verb (denotes an action). *(Phone* is a respectable word with its own meaning in, for example, linguistics.)

> UGLY: After Max finished being charmingly polite to the less than brilliant client, he tore the phone from the wall; then—with a ferocious growl—he threw the instrument of his displeasure across the room.

> GOOD: You can buy a single device that comprises an alarm clock, a facsimile machine, a radio, a coffee maker, a voice-mail system, and a telephone; to create total chaos in your home, you need to disable only one component.

> UGLY: It's not uncommon for people who negotiate via phone calls to get phone ear, a syndrome characterized by constant tinnitus.

> GOOD: Twisted-pair wire was originally the same as voice-grade telephone wire.

UGLY: Lyn kept phoning her publisher, but the automatic messaging system kept erasing her voice mail and beeping at her, causing her to become annoyed.

GOOD: Max telephoned Lyn every night while he was traveling on business, causing her to cease to be annoyed, at least for the duration of the telephone call.

GOOD: Lyn stayed up late waiting until she could place a telephone call[315] to Ladakh, only to be told that no telephones existed in the monastery; Lyn was once again annoyed.

GOOD: Peter did finally pick up Lyn's mangled voice mail, but then he forgot to call Lyn because he was trapped in interminable meetings all day; Lyn had a difficult day.

You should write *airplane* when you are referring to a vehicle for aerial transportation; you should write *plane* when you are referring to a two-dimensional entity or a shop tool.

UGLY: To Lyn's alarm, the plane began to rattle and shake during the final descent.

GOOD: Max understands rational approximations of real numbers, but the task of approximating planes leaves him puzzled.

GOOD: The airplane from which Max attempted to log in was not the one that rattled and shook.

SPLENDID: Soren used finger planes—tiny brass planes that you wear on your fingertips—to shape the planes of the wings of the model airplane that he built.

---

315. Although it is fine to write *telephone* as a verb, there are other phrases that you can use.

You should *use standard English spelling* in formal writing.

BAD: BB wriggled thru the door (which had been left just 2 inches ajar), and then got locked in the closet all nite long (which she thought was not all rite).

GOOD: Red squeezed through the cat door (which had been made for mere-mortal–sized domestic cats), and then stayed out hunting under the moon all night long (which he thought was decidedly all right).

BAD: Max felt liteheaded after giving his presentation, during which, among other catastrophes, the litebulb on his projector had blown out.

GOOD: Lyn felt lighthearted whenever she received a lengthy kiss from Max, during which expression of affection she could not breath.

GOOD: Max and Brian discussed bringing to market a lite[316] version of their software.

 THE PRINCIPLE FOR LUCID WRITING here is that you should avoid truncating words in formal writing. When you are writing lite reading matter, such as silly memos to amuse your lab group, you can do as you wish.

---

316. The word *lite* has taken on a meaning in English different from that of *light*. When you wish to denote, with some cynicism, a version that has fewer calories, less fat, or fewer features, you can use *lite*.

132 Percent

THERE ARE A few simple rules governing the use of *percentages* that allow you to write a document that is consistent both with itself and with those of other writers.

 You should spell out *percent* in most contexts, rather than using the percent sign (%). The exceptions are that you should use the sign in tables, you can use the sign in figures and captions, and you can use the sign in publications (or even individual papers) in disciplines that use percentages frequently, such as statistics.

> UGLY: Jim said that 40% of their clients had no taste, and a different 40% failed to pay their bills.

> GOOD: Lauralee said that 100 percent of their clients had at least one redeeming feature, and that 0 percent had one redemptive factor.

You should always use numerals, rather than spelled-out numbers, with *percent* (unless, of course, the number is at the beginning of a sentence following a period or colon).

> BAD: Approximately two percent of people living in the United States know what a scroll bar is; of the rest, twenty-six percent think that papyrus is involved.

> GOOD: Approximately 87 percent of people living in Silicon Valley know what a GUI is; of those, 67 percent know how to pronounce the acronym.

You should always write *percent* as one word; you should not write *per cent*.

> BAD: Random errors in our direct measurements of the quantity of interest accounted for 50 per cent of the weird data points.

> GOOD: Nearly 90 percent of uncertainties in empirical quantities arise from the following sources: statistical variation, subjective judgment, linguistic imprecision, variability, inherent randomness, disagreement, and approximation.

 You should not repeat *percent* when you give a range. You should, however, repeat the percentage sign if you are using it.

> UGLY: Soren has colds on 5 percent to 10 percent of days, depending on the season.

> GOOD: Soren can repair 40 to 60 percent of the problems that Lyn's car has, depending on in what mood he is.

> GOOD: Lyn sees Soren 80% to 90% less often than she did when they were both carefree bachelors.

You should remember that a percentage is singular.

> BAD: A large percentage of security breaches result from idiotic choices for passwords.

> GOOD: A small percentage of trap doors is placed inadvertently by program and system developers.

When you report studies on *samples,* you can give both the number and the percent of individuals in the sample that have the attribute that you are describing. Give one first, then give the other in parentheses. The order is up to you, but you should maintain the same order throughout your document.

BAD: Forty percent (4) of the men in the study sample reported difficulty using the interface; 3 (30 percent) of the men had no clue how to use a keyboard.

GOOD: Sixty percent (12) of the women, and 80[317] percent (14) of the children under 12 years of age, reported difficulty navigating the men's interfaces.

You should hyphenate a number plus *percent* when the two together constitute a compound adjective (a modifier comprising two words).

UGLY: Max estimated that there was a 10 percent chance that he would be home before the sun set, if only he could manage a 4 percent reduction in his workload for that afternoon.

GOOD: Lyn used a 10-percent screen when she designed the boxes for her book, and 20-, 50-, and 80-percent screens for the icons that highlighted the principles.

 THE PRINCIPLE FOR LUCID WRITING here is that you should know 100 percent of the few simple rules governing use of *percent*.

---

317. The *sixty* is spelled out because it occurs at the beginning of a sentence; the 80 is set in numerals because (1) you should use numerals for all numbers greater than nine, unless you have a reason not to do so; and (2) you should use numerals with *percent*, unless you have a reason not to do so.

133 Object, Modifier, Activity

THERE ARE conventions that allow you to distinguish clearly for your reader terms that can be used to describe an object (noun), to modify another word (adjective or predicate adjective), or to describe an activity (verb). (Such terms usually comprise a word plus a preposition.)

Use of term	Set as	Example
noun	one word	login
adjective	one hyphenated word	log-in
verb	two words	log in
(predicate adjective	two words	log in)

Consider these examples:

GOOD: At login, the system requests the user's password.

GOOD: The log-in file had been corrupted.

GOOD: Max once tried to log in from an airplane.

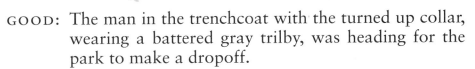

GOOD: The man in the trenchcoat with the turned up collar, wearing a battered gray trilby, was heading for the park to make a dropoff.

GOOD: The park bench with the artistic pigeon droppings was a drop-off point for the undercover mafia chili-pepper thieves.

GOOD: The man hurried along, hoping to drop off the jala-peños before anyone became suspicious.

GOOD: Betsy never forgets to put on her makeup before she leaves the house.

GOOD: Lyn bestowed on Max a make-up kiss.

GOOD: Max then agreed ardently that they should make up immediately.

GOOD: The turnoff from the highway looked hideously dark and forbidding.

GOOD: Nonetheless, Max begged Lyn to take advantage of the turn-off exit before they ran out of gas.

GOOD: "Okay," Lyn acquiesced, "I'll turn off here."

GOOD: "Wow, Judy!" exclaimed Steve, "That new hat is a knockout."

GOOD: Judy meant to give Steve an enthusiastic hug, but she spun around so energetically that she accidentally delivered a knock-out punch.

GOOD: "Oh, gosh, Steve," sobbed Judy, gathering Steve's inert form in her arms, "I didn't intend to knock you out literally as well!"

GOOD: You should make a backup of your system disk.

GOOD: You can use floppy disks as the back-up medium.

GOOD: You should back up your files frequently.

GOOD: You should keep in touch with your user base, so that you can do followup.

GOOD: For example, you can send out follow-up questionnaires, asking whether users approve of added features in a new release, or asking what features they would like to see in the next release.

GOOD: You can also use electronic mail to follow up on suggestions and complaints.

GOOD: Have you any idea what a failsafe might be? Is it a secure place to store your failures?

GOOD: The scientists designed a fail-safe program.

GOOD: Critical applications should fail safe.

In this example, fail safe *is the activity, or verb, form.*

GOOD: That is, critical applications should be fail safe.

In this example, fail safe *is a modifier, or adjective. Note that modifiers that follow their objects (predicate adjectives) are set as two, unhyphenated words.*

GOOD: As soon as the touchdown was scored, the blastoff of the rocket commenced with the announcer beginning to count down from 10.

GOOD: Player 43 made the touchdown pass,[318] while the blast-off countdown approached zero.

---

318. I am informed by Ronald Barry that *touchdown* is now the correct styling of the adjective, when applied to U.S. football.

GOOD: When the team jet touches down at home, the city will blast off a spectacular display of fireworks and the people will give a count-down cheer.

Note that you should not close up to one word terms (be they hyphenated or unhyphenated) that legitimately comprise two words; you should especially not close or hyphenate terms that comprise a modifier and an activity in the *ing* form.

BAD: Can you explain the ways in which decisionmaking and bungeejumping require similar skills?

GOOD: Have you noted that car driving and bicycle riding can conflict on the roads in Woodside?

BAD: Many people in this country do not have access to good healthcare.

GOOD: Health care was a major concern for Hilary.

BAD: When the lights went out, the frontend exploded.

GOOD: A good front end helps to sell the product.

Note that there are exceptions, because words that start off as two words sometimes get so used to each other that they join up as one. Here are examples, both directly and indirectly related to the preceding discussion.

GOOD: Although Lyn was reputed to be a whizbang copy editor, she was a lousy proofreader; copy editing and proofreading are two different skills.

GOOD: Note that *breathtaking* and *stunning* both imply that the beholder's experience has physical (and not necessarily pleasant) components.

GOOD: Intext lists are those that are set intext; that is, they are set in the text line, rather than displayed.

GOOD: Max and Lyn settled lightheartedly into the armchair together, spilling over slightly onto the armrests.

THE PRINCIPLES FOR LUCID WRITING here are that, when a term can take on any of three roles, you should set verbs as two words, adjectives as one hyphenated word, and nouns as one word: To shake down a target, a shakedown artist orchestrates a shakedown.[319]

319. She may thus set you up to be upset about the set-up situation.

134 Rewords

$\mathcal{S}$ ET REWORDS—that is, words beginning with *re*—as single, unhyphenated words.

> BAD: Max realized that they would have to re-design the software from scratch.

> GOOD: The automatic porting program saved them from having to reimplement the entire interface.

> BAD: The airline refused to re-estimate the landing time, even though the airplane[320] took off 4 hours late.

> GOOD: Reentry can be tough when you have been away for a long time.

> BAD: Marty was distraught because Paul had pointed out that he would have to re-do both theorems.

> GOOD: Marty then calmed down, and decided instead to reevaluate whether his energy would be expended more effectively elsewhere.

---

320. In passing, note that you should use *airplane*, rather than *plane*, especially if your text mentions the planes of geometry, for example. I have seen excessive confusion resulting from use of *plane* for *airplane* in such contexts.

> BAD: Richard took Betsy and Madeline to a re-release of *Casablanca*.

> GOOD: Richard was helping Lyn to rerecord her outgoing message, but he dropped a pot and sent Lyn into hysterics while the tape was running.

> BAD: Steve and Judy were having trouble re-conciling their differences, so they hiked up a glacier so fast that they were too tired to argue when they got to the top.[321]

> GOOD: Steve and Judy recycle everything in sight.

Note that, to avoid confusion, you may wish to make an exception for the following case:[322]

> GOOD: Do you resent that I re-sent your electronic mail to my boss?

THE PRINCIPLE FOR LUCID WRITING here is that you should close up all rewords to avoid rewriting them later.

---

321. The technique provides adrenaline as needed for daunting physical challenges and other emergency situations.

322. Suggested by Ronald Barry.

135 Further Versus Farther

THE WORD *farther* refers to distance, whereas *further* refers to all other extents or degrees, and can also denote an activity.

BAD: During 1995, Michael and Shellee lived further away than Lyn would like.

BAD: To Lyn's mortification, Shellee was farther along in her exercise program than Lyn was.

GOOD: When he was on form, Michael could bicycle farther than Lyn and Shellee could walk.

GOOD: Michael and Shellee were hoping for further opportunities to dance to Dr. Loco's Rockin' Jalapeño band.

GOOD: During a scrumptious Persian feast, Shellee furthered Lyn's interest in Chinese herbal medicine.

BAD: "These points are further out than two standard deviations," exclaimed Max.

BAD: "One farther point I would like to make," said Bob, "is that the graphing tool isn't working."

GOOD: Max tried moving the variable's value farther out along the x axis, but the curve was still distinctly odd.

GOOD: "Let's have further discussion after we break for lunch," suggested Max in an affable tone, "so that I do not become further annoyed."

GOOD: Thus did Max further Bob's education in diplomacy.

BAD: Red wished that BB would stay further away from him when he was hunting.

BAD: Red farther wished that BB would stop yowling and meowing incessantly.

GOOD: BB realized that the ground was considerably farther down from the branch than she had anticipated.

GOOD: BB decided that further planning was required, before she was further inconvenienced.

BAD: "This icon," Lyn explained impatiently, "should be much further down on the page."

BAD: "And farthermore," Lyn continued her litany, "I have lost the example for *farther versus further.*"

GOOD: "Well," reasoned Jan, "don't write any further material into you get your brain plugged back in."

GOOD: "You're right," Lyn nodded. "I'll just sit a little farther away so that I can stop driving you batty."

SPLENDID: "We can go further into the issue," Jan comforted, "after you have calmed down further."

BAD: Powerful transmitters let you move the receiver further away than was previously possible.

BAD: A farther advantage is that there are no wires over which to trip.

GOOD: A wide-area network can handle nodes that are farther away from one another than are those in a local-area network.

GOOD: Networking developed further after introduction of fiber-optic technology, and now is used in automated further-education systems.

THE PRINCIPLE FOR LUCID WRITING here is that you should use *farther* to denote distances, and, further, you should use *further* to denote any other dimension of increase.

136 Pronouns for Recipients

PAY ATTENTION to which form of a pronoun you use in a given sentence. If the creature or object denoted by the pronoun is the recipient of the action, rather than being the agent, you need a recipient (an objective) pronoun.

Max   handed   Red   to Lyn

He (agent) handed (action) him (recipient) to her (recipient).

Note that I am not suggesting that the second sentence would be acceptable, given the plethora of male pronouns.

Agent	Recipient
I	me
he	him
she	her
they	them
we	us
who	whom
you	you

The rules governing pronouns are complex; if you find yourself in doubt about a given pronoun, look up the answer to your question in a good style manual.[323] Here, we shall look at examples that will help you to begin to tune your ear (which is always a useful exercise).

---

323. Your best bet, because it is accepted by most publishers, is the most recent edition of the *Chicago Manual of Style,* Chicago: University of Chicago Press.

BAD: Between you and I, it is not always easy to avoid using the terminology of English grammar when you are explaining the finer points of that subject.

GOOD: Between you and me, it is sometimes difficult to think of any but the most trivial and boring examples, yet I believe that it is important to break the cardinal rule that the content of the examples should not distract the reader from the point of them.[324]

BAD: It's me who drew the ternary-tree representation of the general tree.

GOOD: It is I who ordered postorder traversal of the binary-tree representation of the general tree.

BAD: You want to talk to Lyn? This is her.[325]

GOOD: No—that cougar is not Lyn! The woman standing near the edge of the picture is she.

BAD: To who did you give the leftover chocolate cake?

Or, worse yet, who did you give the leftover chocolate truffles to?

GOOD: I am glad to hear that you gave it to Brendan, who will certainly appreciate it, given that he bicycles a few hundred miles each day.

---

324. If you enjoy reading this book, you may absorb the knowledge in it without realizing what you have done (until you next set out to write).

325. The alert reader—that means you—will note the unreferenced *this*.

BAD: I shall severely whip with a strand of silk whomever is caught stealing my chocolate-chip cookies in the middle of the night.

GOOD: In fairness to those masochists who are shy, I shall not whip anyone who begs me to do so.[326]

BAD: After he discovered several major dents in his back, Max asked Lyn just whom had been eating hard pretzels in bed again.

GOOD: Lyn replied that he knew perfectly well who, as it had been he who devoured the entire box, and he never had learned the art of keeping the crumbs in his hands rather than all over the sheets.

BAD: Do you believe that you are as good as them?

GOOD: They believe that they are as good as I am.

BAD: The hysteretic differentiators were offered to all the girls except she.

GOOD: All the girls except her were awarded spanking-new hysterical differentiators.

---

326. The assertion does not imply a promise to whip anyone who keeps quiet.

BAD: Lyn did not know who (or what) she had tripped over in the dark.

GOOD: Max knew for whom (and why) he was bringing home a colossal bouquet of deep red roses.

THE PRINCIPLE FOR LUCID WRITING here is that you should think about whether the creature or object that you wish to denote by a pronoun is an agent; if it or he or she is not, use *it* or *him* or *her*.

137 *Authorship on Research Articles*

I̲T IS GOOD practice to agree with your coworkers in advance how you will handle authorship on papers that result from a research project. There are various ways to handle assignment of authorship, not to mention assignment of the various responsibilities involved in carrying out a project. That everyone on the research team agrees in advance about the responsibilities and rewards is what is important.

I suggest that you construct a list of roles, and that, during your initial project meeting, you assign roles to people. The resultant pairings should be circulated to all team members, and should be updated periodically if and when roles or assignments change.

I offer here one of many possible approaches to authorship.[327] I suggest that you not so much adopt or even adapt it, as use it as a springboard for open discussion with your coworkers. It is, obviously, designed to cover only one general area of research.

Authorship generally reflects a *primary contribution* to the project. One way that you could classify the primary contributions is as follows:

---

327. Gregory Provan and Max Henrion contributed to my thinking about this matter.

1. Generation of original ideas:
   - Identification of issue and inspiration for approach
   - Development of theory and proofs
   - Design of algorithms
   - Design of user interfaces
   - Design of experiments

2. Project oversight and management of execution (getting the work done, which may involve doing the work yourself, or may involve delegating it to other people whom you then have to supervise):
   - Implementation of software
   - Running of software
   - Performance of experiments
   - Collection of data
   - Running of statistical analyses
   - Formatting of papers
   - Identification and generation of references

3. Responsibility for the written report (which may, again, mean doing it yourself, or delegating and supervising):
   - Generation of an outline
   - Writing of abstract, introduction, and background
   - Review of related literature
   - Description of research methods
   - Interpretation of data-analysis results
   - Design of figures
   - Summarization of conclusions
   - Suggestions for future work

If one person both has the *original ideas* and takes on *management,* that person is responsible for the paper; if two people fulfill these roles, usually the person *managing the project* is responsible. The person who is responsible for the written document is the *first author.* If the person on whose ideas the research is based is not the first author, that person is the *last author.* Note that the first author often, but not always, does the writing. It may well be that the person (or people) who takes the roles of ideas and management is not a good writer.

Authorship on a paper is not an acknowledgment of writing skills — it is an indication of research performed. It is perfectly acceptable to obtain the help of a technical writer or editor, including the help of someone who has worked on the project in another capacity. If you make the determination of what goes into the paper and what does not, what gets covered and what does not, which paragraphs are acceptable and which are not, how the topic should be introduced, what the conclusions should be, which figures should be used, who should be cited, and so on, then you are still first author, even if you do not construct the sentences.

Given that the *last author* is the person on whose ideas the project is based, being principal investigator on a grant, for example, or otherwise designing the research, may well be regarded as sufficient. Similarly, being principal investigator and overseeing the writing is sufficient for first authorship.

Other authors are assumed to have contributed substantially to the design or management of the research. They may be listed alphabetically, or by a weighting scheme designed to reflect contribution. The people who carry out the work, but who do not think up the ideas or manage the project, are not necessarily entitled to any authorship. Authorship of a paper is taken to indicate that you are a researcher on the project (or that you did theoretical work), rather than that you carried out work (or that you did practical work). People who only carry out work (be they editors, technicians, programmers, statisticians, librarians, or computer scientists) are properly mentioned in the acknowledgments section. If the first author believes, or the first and last authors agree, that a person who has (in this sense) carried out work has, in the process, also contributed intellectually and substantially, then she or they may decide to put that person's name on the paper. You should determine in advance what it means to contribute intellectually and substantially.

THE PRINCIPLE FOR PEACEFUL COWRITING here is that you and your coworkers should discuss in advance, and should agree to, the various roles and responsibilities that you will each have in carrying out a research project and reporting the results.

Courtesy Dona DeP. Oppenheim

138 Respectively

T HE USEFUL WORD *respectively* indicates what object or creature belongs with what other object or creature.

You should use *respectively* when you write about *n* items, each of which is associated with one of another *n* items. Without *respectively*, the first *n* items are assumed to be associated with all *n* of the second set of items.

BAD: Marina, Nicola, Dona, Helen, and Lyn are honorary mothers to Savannah, Tabitha, Woozle, Cleo, and Red and BB.

The example sentence lies; Marina is honorary mother to only Savannah, Nicola is honorary mother to only Tabitha, Dona is honorary mother to only Woozle,[328] Helen is honorary mother to only Cleo, and Lyn is honorary mother to only Red and BB.

BAD: Felines and humans use mice as prey and as pointing devices.[329]

As a counterexample, Red is not yet computer literate, and Lyn does not yet fully appreciate the thrill of mouse hunting.

---

328. Dona is also real mother to Lyn.

329. For most readers of this book, this sentence would not require *respectively*; for the population at large, the addition of *respectively* might be insufficient to clarify meaning. This example highlights the importance of having nose.

GOOD: Steve, Michael, Richard, and Misha are married to Judy, Shellee, Betsy, and Holly, respectively.

Without respectively, *the sentence would describe a communal marriage of eight members.*

You should place a comma before *respectively* when you set *respectively* after the terms to which it applies.

BAD: Soren and Lois know nothing about accounting and motion physics respectively; the negation of this assertion, however, is not true.

Respectively *is set after* accounting and motion physics; *it should be preceded by a comma.*

GOOD: Steve and Judy know about high-energy physics and breast feeding as a method of birth control, respectively; the negation of this assertion has an unknown truth value.

Respectively *is set after* high-energy physics and breast feeding as a method of birth control; *it is correctly preceded by a comma.*

BAD: Did you know that the original Unix interface and the Macintosh interface are command line and graphical respectively?

GOOD: Have you considered that impact and laser printers have the disadvantages of being noisy and expensive, respectively?

You should not use *respectively* when the meaning that it would add to the sentence is already implied clearly, either from context or from knowledge that your reader can be assumed to have. Such sentences tend to be clumsy either way, however.

UGLY?: Interrogator and suspect ask and answer questions, respectively.

The classification is questionable, depending on whether you believe that the roles might be reversed.

GOOD: People with sweet teeth and people with a yen for savories prefer chocolate or sharp cheese for the final course of the meal.[330]

SPLENDID: Birds gotta fly, and bees gotta buzz.[331]

UGLY?: Max and Lyn went to work and groomed the cats, respectively.

GOOD: Max and Lyn are a decision analyst and a writer.

SPLENDID: Max was finally able to take off time for a holiday, but Lyn was unable to accompany him because she had to work on her book.[332]

THE PRINCIPLE FOR LUCID WRITING here is that you should use *respectively* to indicate that the relationships described are between members of groups in your sentences, rather than between the groups themselves. Max and Lyn are a man and a woman, respectively.

---

330. People with civilized taste buds naturally have the sweet *and* the savory, in sequential order.

331. Placing the items with their companions avoids the problem entirely.

332. See previous note.

139 Possessives

POSSESSIVES are words that indicate that one creature or object belongs to another. Knowing how to form them will save you from a few common, embarrassing errors.

 You should form simple possessives of singular creatures by adding apostrophe s ('s).

> BAD: Maxs habit of stroking his right eyebrow vigorously by crooking his left arm over his head and dangling his left forefinger over his forehead was leading to marked facial asymmetry.

> GOOD: Lyn's habit of lighting 20 to 40 candles every evening by first lighting a small red prayer candle and then using it to light all the other floating, column, taper, and novelty candles was having a marked influence on wax futures.

You should form simple possessives of plural creatures whose names end in s by adding an apostrophe.

> BAD: The computers's operating systems's processes were hung.

> GOOD: The page-replacement algorithms' code turned up in the trashcan.

> GOOD: The cacti's spines left holes in the boys' trousers.

> *When the plural does not end in s, add apostrophe s.*

When you are writing about two creatures who share ownership, you should form the possessive with only the *second* one.

> BAD: In a backyard in Palo Alto, Holly's and Misha's chickens and rabbits had shared their feed with the local rat brigade.

> GOOD: Steve and Judy's postcard reported that the couple had spent $1\frac{1}{2}$ hours lying on a beach in Italy slathering their hot bodies with fragrant oil.

 When the creature who is the owner has a name ending in the sound *eez,* you should use only the apostrophe, just as you would not pronounce the possessive *eezes.*

> BAD: Cleanthes's stoicism is as well known as is Cleisthenes's diplomacy.

> GOOD: Circes' sorcery did not make all that notable a change in the state of Odysseus's men.

You should set the possessive form of a pronoun without the apostrophe or the s.

> BAD: His's machine be broke.

> GOOD: Her machine runs faster than their machine does.

When the pronoun follows the item owned, you should use the possessive form of the pronoun without the apostrophe, but with the s when appropriate.

> BAD: The nuclear-waste problem is your's.

> GOOD: The industrial cleanup problem is hers.

> BAD: The oil-spill report is their's.

> GOOD: The environmental-impact statement is ours.

BAD: Rambo, a maniacal German Shepherd that attempts to bite Max when Max runs up the road, is not his's.

GOOD: Jackie and Woozle, the fluffy fat cats, are hers.

THE PRINCIPLE FOR LUCID WRITING here is that you should generally, but not always, use apostrophe s ('s) to indicate that an object is someone's.

Image based on a photograph copyright © Paul Fusco/Magnum Photos

140 Clichés, Jargon, and Euphemisms

CLICHÉS AND JARGON, and idiotic euphemisms, degrade your writing. *Clichés* are always originally particularly apt phrasings that become tired through overuse, until they are so thin as to have lost all meaning, descending into the realm of the hackneyed and trite. *Jargon* is verbiage that obfuscates rather than clarifies; you should not confuse it with the highly precise technical terms adopted by a given discipline. Many *euphemisms* are jargon terms. Our political language is rife with all three pests, as is our advertising language.

> UGLY: Each department was asked to respond to the latest productivity challenge.[333]
>
> GOOD: Each department was notified of the latest 10-percent budget cut.

> UGLY: The senator proposed paying for the new housing initiative via revenue enhancement.[334]
>
> GOOD: The mandated changes will be paid for by increased taxes.

---

333. Contributed by Max Henrion; the meaning of the euphemism is explained in the good example in the pair.

334. Here, too, the meaning is explained in the good example of the pair.

UGLY: Both sides in the conflict suffered collateral damage, as well as friendly fire that failed flagrantly to encourage friendship.

UGLY: The possibility of capital punishment led inevitably to a hung jury, despite the hanging judge.

UGLY: When you get stuck writing a document, try playing with your own euphemisms.

UGLY: Most new authors tell their publishers that they cherish a firmly held belief that they have written readable books.[335]

UGLY: The new powerful software sported an innovative design and unlimited functionality.

UGLY: The twentieth century has seen an explosion of research in computer applications.

UGLY: The company announced that it would undertake a major smart-sizing effort over the next year.[336]

UGLY: The latest messaging technology allows you to communicate effectively with your workgroup partners, and to open conversations to explore new directions and to avoid breakdowns.

UGLY: The loved one was laid to rest amid great pomp and circumlocution.

UGLY: Contemplating the vast expanse of the Pacific ocean from Windy Hill is an ataractic undertaking.

UGLY: We could have avoided this snafu if we had remembered GIGO.

---

335. Based on a suggestion by Peter Gordon.
336. Based on a suggestion by Suresh Chanmugam.

UGLY: Our software is based on a new paradigm in spread-sheets, putting at the tip of your fingers the vast power of computerized expertise to guide your business decision forecasting.

UGLY: Len tried to engage Max in a constructive dialog to formulate creative options for their mutual benefit, but Max had had a long hard day at the office and was too exhausted to see straight.

UGLY: Tom explained to Max that, to leverage the targeted clients as they prepared to go to market, the company would have to develop a more highly polished business face.

UGLY: Politicians are usually cautiously optimistic, whereas advertisers entertain no ifs, ands, or buts.

UGLY: Lyn's heart was bursting with joy, her feet barely touched the ground, her heart was singing, her toes were tingling, her heart was worn on her sleeve; her feet, however, were made of clay.

Proverbs—popular, pithy sayings—can be used as clichés also.

UGLY: When you are considering coauthoring a book, remember that too many cooks spoil the broth; then again, many hands make light work.

UGLY: Fools rush in where angels fear to tread; but, nothing ventured, nothing gained.

UGLY: Great minds run in the same channels, and fools think alike.

UGLY: People who are penny wise and pound foolish appreciate that a penny saved is a penny earned.

UGLY: Marry in haste; repent in leisure; but remember that she who hesitates is lost.

THE PRINCIPLE FOR LUCID WRITING here is that you should render unusual mots and gnomes, rather than appropriating the consuetudinary argot of other scriveners' missives.

141 Design Elements and Eye

THE CONCEPT OF a design element is critical in the document-production process. A *design element* is a chunk of text that looks the same as other similar chunks. Thus, *base text* is one design element, a *level 3 head* is another, and a *numbered displayed list* is a third.

If your manuscript is being published by a journal, magazine, newspaper, book publisher, or similar publisher, the design specifications will be developed by a designer. *Design specifications* indicate how a given design element should be set; for example, they might say that the base text should be 14-point Sabon, with 2-point leading (space between lines), no indentations at the beginnings of paragraphs, and so on.

If your manuscript will be set according to a designer's specifications, then *your primary job is to prepare your manuscript such that another person can see easily how many and what kind of design elements you have.*

You should use a minimum number of design elements. For example, you should probably have only one kind of bullet displayed list, and only one kind of numbered displayed list—so you should not set 12 different kinds of displayed lists. Do not draw numerous different boxes, with different rule weights and shapes, around words, definitions, theorems, and so on; rather, leave it to the designer to decide how such design elements should be handled. Your job is to make clear that this chunk is a theorem, this chunk is a definition, and so on. Beyond that, the less you do, the less undoing your publisher will have to do.

An excellent approach that you can use is to make up a list of design elements, and to use that list as you write. Your publisher will love you if you submit a list of design elements, called a *design memorandum,* with your book manuscript. To give you an idea of what entries such a list might have, I show you an initial design memorandum for this book in Box 141.1. Be aware that the design elements for this book are particularly numerous because of the nature of the material: I have shown you an example of nearly every design element, so my designer[337] was faced with a daunting task! You can also *key* your manuscript for your publisher; that is, you can indicate on a sample chapter, using a simple letter code what is which element (key numbered lists as such, chapter titles as such, figure captions as such, and so on). If you submit your manuscript electronically, such keying will be embedded already in your file, in the form of the environments that you have defined.

If your manuscript will be set with your own design, then you should read several of the many books available on typography, composition (here, how you put together words and text, for example), and layout for book (or brochure, or manual, or whatever) design.

Writing design specifications requires *eye,* a skill that is similar to *ear* and *mouth.* Certain talented people have innate eye; other people require years of training to develop eye. Be aware, in any event, that eye is no simple skill. My primary advice to you is to *keep it simple and consistent.* You should make up a limited list of design elements, and should design all the elements to go with one another.

---

337. My designer, after an initial unsuccessful foray, turned out to be me and Joseph Norman.

THE PRINCIPLES FOR LUCID WRITING here are that you should think about the design elements in your manuscript before you begin writing; should develop and maintain a design memorandum; and should use a few, consistently set elements.

Box 141.1 Design memorandum for *BUGS in Writing*.

TO: Peter Gordon

FR: Lyn Dupré

RE: Notes on Design, *BUGS in Writing*

ON: 1 July 1994

The type should be large and simple, and the general look and feel should be open and inviting — fun even! — like a children's book.

Each segment should be preceded by a page break.

There should be plenty of white space around the displayed examples.

There is only one type of level 1 head or chapter heading — it's a mix of both (the segment heads). Should be cap/lc.

The examples are set in groups sometimes, with extra space between to show grouping. So there should be one spec for interexample spacing within a group, and one for extra space between groups.

Many of the examples are annotated. The annotation follows the example and is preceded by a line break. It contains at least one complete sentence, ends in a period, and is set in italic type.

The subpoints within a segment should *not* be interpreted as bullet-list entries. Rather, we need an icon or dingbat to mark separate points. A cat would be suitable, or a cat's paw. (Please note that, in general, cats would be a highly appropriate item to drag in — by the tail if need be — to the design.) The effect should be that of a thick arrow or pointing hand. Not all segments have subpoints.

Each segment ends with "The principle(s) of lucid writing here is (are)...." This portion of text should be set off, by a box, screening, icon (cat on a CRT monitor?), or alternative type. The effect should be that the principle stands alone, rather than being just the final text in the segment.

We need specs for the following elements:

- About the Author backmatter; head and text
- BUGS list; explanation of the classification system, to be set periodically in book where segment breaks leave more than one-half of a page of whitespace
- Bullet list
- Bullet list extract
- Boxed text (numbered and captioned, with source notes)
- Bullet list within box

- Bullet list within example
- Bullet list within footnote

- Chapter/segment opener (number, title, some kind of opener)
- Code segment (set in monospace typeface) within box, to be set as a numbered and captioned program
- Code segment (set in monoface typeface) within example
- Code, single words (set in a monospace typeface), in examples

- Copyright page

- Definition (probably should be boxed; can look like theorem)
- Equation, displayed, within example
- Equation, displayed and numbered, within example

- Equation that comprises an example

- Example (BUGS); hang BUGS, align with right justify on colon, text should be indented from both margins
- Example (BUGS) annotation
- Example (BUGS) continuation (second paragraph or other continuation text)
- Example (BUGS) group spacing (between sets of examples or, occasionally, before a single example that is not grouped)
- Example, textbook-style, with *question* and *answer* heads and text (probably should be boxed for clarity)
- Exercise, textbook-style, with lettered displayed parts list, also within example

- Exercise, textbook-style, numbered but no subparts, within example

- Extract within an example; must be clearly different from regular example text; use narrow column, tight leading, Times Roman or other different serif face, small font

- Extract within text; see preceding entry
- Figures — placement, enclosing boxes, labels, line weights, and so on
- Figure captions and source notes

- Figure labels (two levels)
- Footnotes (must be set at page bottoms, must not wrap to next page, must have rule above)
- Frontmatter head (dedication, disclaimer, foreword, required reading)
- Frontmatter text
- Icon for subpoints (cat's paw?); should hang in margin
- Index by Category: Alphabetical by subheads (10 categories), with alphabetical entries of segment names, segment numbers, and page numbers. Thus, need specs for heads and for entries.

- Index of Principles: Alphabetical by segment title, with segment and page number, annotated with "Principle: ———."
- Multicolumn list
- Multicolumn list head
- Multicolumn list within example — head; set centered over column
- Multicolumn list within example; set with column entries centered
- Numbered list

- Numbered list extract within example
- Numbered list within example

- Outline list (bullets embedded within numbered list) in an example
- Outline list (bullets embedded within numbered list) in text

- Paragraph divider, to divide sections of the *Read Me: Ear* segment; should have effect of a horizontal rule (cat paws? cat with long tail?)
- Principle — should be boxed or otherwise definitely set off from text, to stand alone; should have graphical treatment of some kind
- Slides: These elements should resemble overhead-projection transparencies or slides. Some have graphics, some are all text. All must be enclosed in a heavy box. Many different type sizes are used, but typeface should be drastically different from that of base text. Will need to be laid out individually.
- Table of Contents
- Table with number and title; column heads (with spanners), body (contains side heads), footnotes, source notes, and legend
- Theorem with proof (probably should be boxed); needs clear differentiation from base text; will be used also for lemma
- Theorem without proof
- Title page; two, one straight, one with numerous subtitles

- Unnumbered list within example

- Unnumbered list within footnote
- Warning page "STOP! Do not read further...."

Note: In final example in segment on human-computer interaction, I use many different faces and type styles; must be set pretty much as shown.

Note: No bibliography or reference sections in text.

142 Word Match

ANY TWO WORDS or phrases that you use in conjunction with each other (most often a verb and a noun, but you could also be misdescribing an entire clause) should *match* each other. Consider these examples:

UGLY: We had to expend substantial time reengineering the system after the Beta test version flopped.

GOOD: We had to spend substantial time comforting the distraught systems engineers.

GOOD: We had to expend substantial energy moving the furniture when everyone switched offices.

The point here is that you can either *spend time* or *expend energy*, but you cannot—or at any rate should not—*expend time*.

UGLY: It is difficult to convince a cat to use a cat door; the answer to this question is to leave the door open at first, and to put cat food on the side on which the cat is not.

No question was asked. Reference to nonexistent questions is a common mistake made by authors of research articles.

GOOD: How do you convince a cat to use a toothbrush? The answer to this question is that you teach yourself and your cat to cooperate in dental hygiene, you wrap your finger in gauze, and you purchase malt-flavored toothpaste.

GOOD: The difficulty that you encounter in teaching your cat to use a flush toilet may be considerable. The solution to this problem is to put a piece of cardboard covered with kitty litter between the seat and the bowl, to punch a small hole in the cardboard, and to enlarge the hole slightly each day. Be careful not to use this technique with a small kitten, who might fall in the bowl and drown.

GOOD: It is difficult to convince a cat that the kitchen counter is off limits. You can remedy this difficulty by placing precariously on the counter many old pots, such that they will fall with a deafening crash if only slightly disturbed.[338]

UGLY: Moisés asked Tom to blow a tune on the piano.

GOOD: Lyn asked Moisés to blow his sax for her.

GOOD: Greg played his guitar for Nicola; it had been so long since he had played that he hoped he would not blow it completely.

UGLY: The weather was raining[339] on Sunday afternoon during the barbecue; Martin, being German and inured to rain, cooked the sausages anyway.

GOOD: It was raining on Monday afternoon during Lyn's weekly visit; Adelheid put Suzannah in the stroller with the canopy, and the two women went for a walk anyway.

---

338. Contributed by BB, who still has nightmares. The discovery of this pedagogic technique was accidental.

339. Weather never rains.

GOOD: The weather was wet on Tuesday afternoon during the picnic; Suzannah enjoyed mashing cookies into lumps of dough and then eating them anyway.

UGLY: "Let's solve this issue now," Malcolm suggested.

GOOD: "I'd rather resolve the issue," Brendan piped up.

GOOD: "It might be more productive to solve the problem," suggested Max, "because issues are tricky to define."

UGLY: We can deduce from these results the hypothesis[340] that all men are repressed by their mothers, after which they must find a woman to derepress them.

GOOD: We can formulate the hypothesis that women invented this system as a way of keeping men dependent.

GOOD: We can draw the conclusion that the strongest marriages are formed in the most repressed cultures.

UGLY: Max called Lyn because he was tired of working, and wondered whether there was a play that they could listen to[341] that evening.

GOOD: Lyn said they could see *Ain't Misbehaving* or *M. Butterfly*.

---

340. It might be dangerous to deduce hypotheses.

341. It would, admittedly, be possible to listen to a play; indeed, a large portion of the audience does listen if the play is good and the actors' diction acceptable. However, most people in most circumstances ask other people about seeing a play.

THE PRINCIPLE FOR LUCID WRITING here is that you should pay attention to the contents of your words, so that you do not place the words in untenable positions. Do not expend time learning this principle; rather, spend time doing so.

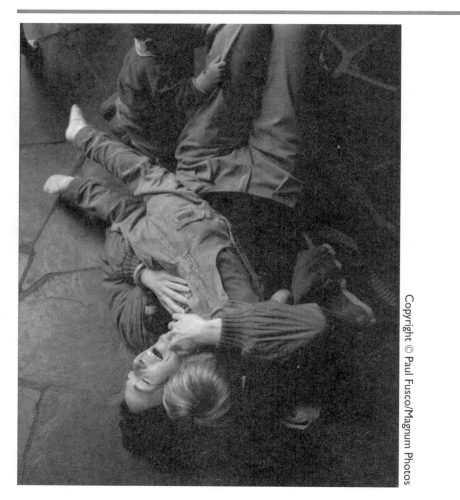

143 Sex Versus Gender

WHETHER A PERSON is male or female is a question of *gender*; whether a couple has undertaken the most common way to conceive a child or to resolve an argument is a question of *sex*.

UGLY: The researchers investigated whether there were any sex-related differences in how users hold a joystick.

GOOD: The study demonstrated a strong correlation between gender and use of electronic mail to establish a social-support network.

UGLY: In English, the battle of the sexes is fought over gender.[342]

GOOD: When Helen taught introductory French, she guaranteed attendance at the following class by promising to explain all about sex; the students were somewhat disappointed when Helen presented a lively discussion of gender.[343]

---

342. Contributed by Peter Gordon. The classification of this sentence is dubious. *Sexes* should be *genders*; however, *battle of the sexes* is a common phrase, whereas *battle of the genders* is not. The use of *gender* is correct.

343. Contributed by Helen Goldstein.

UGLY: It is difficult to imagine a society that does not propagate sex-related role models.[344]

GOOD: People who believe themselves to suffer from gender mismatch face an arduous undertaking, comprising counseling, medication, and sometimes surgery.

SPLENDID: Gender identity and sexual identity are distinct; a transsexual may be either hetero- or homosexual.

UGLY: When Holly still lived in Palo Alto, she was horrified to discover that the chick that she had adopted turned out to be a rooster; the bird's sex would be noted all too soon by the neighbors of her suburban house.

GOOD: Holly found a good home for the rooster, and everyone relaxed after the case of mistaken gender.

GOOD: It can be difficult to sex a baby chick.

UGLY: Betsy had amniocentesis to determine the sex of the child that she was expecting.

GOOD: Richard told Betsy that he had no preferences regarding the gender of their child, because he would be ecstatic whether they had a girl or a boy.

GOOD: We shall avoid mentioning sex in this context.

---

344. It would be entirely possible for this sentence to receive the classification good; the sentence might be ugly in the context of, say, a tract on why there are few women in electrical engineering, but good (vis-à-vis style) in the context of a Masters and Johnson–style study.

UGLY: Red and BB thought that they were exemplars of different sexes.

GOOD: Max and Lyn are delighted to be of different genders.

THE PRINCIPLE FOR LUCID WRITING here is that you should distinguish males and females by gender.[345]

---

345. Whom you distinguish by sex is a private matter.

144 Awhile

T HE WORD *awhile* means for a while; thus, the phrase *for awhile* is redundant. In most cases, you should write *a while*.

BAD: Lyn was in the neighborhood, so she decided to drop in at Doug's office for a while to see how his guidelines were coming along, and generally to interrupt his busy schedule.

GOOD: "Why don't you kick off your shoes and stay awhile?" Doug suggested.

GOOD: "Well, I have to leave in a short while," answered Lyn as she flopped into the chair, "but I'll help you to drink that chocolate soda first."

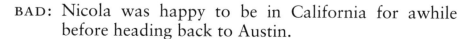

BAD: The program takes awhile to do the calculations and to graph the results.

GOOD: It may take a while to load the software, given that there are 18 floppy disks; just set aside a weekend if you want to undertake the job.

BAD: Nicola was happy to be in California for awhile before heading back to Austin.

GOOD: Greg was happy to be going to the East Coast for a while, before returning to his work in the Bay Area.

BAD: Learning to write error-free prose will take you awhile.

GOOD: A while back, you had to prepare your manuscript on a manual typewriter, and a spell checker was a good friend on whom you imposed.

GOOD: It's been a long while since I have used that white gunk in a little bottle to correct typos.

THE PRINCIPLES FOR LUCID WRITING here are that *awhile* means *for a while,* and that you should usually write *a while*. If you invite a literate person to stay for awhile, she may prefer to leave.

145 Footnotes

FOOTNOTES ALLOW you to add to your document information that is relevant but that does not belong in the text line. You can think of them as optional reading.[346]

 You should write full sentences[347] for your footnotes, rather than setting only fragments.[348, 349]

> UGLY: An LFU[350] page-replacement algorithm keeps a counter of the number of references that have been made to each page.[351]
>
> GOOD: Most frequently used (MFU)[352] replacement assumes that the page with the smallest count was just brought in and has not yet been used.

 In a document that contains few citations, you can provide references in footnotes.

---

346. They are optional reading only insofar as the base text is required reading.

347. This footnote is an example of a full sentence.

348. an example of a fragment

349. A fragment is any string of words that does not constitute a sentence.

350. least recently used

351. Note that the previous footnote is classified as ugly for three reasons: (1) it is not a sentence, (2) it defines a term, and (3) the letters do not match the term.

352. Define your abbreviations in the base text, rather than in footnotes.

GOOD: *Mors certa; amor incerta est.*[353]

GOOD: *Solum certum nihil esse certi.*[354]

 For any references to personal communications, you should use footnotes.[355, 356]

GOOD: Ingo now requires that all the workers at his startup take 6-week vacations every year.[357]

GOOD: Cats should be fluffy, but not prose.[358]

You should use footnotes to provide information, or to hold a discussion, that is tangential to the material in the text. For example, a footnote may provide further information that is interesting but not critical. It may explain notions with which only a small percentage of your readers are not familiar. It may hold asides, or jokes, or bits of arcane erudition. Do not use footnotes to provide information that is critical to your discussion, however.

---

353. Dupré L, Henrion M. The most difficult decision of all. *Home Decision Analysis,* 6(5): 41–527, 1995.

354. If you are interested in learning about uncertainty, I recommend that you read the following: Henrion M. *Uncertainty: A Guide to Dealing with Uncertainty in Quantitative Risk and Policy Analysis.* New York: Cambridge University Press, 1990.

355. You can cite personal communications as name and date, or as number, and tell the reader in the reference section that your information is from a personal communication, rather than from a published study. However, you slightly mislead your reader, because, from the text, she cannot tell that the reference carries only the weight of a personal communication.

356. Note that such footnotes do not need to contain the words *personal communication,* as demonstrated by the second example.

357. Personal communication, John Egar, Woodside, CA, 1994.

358. This sentiment was expressed to me by Max Henrion, who apparently believes that cats should not be prose.

BAD: Lyn used fluffy logic[359] to figure out why people always remember their first loves fondly.[360, 361]

GOOD: Demos[362] was developed by[363] Max Henrion[364] at Carnegie Mellon[365] University.[366]

GOOD: Lyn[367] interrupted[368] her work[369] to feed Red[370] and BB.[371]

---

359. Fluffy logic is a system of deductive reasoning developed over many years in the Dupré–Henrion household. It requires facile argument, slippery hypotheses, dubious assertions, and an array of other proven techniques.

360. Lyn's theory is that, because a person has never been disappointed by a lover, she is not yet crazed; after the first love, however, the person is totally nuts, so the relationship is considerably more difficult. Thus, the person remembers the first relationship as a peaceful idyll, forgetting that it ended painfully.

361. Max destroyed this theory, pointing out that it requires acceptance of the assertion that neither Max nor Lyn was nuts before encountering a first love.

362. Demos is an acronym for Decision Modeling System; note that the name follows British convention, whereby only the first letter is capped in an acronym that is pronounced as a word, rather than as a series of letters.

363. That is, Dr. Henrion both created the software and did the initial development of it.

364. Dr. Henrion is English, so he has a marginal excuse for adopting British usage.

365. The name was formerly hyphenated, but was officially changed by the university to two words, several years ago.

366. Dr. Henrion supervised the Demos development project while he was on the faculty at Carnegie Mellon University.

367. Lyn has been driving Max to distraction, because she keeps popping out of bed in the middle of the night to add another example to her book.

368. That is, Lyn stopped working and climbed upstairs.

369. Lyn works as a writer and artist.

370. Red eats frequently, which is not surprising, given that he is larger by a factor of 2 than most Siamese cats.

371. BB is considerably smaller than Red, and thus requires less sustenance.

THE PRINCIPLE FOR LUCID WRITING here is that footnotes provide a useful vehicle for introducing into your text material that belongs in the text, but that also does not belong there.[372]

372. You have presumably noticed that this book contains myriad footnotes. To get an idea of the kind of material that is appropriate for footnotes, you can simply skim through those footnotes.

146 Mouth

E AR IS ONE OF the characteristics that good writers develop; ear allows writers to hear what is correct. Another such characteristic is *nose,* which allows writers to identify their audiences accurately. *Eye* allows writers to lay out their work pleasingly, to combine graphics and text intelligently, and to use the design elements most suited to the exposition. *Mouth* is the means by which an author speaks; the amount of mouth that an author has is the degree to which she permits herself to be visible to her audience.

When you write, you control the degree to which your personal views, biases, and beliefs come through the words. Most expository writing in this country gives you little clue about the feelings or perceptions of the authors. Yet, writing that has a perspective, that has a definite person driving it—writing that has mouth—is considerably more interesting.

In the years that I have worked as an editor, I have read hundreds of textbooks; I discovered long ago that the subject matter has almost nothing to do with how interesting a book is. Rather, it is the author's enthusiasm for her subject—or lack thereof—that determines the interest generated in the reader. Good expository writers are good teachers. They get excited about their subjects. The most eloquent of them convey respect, wonder, and awe for their discipline. They want to tell you about an idea or fact because it is neat, miraculous, absurd, or otherwise worth your attention. An author who just drones along presenting fact after fact, or argument after argument, even if the discussion is faultlessly structured and the syntax is perfect, produces soporific prose.

In contrast, the writer who tells you what her biases and opinions are is taking a risk. You may hate her or love her, but you will certainly have a reaction to her. I encourage you to take that risk, to tell your reader who you are and what your relationship to your subject matter is. Develop mouth.

When you introduce your topic, whether you are describing a system, program, or clinical research study, always *begin by justifying your work in terms of a big picture.* You must have a general hypothesis about the world that implies that your work is valuable. You may have implemented an algorithm for scheduling complex, interdependent tasks. Instead of beginning by talking about your algorithm, however, talk about what scheduling is, why it might be important to have a computer-based tool to assist humans in doing it, what are the research problems that need to be solved to create such a tool, and how your work contributes to those solutions. I call this the general hypothesis of your work, and you may not be able to talk about testing it directly; rather, you may appeal to literature or even common sense to convince your reader to accept the general hypothesis. The local hypothesis is that your algorithm will perform in a certain way, and that is the hypothesis that you prove or disprove in your paper.

By thinking in terms of hypotheses, or in terms of justifications, you show your reader why you believe that your work is important, and to what you believe it contributes. You begin to develop mouth.

You should *express appropriate, relevant opinions.* Such opinions should have to do directly with the work that you are reporting. If you think that a school of thought is misguided, and can explain cogently why you think so, and you will not be committing political suicide if you do so, then, provided that such an opinion is directly relevant to justifying your own work, say so. I am certainly not suggesting that you enter into a political harangue in the middle of an article on object-oriented programming. I am not suggesting that you jump on your hobby horses, onto soap boxes, or into pulpits whenever you get the chance. I am suggesting that you not hide your opinions, which you as an expert have developed thoughtfully.

You should *loosen up*. This aspect of mouth is the key component, yet it is extremely difficult to describe. I think that you first have to develop a comfortable relationship with written language; that is, you first have to develop ear. Ear permits you to apprehend, intuitively, whether a given construction is right. Once you have ear, you can forget the principles that allowed you to develop it; you can simply listen. And, once you have ear, you will stop worrying about the medium that you are using to express yourself, and will instead be able to concentrate on what you want to express. This book will help you to develop ear.

Next, you have to be comfortable with your topic. You have to be able to think clearly about what you want to communicate, or you will not be able to convey your thoughts to your reader.

Finally, you have to let the writing flow out of you, to let your mind leap ahead with thoughts. Certain people write more easily if they do not worry about structure and organization when they start, but rather just get the ideas down on paper, to be organized later. I have often worked with authors who are daunted by writing, but who are articulate and fluent in describing the topic about which they want to write. An author will present me with a mangled, indecipherable paragraph; I will ask, "What does this paragraph mean?" and will receive a cogent, well-organized answer. Then, I say, "So why didn't you write *that?*" In other words, writing as though you are talking to someone you know can help you to let your writing flow.

You should *use examples wisely and frequently*. Avoid writing boring, pedestrian, or vague examples. Make them specific, but do not be afraid to make them whimsical or otherwise interesting. Examples provide an excellent way to tie your discussion to reality, to tell your audience what the point of your work is. Whenever you write or talk about theory and concept, tie your ideas to numerous real-world examples.

THE PRINCIPLE FOR LUCID WRITING here is that this book has plenty of mouth.

147 Boxes

Boxes are simply *containers for text* that has been enclosed and set off from the main body of text. Most popular news magazines use boxes extensively, so leaf through one of them if you are not clear about what a box is.

Boxes add to your repertoire of ways to present information to your reader; most writers do not know about boxes and thus do not use them. The next time you find that you want to include a block of text, but you cannot find a way to integrate it into the flow of your discussion, ask yourself whether a box would be appropriate.

You should use boxes to delineate text that is *tangential* to the primary subject, or text that you are using as you would a figure. You can place quoted text or your own text in a box, and the text can comprise a displayed list.

If, for example, you are writing a long history of a workaholic, you might include boxed text that describes the epidemiology of workaholism, or that analyzes workaholism as a cross-cultural phenomenon. If you are writing about the history of computers in biology, you might have boxed text giving brief personal histories of well-known scientists in the field. If you are writing about stress reduction, you might have a box that contains a list of steps for a relaxation and visualization exercise.

You can also use boxes to reproduce text from other sources, such as advertisements or leaflets (be sure to get permission!).

You should give each box a *number*. If you are using double numbers (for example, 7.3, 14.2) for 1 heads, figures, and tables—in a book chapter, say, or in a proposal—then use double numbers for boxes as well. Otherwise, use single numbers (for example, 3, 18).

> GOOD: Box 12.8 reproduces a lively discussion held early one morning in the bedroom suite of the Dupré–Henrion household.

You should give each box a *title*. You can use complete sentences if you wish, although it is more common to tag—that is, to name—the box. You should generally set the title with the first letter capitalized and with a period at the end, as you would a figure caption. If your design calls for it, however, you can use mixed capital and lowercase letters, and no period, as in the second of the following examples:

> GOOD: Box 4.3 Tips for cleaning a cat's tail.
> GOOD: Box 6.7 Carver Mead: A Scientist with Many Hats
> GOOD: Box 9.4 Advertisement for snake oil, for posting on fences, circa 1907.

You should include a *credit line* and should obtain *permission* if you use material from a source other than your own mind. You obtain permission from the holder of the copyright—the holder is usually, but not always, the publisher. The permission will probably specify how the credit line should be set. If it does not, you can write a credit line like these examples (note that you should give the page number from which you took the material):

> GOOD: *Source:* Adapted with permission from Dupré, L. *Bellybutton Lint and Inspiration: A Guide to Peculiar Meditation.* Woodside, CA: Homespun Press, 1963; page 45.

GOOD: *Source:* From Henrion, M. Risk-aversive behaviors and personal bias. *Journal of Public-Policy Nonsense,* 7:85, 1994; used with permission.

GOOD: *Source:* Data collected by Red and BB, who stood night watch on the egress from the space between the ceiling and the floor and recorded the number and types of creatures using the pathway.

GOOD: *Source:* Based on a painting by Lauralee Alben.

GOOD: *Source:* Package insert, Healthy Child Snake Oil, packaged by Gullible Sales, Inc., Oaktree, CA, copyright © 1821.[373]

GOOD: *Source:* Arnold B, Henrion M. Launch without tears. *Silicon Journal,* 3(7):94.

You should *call out* each box. That is, you should mention the box by number in the text, at least once.

GOOD: Box 7 details the evolution of the personal computer.

GOOD: Boxes 147.1 and 147.2 provide examples for the material in this segment.

GOOD: A checklist for diagnosing failure of a program to perform as expected is given in Box 5.6.

GOOD: Box 147.1 examines how you can encourage your reader to care about your topic.

GOOD: The concept of a transporting angel is central to certain pages in this book; if you have been wondering what a transporting angel is, you can satisfy your curiosity by reading Box 147.2.

GOOD: The principle for lucid writing in this segment is contained within Box 147.3.

---

373. What is incongruous in this example?

Box 147.1 Why should your reader care?

When you write an article, you should remember that the critical information to give your reader right away is an answer to the question "Why should I spend my time reading your article?" You should answer that question in both the abstract and the introductory paragraphs.

Frequently, authors are so wrapped up in their research projects that they forget that their readers do not yet know the subject of their endeavors. They jump right in by telling the reader the details of the study, without ever explaining what it is they have done, within the general picture, and why they have done it. For example, if you write that you have been studying voice input for progress notes in a medical-records system, and immediately start talking about the problems of parsing natural language, you have neglected to tell me the critical information that, for example, previous experience has shown that physicians are reluctant to do keyboard entry of patient data, or that coding schemes often fail to capture the nuances of natural-language notes, or that the single greatest obstacle to widespread implementation of such systems has been lack of physician-user acceptance. Even more important, then, you need to tell me why I should think it a shame that such systems have not been entirely successful to date, and why I should be excited that you are doing research that will reverse that situation.

In a proposal, failure to say clearly what you are offering and why your reader should be interested — what the big picture is and where your work fits into it — is a fatal flaw. If your reader has to wade through 30 pages of detail before saying, "Aha, so *that's* what they are proposing to do!" then you do not deserve the money. Say immediately precisely what you propose to do, and why your results will be valuable. Remember that you should, of course, tailor your presentation to the agency from which you hope to obtain funding.

In your conclusions, return to your justification. In an article, show that you have, after all, accomplished the work that you promised to report when you began. In a proposal, quickly sum up how you have covered the ground and have justified your planned research.

Box 147.2 Transporting angels.

A transporting angel is a person who enters your life to lift you to a new and higher level of consciousness or awareness. The process of being transported is frequently severely uncomfortable, and the transportee often perceives the transporter as a devil sent especially to irritate, frustrate, infuriate, or otherwise torture. Thus, the person lucky enough to be visited by a transporting angel often struggles mightily, ranting and raving and generally making a major fuss, and thus drastically complicating the transporting angel's job. In calm moments, however, the transportee nurtures a growing suspicion that all this discomfort is serving a critical purpose, and that the destination is well worth all the jouncing and jiggling of the journey.

Source: From Dupré L, Henrion M. *Kicking and Screaming All the Way.* Woodside, CA: Enlightenment Press, 1994, page 1. Used with permission.

Box 147.3 Principle for lucid writing.

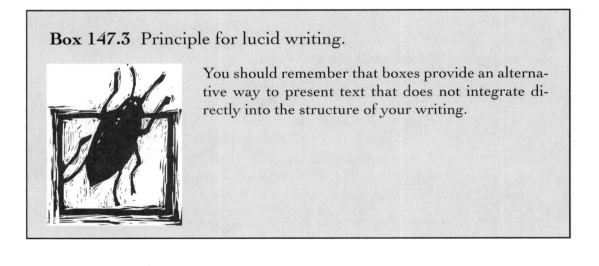

You should remember that boxes provide an alternative way to present text that does not integrate directly into the structure of your writing.

148 Exercises, Examples, and Questions

YOU SHOULD USE *exercises, examples,* and *questions for discussion* in expository material such as a textbook, or as handouts to go with a presentation, or in any situation in which you are teaching people and want to give your audience a chance to rehearse what it is learning.

- An *exercise* is a question that asks the reader for an answer that can be gleaned directly from the text. In other words, if you read (and understand) the text, you can answer the question correctly. Furthermore, an exercise has at least one correct answer, and defines precisely what constitutes a correct answer.

- An *example* is an exercise plus an answer. Examples are often set within sections, rather than at the end. You can use examples to show your reader how to apply a technique that you have described in your text.

- A *question for discussion* is a question that asks the reader for an answer that may not be covered directly in the text, or that may have more than one (or even no) correct answer. You should use questions for discussion to explore ideas that are tangential to your text, or to encourage your reader to think about (and to discuss) ideas that are presented in your text.

- You can also use *essay questions*—which are the same as questions for discussion, except that they request the reader to write down her ideas, rather than to discuss those ideas with her peers.

- You can also use *practice exercises,* which ask the reader to undertake a practical experiment or study, such as using a computer, observing a cat at play, writing a program, or interviewing an expert.

You should *number* all exercises, examples, and questions for discussion. Use double (3.6, 19.2) or single (5, 17) numbers, depending on whether you are using double or single numbers for your level 1 heads. (If the exercises, examples, and questions for discussion are presented in a handout, use single numbers, of course.) You should capitalize the words *Exercise, Example,* and *Question* when—and *only* when[374]—you set them in running text *with a number,* just as you would *Figure, Table, Chapter,* or *Section.*

> GOOD: Example 4.5 shows how to work through the proof that Max and Lyn live together. The examples that follow deal with mundane details.

> GOOD: We leave the proof of this theorem to Exercise 5.6. Several of the exercises in other chapters are even more challenging.

> GOOD: In Question 8, you are asked to explore whether this book qualifies as a work of fiction. The previous seven questions are immaterial.

You should set *exercises* and *questions for discussion* at the *end* of your text (end of the paper or end of the chapter; certain writers set

---

374. You should, of course, capitalize these words when they occur at the beginning of a sentence or in a line that is set in mixed capital and lowercase letters.

them at the end of each section), and should set *examples within* the text at the appropriate place.

 You should *call out* all examples; examples are just like boxes or figures or tables in this regard. That is, you should refer to all examples by number in your text. (You do not need to call out exercises or questions for discussion.)

>   GOOD: Example 5.7 shows how to take an integral.
>   GOOD: Example 4.3 shows how to take a walk.

 You should *distinguish* the question from the answer in your examples. You can use any convention that you like, as long as you are consistent. You also should set an em box (□) to indicate where your example ends and your regular text resumes. Example 148.1 is an example of an example.

When you ask for more than one answer in an exercise, you should *specify the number of answers.* You should never use terms such as *some, a few,* or *several* in this context. Remember, anyone should be able to determine without doubt whether a given scribble constitutes an answer to an exercise. If you ask for *some examples* and your reader provides two, has she completed the exercise satisfactorily?

>   UGLY: Name a few babies mentioned in this book.[375]
>   GOOD: Name five cats mentioned in this book.

>   UGLY: What are some of the activities that Max undertakes
>   on a normal workday?
>   GOOD: What are four activities that Lyn enjoys?

---

375. Note that *baby* is not defined; is a 1-year-old-child a baby?

Example 148.1 Calculation of productivity in a workweek.

Question: During the next week, Max must write one outline for a manual; two white papers; three SBIR proposals, phase I; seven letters to colleagues; one review of a paper for a journal; 495 electronic-mail messages; an update of the design specifications; three reviews of employees; one job specification; two advertisements; and one love letter. He must also attend a 2-day conference in Washington where he is conducting an all-day tutorial; make 184 telephone calls; return 734 telephone calls; attend two staff meetings; supervise 12 employees; soothe one disgruntled client; learn how to calculate the maximum expected belch (MEB) of a waste-storage tank; eat seven power breakfasts; take Lyn out to dinner and a movie twice; and sleep 7 nights.

The time (in hours) allotted to each task (performed once) is as follows:

outline of manual	102.0
white paper	9.0
SBIR proposal	114.0
letters	0.5
paper review	4.0
electronic mail	0.05
designs specs	10.0
employee review	1.0
job specs	1.0
advertisement	3.0
love letter	40.0
conference	48.0
telephone call	0.2
staff meeting	1.0
supervise employee	0.5
soothe client	5.0
MEB	9.0
power breakfast	1.0
Lyn	14.0
sleep	8.0

Calculate the percentage of tasks that Max will accomplish, or suggest an alternative solution.

Answer: Max should soak in a hot tub with Lyn under the redwoods and stars on a clear cold night. Then, he should sleep for 12 hours. Finally, he should rewrite his to-do list. □

UGLY: Describe several of the creatures brought home by Red.

GOOD: Identify four people who work (or have worked) at one of Max's companies.

UGLY: Name some of Lyn's friends.

GOOD: Name six people who are acknowledged as contributing material to Lyn's book.

When you ask a *yes-or-no question* in an exercise, you should almost always request that your reader *explain* her answer. Otherwise, simply *yes* or *no* will constitute a correct answer. Be careful not to give away the answer (yes or no) when you ask for the explanation. Do not write *if so* when you mean *if you do* (or *if it is, if she would,* and so on), and do not write *if not* when you mean *if you do not.*

UGLY: Do you think that Gelareh was sleep deprived during most of 1994?

UGLY: Do you think that Malcolm was particularly busy during most of 1994? Explain why he was.

You have just told your reader that the correct answer to the question is yes.

UGLY: Do you think that Alyssa is a lucky kid? If so, describe in what ways she is. If not, explain what would have to change for you to think of her as lucky.

This example would be classified as good were it not for the if so *and* if not.

GOOD: Do you think that Malcolm is tolerant? Explain your answer.

UGLY: Do you know where Mike and Shellee are living?

UGLY: Do you know where Red and BB are living? If you do, describe where.

GOOD: Do you know where Steve and Judy are living? If you do, describe where. If you do not, describe how you might find out. Be careful that your answer applies to the time that you write it, rather than to the time that this book was written.

UGLY: Do you think that reading this book will improve your ability to give presentations?

GOOD: Do you think that reading this book will improve your social life? Explain your answer, giving an example of each point you make.

UGLY: Can you name one of Carver's scientific hats?

GOOD: Can you guess in what kind of institution Eddie works? What is your best guess?[376]

 When you ask several related questions within one exercise, you should use *lettered parts*. You can phrase the question as a complete sentence, followed by fragment parts. In this case, you can set the parts as a list within the text line, or you can use a displayed list.

GOOD: Identify the jobs of the following people: (a) Devon, (b) Brian, (c) Boris, (d) Suresh, (e) Len, (f) Max.

---

376. *Best guess* is one of the few phrases where *best* is not objectionably fuzzy.

GOOD: What kind of creatures are the following?

> a. Swix and Spud
> b. Reno and Cosmo
> c. Sophia and Madeline
> d. Bugs and Wiggles
> e. Blackford and Ursula
> f. Gabe and Bruno
> g. Buttercup and Hershey
> h. Red and BB

GOOD: Name the partners of the following creatures:

> a. Misha
> b. Lauralee
> c. Lois
> d. Marina
> e. Sachiko
> f. Moisés
> g. Max
> h. Red

Alternatively, you can phrase the question as a subject area or topic, followed by parts that comprise one or more sentences. (Note that you should not, however, jump into a part without providing an introductory sentence.) In this case, you should use a displayed list for the parts.

GOOD: This book has introduced you to a number of characters, and has taught you about their lives. Let's see how carefully you have been reading.

> a. What are three of Joe's areas of expertise?
> b. What big event did Brian and Adrienne celebrate in 1994?
> c. What academic advancement did Sally receive in 1994?
> d. What is one of Marina's many talents?

e. What is one skill at which both Darlene and Thérèse excel?

f. For what four animals does Holly have a marked fondness? Note: Consider *Homo sapiens* as well as other species.[377]

g. What three careers could Egar pursue?

h. What favor did Soren used to do for Lyn, and why did he stop?

i. Why do Al and Mary have two houses?

j. What kind of instrument does Tom play?

k. How many Gregs, Homers, and Brians are there in this book?

l. What is the genealogy of a guardian angel?

You can also use lettered parts to create *multiple-choice questions*. Note that you can use commas, semicolons, or periods between the parts if you set the parts as an intext list—the only rule to follow is that you should be consistent. If you use periods, begin each entry with a capital letter; otherwise, begin each with a lowercase letter unless you have good reason to use a capital letter.[378] You can also use a displayed list for this purpose (with end periods if the entries are sentences, and with no end punctuation otherwise).

GOOD: How often does Max work? (a) Every day. (b) Five days per week. (c) Only on weekends. (d) Rarely.

GOOD: What made Brian cry? (a) the release date, (b) the bugs, (c) the onions, (d) all the preceding answers.

GOOD: Who is Lyn's transporting angel?

    a. Max
    b. Red
    c. BART

---

377. Based on a suggestion by Misha Pavel.

378. A good reason might be, for example, that the word is a proper name.

When you write a *question for discussion*, you should *suggest* the type of answer that you wish the reader to explore.

GOOD: Do you think that Max and Lyn will be together 20 years from now? Provide explicit evidence for your opinion, citing relevant articles and research studies. Draw a relational diagram to emphasize the causal links that you perceive. Interview several couples, and enter your data into a relational database. Perform multiple regression analyses on cells. Apply the results to answer this question.

GOOD: Do you think that the methods used to present material in this book are helpful? Explain the ways in which you think they are, and the ways in which you think they are not. Identify problems that you had in understanding the principles. Discuss the extent to which you found the material relevant to your own writing. In what ways do you think the book could be changed to serve you more effectively? What other information would you like to see in it?

Please send your conclusions to

dupre@awl.com

 THE PRINCIPLE FOR LUCID WRITING here is that you should use the pedagogic tools of exercises, examples, and questions for discussion whenever appropriate. Should you be certain that your exercises identify precisely what will constitute a correct answer? Explain your response, citing evidence to support your opinions.

149 Writer's Block

Writer's block is a peculiarly nasty syndrome that can strike any author. It is the author's equivalent of stage fright—a loss of nerve that can paralyze you.

If you are planning to write—or even are already writing—a document that sums up your knowledge on a topic, then you are exposing yourself in two ways. First, you are making public your expertise, and thus are inviting other experts to point out to you what is wrong. Second, you are making public your writing ability, about which you may feel less than confident. This book will help you to lay to rest the second form of insecurity. It will also tune up your ear, so that soon you will not even have to think about what is correct: You will simply know what sounds right.

The first fear—of holding up your expertise for public scrutiny—has little to do with your knowledge of your subject. For example, most graduate students go along happily researching their thesis topics for years, without too much doubt that what they are doing is important. Faced with the task of writing a dissertation, however, they may suddenly think, "Gosh, what if no one cares about this stuff? What if everyone thinks it trivial, or misguided?"

In addition, doctoral candidates often face the peculiar problem that, when their dissertations are written and their degrees earned, they will have to switch from being students to holding jobs. Major career changes are stressful; it is much easier not to write, and thus not to finish.

A third, related problem that you as a writer may have springs from the idea that you are putting down everything you know about the topic under discussion, yet you do not want to stop learning, to draw the line, and to publish. "By the time that the article comes out," you think, "I will know so much more, I will be ashamed of what I said." You can use this notion to procrastinate for years, at which point you can argue that your results are too old to be of interest.

This loss of nerve can have devastating effects if one of your primary tasks in life at the moment is to write a document. You may be convinced that you are a worthless dodo, incapable of writing a single sentence that anyone else would want to read.

My publisher suggested to me the seed of the idea for this book more than 7 years before I sat down to write. I was always busy with other projects, and was not sure how I wanted to approach this one, so we let it slide for a long time, until the right moment came and we developed and signed a contract. Having felt complete confidence all those years that, when the time came, I would simply write effortlessly,[379] I now woke up one morning convinced that I knew nothing about the discipline I had pursued for the previous 25 years. Write a book? Hah! No way!

Instead, I went into a down state, avoiding talking or even thinking about the project for fully 4 months. I was not my usual laughing self; I was all yin, I was gestating, I was composting. Yet, you are reading the book that I wrote—and had considerable fun writing.

There are numerous ways to overcome writer's block. I was helped by three people's input, which I shall pass along to you. First, Kirk McKusick told me that he writes by jotting down not so much an outline as a set of pieces that will eventually constitute the document.

379. Writing is *always* plain hard work for everyone, of course, but certain of us are fortunate in that the work flows easily—most of the time.

He starts by choosing the simplest, easiest-to-write-about piece; in any given writing session, he continues to choose whatever topic looks most inviting. By the time that only nasty bits are left, he has sufficient mass behind him to carry him through on momentum. I used this technique, and for me it made the task vastly simpler. This book is particularly well suited to a divide-and-conquer approach, but any document has components, and there is no absolute reason why you should write those components in sequential order.

Second, Carver Mead related to me an incident that had permitted him to overcome his own grueling case of writer's block. One weekend while he was in Jack London park, he approached a museum case in which sat an original London manuscript. London was one of Carver's boyhood idols. The placard on the manuscript read, "Jack London wrote 2 pages per day." Carver stood, transfixed; "Hey," he thought, "I can do *that!*"

In both cases, you are breaking down the single overwhelming or seemingly impossible task into a series of doable subtasks. You are committing to undertake an action that you are capable of carrying out, even if you are not feeling like Superwriter. The approach works for just about any problem that you need to solve, and it works for overcoming writer's block.

Third, Eddie Herskovits offered to receive and to read, via electronic mail, my first few paragraphs. I made a deal with him: I promised to send him one paragraph on a particular day. I kept that promise, so I then promised to send him another paragraph on the next day. After a few days, I felt sufficiently confident that I committed to sending a paragraph on each day for the next week. By the time that week was out, the block had melted, and I was getting interested in what I was writing. A few weeks later, I was glued to my terminal, and my only problems were trying to catch up with all the ideas that were whizzing through my head, and remembering to eat.

In this case, the trick was to make a commitment to someone other than myself. It worked for me, and it will work for anyone who has an abject terror of missing deadlines or of breaking promises. For people who are a bit more laid back about life, it may not be effective. It may give you an idea, however, of what trick might work for you. You almost have to sneak up behind yourself, to start writing without realizing or admitting that that is what you are doing.

In summary, if you unexpectedly freeze up when faced with a writing task, keep telling yourself that you are experiencing a normal form of incapacitation. Get all your friends to tell you that your self-doubts are nonsense, and that they have unflinching faith in your abilities. Ask them to refuse to hear you when you moan that you will never write a word, and to tell you instead how confident they are that you will find your voice again. Then, break down your writing task into small pieces, using whatever technique appeals to you. Start writing. Tell yourself that it does not matter if what you write is garbage; all that matters is that you write words, or close approximations to words. If you cannot write words, then draw pictures. Once you have got through the block, you can edit and rewrite, and that task will be simple. Keep writing, even if only a few sentences at each sessions. It works.

THE PRINCIPLE FOR LUCID WRITING here is... the principle for lucid writing here is... the principle for lucid writing here is....

150 Acknowledgments

By THE TIME that they publish a document, most authors are indebted to numerous people and organizations. The acknowledgments section is the proper arena for expressing gratitude to all those people.

In journal articles and other short documents, the acknowledgments section is usually set at the end of the text, before the references. Certain journals set the funding sources as a footnote to the title or to the authors' names. In books, the acknowledgments section is usually set at the end of the front matter, either after the preface or as a subsection within the preface.

You should spell *acknowledgments* with only two *e*s. The alternative, *acknowledgements,* is the British spelling and is incorrect for manuscripts published in the United States. This spelling error is the second most common of those you will see in heads.[380]

> BAD: I forgot to update the acknowledgements!
>
> GOOD: The acknowledgments section gives you an opportunity to express a fraction of your indebtedness and appreciation.

---

380. The most common is *forward* to mean *foreword.*

You should mention in your acknowledgments people who contributed to the work reported in your document, but whose contribution does not merit joint authorship. For example, you might acknowledge people who discussed ideas with you, who performed laboratory work, who programmed a module, who helped you with the statistical analysis, who helped you with the artwork, or who edited your text. In books, authors often mention friends and family whose patience and support, or cold collations, for example, made the writing project bearable. Such acknowledgments are not usually appropriate for articles.

You can either identify the feeling or attitude that you have toward the contributor (gratitude, indebtedness, and so on), or you can simply list the contribution and assume that your reader will understand how you feel.

GOOD: Peter Gordon, my sponsoring editor and guardian angel, provided during this book's development, among other gems, the initial concept, technical expertise, substantial hand holding, and a pair of jade balls. The prodigious feats that Peter performed to enable the publication of this book put Hercules to shame. Then, not yet flattened by his labors, Peter tried to market the book as well, thus going above and beyond the call of … Lyn howling.

GOOD: Helen Goldstein exercised her usual astonishingly diplomacy while shepherding this book through untold obstacles on the shoals of production and marketing. Helen's unstinting attention to detail, her tolerance for my incessant demands, and her appreciation of kittens deserve several stars from heaven.

GOOD: Richard Adamo has remained a staunch friend, shoring me up when I doubted myself, encouraging me to reject mediocrity at the expense of peace;

Richard's quirky, razor-edged wit made me burst into giggles amid the book's most trying moments.

GOOD: My friend Marina Nims read various drafts of the manuscript and pointed out, with her usual gentleness and kindness, places where my prose was opaque. Marina is always an excellent listener, knowing precisely when to laugh and when to groan.

GOOD: Ann Barry's optimism, encouragement, and endless support allowed me to discover my voice.

GOOD: Patrick Henry Winston gave me early, fast, winsome, and thoughtful feedback, my responses to which substantially increased clarity and reduced idiocy in the text.

GOOD: I had the good fortune of getting to know Joseph Norman at the critical moment when this book most needed him. He dove into the project of design, bringing to the task precision and attention to detail. Joe lost many nights of sleep during the final countdown, unselfishly dedicating himself to excellence. I, the book, and you all benefit from his contributions.

GOOD: When the fire-breathing dragons threatened me with a disaster unimaginable, Adrienne Esztergar (hitherto oblivious to my appropriation of her name for a character in this book) strapped on her shimmering armor, drew her mouse from its sheath, leaped astride her page-layout software, and came to my rescue. She took over page makeup without turning a hair when mine had already been torn out and thrown on the floor, and calmly decimated the army of problems that we faced. When the dragons regenerated and once again threatened to consume us, Jan Clayton weighed in with unruffled competence, shoring us up and slicing through obstacles to bring the project to completion.

GOOD: I am indebted to John Gamache, whose effervescent creativity and instant grasp of the book's personality allowed him to design the enchanting cover and the icons used in the book, in between the closings on two houses and the resulting move. In addition, he responded to numerous changes and telephone calls with good humor, for which I am deeply grateful.

GOOD: I am profoundly grateful to Max Henrion, who tolerated continual conversational interruptions to allow me to jot notes on any scraps of paper or flesh that I could find, who listened understandingly to my endless perseverations about all events and thoughts related to the project, who provided a serene and creature-filled environment for my work, who served as my primary computer-support staff, and whose patience and pertinacity allowed him to tread the heavily mined path between constructive criticism and inducement to riot. As though that were not enough, Max proofread the manuscript in 2 dizzy days, and broke his hand while arguing with it.

Max has forged with me my first true home—a state of mind and heart, more than a physical location—and in it I have discovered the creativity and energy to bring forth this book.

You should generally use people's full first names, rather than their nicknames (and certainly rather than silly, made-up names), when you acknowledge their *work-related contributions* to your formal written documents. You can use titles if you wish, but be consistent in giving them to everyone or to no one. When you acknowledge friendship or any non–work-related help, you can use whatever form of a person's name you prefer; you can even use only the person's nickname.

UGLY: I cannot express my gratitude to[381] Ron Barry, Ellie Finch, Jenny Knuth, Doug Owens, Dick Rubinstein, and Wyn Snow.

GOOD: Many people gave generously of their time and intellect, reading and patiently correcting various drafts of my manuscript. From each of them I received extraordinarily helpful advice; several of them also contributed examples. Among other reviewers, the following thoughtful people deserve credit for helping me to polish this book: Ronald Barry, Ellen Levy Finch, Jennifer Knuth, Douglas K. Owens, Richard Rubinstein, and Wynter Snow.[382]

UGLY: Greg, my hiking companion, selfishly sprained his ankle midway through the writing process, leaving me to compose alone on my morning adventures.

UGLY: Avi sent regular messages letting me know that he was too busy to think, much less to send a message.

GOOD: Dr. Provan provided the delirium-tremens version of the QMR database.

GOOD: Abraham Silberschatz read a draft of this manuscript while doing 19 other tasks, sending staccato bits of electronic-mail advice.

---

381. The introductory phrase alone would be sufficient to classify this sentence as ugly. If you cannot express your gratitude, then what are you doing when you write the phrase? Similarly, do not say *I cannot thank sufficiently.*

382. Note that, when you want to indicate that you are equally indebted to multiple people, you should list the names alphabetically if you want to signify that the ordering is not by weight of contribution.

UGLY: Drs. Pavel and Jimison's warm hospitality ensured that Max and I were well fed and well rested during a critical small vacation.

Presumably, the visit was not a formal business undertaking, so the titles are out of place.

GOOD: We should all stop by to thank Brian and Adrienne for the fabulous postnuptial barbecue and chocolate party that they threw yesterday, even though they neglected to provide tongs or potholders, and everyone came home with lightly toasted fingers, and several people are still trying to figure out how to charbroil a chocolate cake.

UGLY: Brendan "MacPerson," and Suresh "BugPerson," kept my Mac in a cooperative mood, despite its blatant bipolar disorder.

GOOD: Red and BB Dupré literally sat at my elbow throughout most of my work on the book.

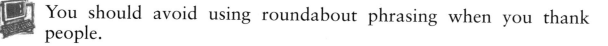 You should avoid using roundabout phrasing when you thank people.

UGLY: I would like to thank Devon Brown for fielding facsimiles and telephone messages.

GOOD: I thank Brian Arnold for guiding me on my first Mac tour, and for showing me that it is possible to appear serene and to remain friendly in the midst of crises and chaos.

UGLY: Thanks are due to the numerous reviewers of this book, who pointed out countless sources of contamination and confusion.[383]

GOOD: The team that Peter and I eventually assembled did a stellar job in producing this book.

UGLY: I would like to express my gratitude to Judy for suggesting the underground version of this book, and a backward R.

GOOD: I am grateful to Steve for taking me out for a night of frenetic dancing, punctuated by confidence-building strokes, just when I needed it most.

UGLY: I cannot begin to thank all the numerous people who contributed.

GOOD: I thank the many people who contributed to this book, both directly, such as by reviewing various drafts, and indirectly, by giving freely of their time and friendship, and thus filling my life with pleasure and joy.

You should acknowledge *sources of funding*, and *providers* of any equipment or other *resources*.

If you are acknowledging grant support, include the grant number. Spell out the name of the funding agency (that is, do not use an ac-

---

383. This example has at least three objectionable characteristics. Can you identify them?

ronym). If you are acknowledging corporate provision of resources, give the company name in full, as well as the location of the company if you wish. If you acknowledge both personal contributions (that is, the people who helped you in your work) and direct sources (the providers of funding and resources), put the contributors first, followed by an extra line break, followed by the sources.

GOOD: Funds for this research were provided by grant #432T9342JK from the Starving Authors Guild.[384]

GOOD: Equipment support for this work was generously donated by Lumina Decision Systems, Inc., of Los Altos, California.

GOOD: Max Henrion supplied nutritional support from numerous Bay Area restaurants; externally forced rest and relaxation in the form of plays, movies, and concerts, as needed; and comic relief at home.

THE PRINCIPLE FOR LUCID WRITING here is that you should include an acknowledgments section where you list contributors to your work and providers of resources, and where you make feeble and inadequate attempts to compensate people for the hours of time and yards of patience that they invested in your work.

I thank you all.

---

384. Dream on.

625

628

Index by Category

Conventions and Standards

Formats for Information

Parsing, Syntax, and Parts of Speech

Punctuation Marks

Terms Often Confused

Terms Often Misused

Terms to Avoid

Terms Tricky to Handle

Types of Documents

Writers' Characteristics

633

Index of Principles

Around   Segment 90   Page 365
Principle: There are sufficient serviceable words that mean *approximately, more or less, about, roughly, generally,* and *in round numbers.* Do not drag *around* into service for this purpose; rather, let it stand for its own perfectly respectable meaning. Draw a circle around this principle to be sure that you understand the idea.

As to Whether   Segment 50   Page 222
Principle: If you find yourself wondering *as to whether* you should use that term, you should simply wonder *whether* instead.

Authorship on Research Articles   Segment 137   Page 563
Principle: You and your coworkers should discuss in advance, and should agree to, the various roles and responsibilities that you will each have in carrying out a research project and reporting the results.

Awhile   Segment 144   Page 588
Principles: *Awhile* means *for a while;* you should usually write *a while.* If you invite a literate person to stay for awhile, she may prefer to leave.

Better, Best, Worst   Segment 69   Page 291
Principle: Avoid using valuative terms that fail to specify the measure that you are applying. Rather than writing better, for example, you should write more concisely.

Blocks: Theorems, Proofs, Lemmas   Segment 47   Page 211
Principles: Set blocks such as theorems and definitions consistently, distinguish the block's various elements, and set off blocks from the base text. □

Boxes   Segment 147   Page 598
Principle: Remember that boxes provide an alternative way to present text that does not integrate directly into the structure of your writing.

Callouts   Segment 62   Page 256
Principle: Provide a callout (for example, see Table 4.2), in numerical order, for every numbered table, figure, program, or box.

Cannot Versus Can Not   Segment 104   Page 427
Principle: Set the negation of *can* as *cannot;* if and only if you intend to negate an activity, rather than an ability, use *can not.* If you cannot remember the principle, you have no choice; if you can not remember the principle (that is, if you can intentionally forget the principle), you do have a choice.

Cap/lc   Segment 83   Page 336
Principle: In cap/lc lines, Set the First Letter Capitalized (with Exceptions), and Set Every Other Word Lead Cap *Unless* that Word Is Not the Final Word and Is a *Preposition, Connective,* or *Article* of Fewer than Five Letters, or Is the *to* in an Infinitive.

Center On   Segment 40   Page 187
Principle: Use more interesting — and perhaps more meaningful — phrases in place of ~~centering on~~ tired, overused terms.

Citations   Segment 65   Page 270
Principles: Always be consistent when you choose among several competing citation styles, and always give your reader pertinent information [Dupré, 1994].

Clichés, Jargon, and Euphemisms   Segment 140   Page 573
Principle: Render unusual mots and gnomes, rather than appropriating the consuetudinary argot of other scriveners' missives.

Code   Segment 94   Page 380
Principles: Set code in a `monospace` — or **otherwise differentiated** — typeface, set comment lines consistently, and set large chunks of code as numbered figures, programs, or boxes.

Colon   Segment 15   Page 60
Principle: Use a colon at the end of a sentence to indicate that further explanation follows: An explanation might consist of a detailed list, for example, or of an example.

Commas   Segment 23   Page 94
Principle: Always include commas that are logical and helpful, such as those indicating pauses, or those after introductory remarks, <u>and</u> also remember not to splice together artificially two sentences simply by inserting a comma and an *and*.

Comparatives   Segment 95   Page 387
Principle: When you use words that imply a comparison, specify precisely what you are comparing with what. Merely making your writing clearer is insufficient; you must make your writing clearer than it used to be, or clearer than your dearest colleague's, or clearer than mud.

Comprise   Segment 75   Page 311
Principle: The whole comprises the parts, whereas the parts constitute or make up the whole. If you ever find yourself using the phrase *is comprised of,* give yourself three lashes with a red pencil and recast your sentence.

Continuous Versus Continual   Segment 128   Page 531
Principle: When you are writing, be on guard against errors continuously, to avoid making mistakes continually.

Contractions   Segment 32   Page 150
Principle: Do not use contractions in formal writing; it's OK to use them in casual writing.

Cross-References   Segment 67   Page 279
Principle: Use liberal, specific, and nonredundant cross-references (see, for example, the top-right corner of page 1190).

Ensure, Assure, Insure   Segment 45   Page 206
Ensure, assure, and *insure* have three distinct and substantially different meanings; use each word correctly to ensure that your readers are assured that you know what you are doing, so that they feel no need to insure themselves against damage from bad prose.

Equals   Segment 57   Page 244
Principle: Do not lose any pieces of the phrases *is equal to, is greater than,* and *is less than;* doing so equals an error.

Equations   Segment 118   Page 469
Principle: After you have mastered a text editor that can handle equations, review the few simple ideas presented in this segment.

Everyone, Someone, No One, None   Segment 14   Page 57
Principle: Everyone should remember that only *no one* and *every one* have a space in them, and someone should remind you that *none* can be singular or plural.

Exclamation Point   Segment 63   Page 261
Principle: Use exclamation points sparingly (except in exceptional circumstances!), and use with them only quotation marks and, occasionally, dashes.

Exercises, Examples, and Questions   Segment 148   Page 603
Principle: Use the pedagogic tools of exercises, examples, and questions for discussion whenever appropriate. Should you be certain that your exercises identify precisely what will constitute a correct answer? Explain your response, citing evidence to support your opinions.

Expected but Nonarriving Agents   Segment 79   Page 325
Principles: To be fair to your reader, <u>you</u> should name whom you intend to name, and should deliver on any promises that you make about the forthcoming identification of an agent.

Feel Versus Think   Segment 84   Page 339
Principle: Limit your use of *feel* to those situations in which a creature is genuinely experiencing a feeling; use *think, believe, argue,* and so on to denote intellectual activity. You may feel confused, but you should not feel that this segment is confusing.

Fewer Versus Less   Segment 129   Page 533
Principle: Distinguish between *fewer* and *less;* making fewer errors will allow you to feel less foolish.

Figure Captions   Segment 43   Page 198
Principle: Write figure captions that provide explanations, legends, and credit lines, and set them consistently. Most important, make sure that each figure caption provides sufficient information that the figure plus the caption stand alone.

Figures   Segment 126   Page 512
Principle: Graphical representation is a powerful and compact form of communication that you should use wisely. Choose figures that are clear, simple, and easy to understand, and that are labeled carefully.

Focus On   Segment 88   Page 357
Principle: Avoid the worn-out *focus on;* instead, ~~focus on~~ turn your attention to more interesting alternatives.

Footnotes   Segment 145   Page 590
Principle: Footnotes[88] provide a useful vehicle for introducing into your text material that belongs in the text, but that also does not belong there.

Foreword and Forward   Segment 46   Page 209
Principle: You should find it astounding that many books open with a section entitled *forward.*

Full Versus Incomplete Infinitives   Segment 30   Page 144
Principle: Use full infinitives in a series in which the infinitives are modified in different ways. That is, to write well, or to avoid mistakes, you should learn when to repeat the *to.*

Further Versus Farther   Segment 135   Page 556
Principle: Use *farther* to denote distances, and, further, use *further* to denote any other dimension of increase.

Fuzzy Words   Segment 36   Page 171
Principle: Always use the most specific, informative term available, so that you maximize the communication per word, all other considerations being equal. In fact, there really is something actually very annoying about truly fuzzy terms.

Gender-Specific Words   Segment 127   Page 525
Principle: To mistress the art of writing, you should avoid using words that are gender specific when the roles that they denote are not gender related.

Half   Segment 119   Page 475
Principle: Use *one-half* to refer to one of two equal parts, and use *moiety* to refer to one of two unequal parts that together constitute a whole. The greater moiety of this book comprises more than one-half of the segments.

Hyphens   Segment 29   Page 134
Principle: Hyphenate most compound-adjectives terms, but do not hyphenate terms that are compound adjectives that follow the noun. A user-friendly book is not user unfriendly.

Impact   Segment 25   Page 108
Principle: Do not use *impact* when you mean *influence* or *effect,* and certainly do not use *impact* when you mean *affect,* because impacting people is incredibly impolite.

Importantly   Segment 101   Page 412
Principle: It is important that you not use *importantly* when you mean *important.* You can write importantly, which is arrogant, or you can write important material.

In Order To   Segment 76   Page 314
Principle: *In order to* is a clumsy phrase that you should avoid using, ~~in order~~ to improve your writing, ~~in order~~ so as to communicate effectively.

Indices Versus Indexes Segment 116 Page 464
Principle: Distinguish between the indexes in, for example, a book, and the indices in, for example, mathematical expressions.

Is Due To Segment 39 Page 185
Principle: Use *is due to* to speak of reparation and perhaps to speak of causes, but do not use it to speak of origination. That much care is due to your reader.

Issue Segment 111 Page 448
Principle: Avoid writing about issues; instead, describe your subject clearly and in sufficient detail. This principle is not an issue.

Italic Type Segment 130 Page 535
Principle: Set in italic type *words that you wish to emphasize, words under discussion, foreign words,* and *variables.*

Its Versus It's Segment 80 Page 328
Principle: It's a good idea to use *its* to mean *belonging to it,* and *it's* to mean *it is.* It is also a good idea to avoid contractions in formal writing.

Key Terms Segment 12 Page 51
Principle: Consider organizing your presentation based on **key terms,** which are terms critical to the meaning of your discussion; defining key terms in text; and using a form of typographical distinction (usually boldface type) to highlight key terms.

Last Segment 87 Page 355
Principles: Use points of ellipsis to indicate missing material in quoted passages and in series,..., and use them with a correctly placed period when appropriate.

Like Versus Such As Segment 27 Page 125
Principle: Always use *like* correctly to refer to likeness, or resemblance. Use *such as* to indicate an example member of a group about which you are speaking. Use *as* to indicate a likeness in activities. If you use your ear as I do, then your prose will be like mine, and you can compose sentences such as this one.

Lists Segment 26 Page 111
Principles: (1) at first blush, you may find the rules associated with the correct use of lists too numerous to swallow (but you should keep chewing); (2) you should learn to distinguish among the various lists; and (3) you should use the type and format that are most suited to the message that you want to communicate.

Literal and Virtual Segment 92 Page 370
Principles: Use *literal* when you mean *not metaphorical* (when you mean *metaphorical,* you should say so), and use *virtual* to mean *not actual* or *nearly.* You should take this principle literally, at virtually all times.

Maybe Versus May Be   Segment 125   Page 509
Principle: Set the term *maybe,* meaning *perhaps,* as one word. Set the verb *may be* as two words.
You may be still learning, but maybe you already know this principle.

Media   Segment 120   Page 479
Principles: Use *media* to denote more than one medium, unless the medium is a human being;
never use *media* as a singular. Either use numerous writing media, or use just one medium.

Missing That   Segment 107   Page 433
Principle: Although it is good practice to leave out unnecessary words, it is not good practice to
leave out words that belong in your sentence.

Missing Words   Segment 70   Page 294
Principle: Use the various ———— that indicate missing words and l——ters.

Motivate   Segment 9   Page   41
Principle: Never attempt to motivate any entity that is not a living creature.

Mouth   Segment 146   Page 594
Principle: This book has plenty of mouth.

Neither Nor   Segment 99   Page 407
Principle: Use *neither* and *nor* together, and take care in placing other words near them, or your
sentence will be neither correct nor clear.

Nonwords   Segment 106   Page 431
Principle: Do not hyphenate nonwords unless the second term consists of multiple words or begins with a capital letter. For example, a non–Dupré system might use a hyphen in *nonwords.*

Nose   Segment 91   Page 367
Principle: Know thine audience.

Not Versus Rather Than   Segment 121   Page 481
Principle: Use *rather than,* rather than *not.*

Note That Versus Notice That   Segment 114   Page 457
Principle: When you wish to flag a sentence for your readers' attention, first ask yourself whether
the content is not sufficient to call attention to itself without your aid; if you determine that the
sentence needs a flashing light, use the simplest choice: *note that.* In addition, notice that *notice
that* has its uses too.

Number Spelling   Segment 24   Page   99
Principle: In general, spell out numbers from one to nine, and use numerals otherwise; there are,
however, many exceptions. This area is one of the few in which ear will get you nowhere; rote
memorization or frequent lookups are your choices.

Number Styles   Segment 34   Page 156
Principle: Learn the rules for styling numbers so that your text will be internally consistent and will be consistent with that of other scientific writers. Keep this book at your elbow, with this page marked, until you know all the rules in this segment.

Object, Modifier, Activity   Segment 133   Page 549
Principles: When a term can take on any of three roles, set verbs as two words, adjectives as one hyphenated word, and nouns as one word: To shake down a target, a shake-down artist orchestrates a shakedown.

Only   Segment 5   Page   23
Principle: Whenever you use *only,* double-check that you intend it to modify only the term that follows it directly.

Oxymorons   Segment 10   Page   43
Principle: Do not use phrases that are internally inconsistent unless you do so knowingly with the intent to amuse your reader. If your words are intelligibly incomprehensible, your reader may become clearly confused.

Parallelism   Segment 85   Page 342
Principles: Enforce parallelism throughout your writing, pay attention to parallelism at each level, and cast all like entities in like form. The three parts of the preceding sentence are parallel.

Parentheses   Segment 37   Page 175
Principle: Use parentheses to enclose asides (or tangential remarks), explanations (for example, of a word), or numbers (or letters) in an intext list.

Passive or Missing Agents   Segment 1   Page   1
Principle: To the extent possible, avoid using passive voice in your writing; instead, use active voice, and name your agents. In addition, by using clauses such as this one, you promise your reader to name an agent: Keep your promise.

Per   Segment 33   Page 154
Principle: Use *per* to mean *for (or on) each*. If you read one segment per day, you will learn a useful principle on each day.

Percent   Segment 132   Page 546
Principle: Know 100 percent of the few simple rules governing use of *percent*.

Persons Versus People   Segment 82   Page 334
Principle: Almost never use *persons;* use *people* when you intend to denote the plural of *person*, and *peoples* when you intend to denote multiple groups of people. People may wish to keep this book near their persons.

Placement of Adverbs   Segment 58   Page 246
Principle: Understand <u>fully</u> the nuances of adverb placement. Unless you have an excellent reason to do otherwise, place adverbs at the ends of the phrases to which they belong. What constitutes an excellent reason, in this case, is a matter of ear and intent.

References   Segment 103   Page 418
Principle: Set your references with compulsive care. If your publisher suggests a style, use it; otherwise, use any you wish, but use it carefully and consistently.

References to Parts   Segment 53   Page 230
Principle: You should generally, as defined in rule 1, omit parentheses around letters or numbers that designate parts, unless, as defined in rule 2, you use parentheses for callouts or for parts of numbered manuscript components.

Remarks Inserted After That   Segment 42   Page 195
Principle: Remember that, when you insert a separate remark after a *that,* you must delimit the remark by placing commas on both sides of it.

Repeated Prepositions   Segment 20   Page 77
Principle: Repeat prepositions to indicate governance of the words that follow, or <u>of</u> entire terms.

Respectively   Segment 138   Page 567
Principle: Use *respectively* to indicate that the relationships described are between members of groups in your sentences, rather than between the groups themselves. Max and Lyn are a man and a woman, respectively.

Rewords   Segment 134   Page 554
Principle: Close up all rewords to avoid rewriting them later.

Sections and Heads   Segment 74   Page 304
Principle: Break up your writing by sections and subsections, each of which you should number. In Section 1, for example, you might set subheads for Sections 1.1 and 1.2, and perhaps for Sections 1.2.1, 1.2.2, and 1.2.3.

Semicolon   Segment 93   Page 374
Principle: Use a semicolon to set off a portion that is itself a complete sentence; furthermore, use a semicolon to separate in-sentence list entries when at least one entry contains a comma.

Sex Versus Gender   Segment 143   Page 585
Principle: Distinguish males and females by gender.

Shall Versus Will   Segment 11   Page 47
Principle: Use *shall* to predict the future (when you are speaking of yourself, alone or with other creatures), and *will* to imply intentionality. I shall teach, and you will learn.

Since   Segment 102   Page 415
Principle: Reserve *since* for time relationships, and use *because* to indicate causal linkages, ~~since~~ because those are the correct uses of the words.

So Called   Segment 113   Page 454
Principle: Avoid using *so called*. If you do use it, do not also use quotation marks or italics to delineate the ~~so-called~~ "term under discussion."

So, So That, Such That   Segment 3   Page   14
Principle: Distinguish among the three terms *so, so that,* and *such that,* so that you write more accurately, such that you use words correctly, so you feel confident when you publish your document.

Solidus   Segment 117   Page 466
Principle: The solidus (forward slash, /) means *and or;* do not use it for equal-weighted pairs or in the redundant term *and/or.*

Split Infinitives   Segment 38   Page 182
Principle: There is never a good excuse for splitting an infinitive: Remember that to even occasionally split infinitives is a sloppy habit.

Spread-Out Phrases   Segment 18   Page   72
Principle: Do not force your reader to backtrack by spreading relevant phrases, over the expanse of your sentence, out.

Style Sheets and Spell Checkers   Segment 124   Page 506
Principles: Develop and maintain a style sheet, and use your spell checker to help you to implement that style sheet.

Tables   Segment 96   Page 391
Principles: Simply remember that tables are highly constrained beasts, and take care to set them correctly and consistently.

Tense   Segment 97   Page 397
Principles: Pay attention to the tense in which you are writing, and do not change tenses without due cause (rather than without having had due cause).

Terms for Human–Computer Interaction   Segment 112   Page 450
Principle: Determine at the outset of your writing project how you will style terms used to describe human–computer interaction. **Use the** *arsenal* of **methods** FOR **differentiating** type, but use it wisely, so that the result is clarification of your meaning.

The Fact That   Segment 66   Page 277
Principle: Avoid using *the fact that;* ~~you should remember the fact that~~-nothing, simply *that,* or a recast will work.

Though   Segment 52   Page 227
Principle: ~~Though~~ Even though the word is common, do not use *though* on its own; use *although* or *even though.*

Titles   Segment 31   Page 146
Principle: Save capital letters for people; do not waste them on titles of office not attached to people. Various presidents are quite different from President Lincoln.

650

Index of Photographs

Copyright © Paul Fusco/Magnum Photos

655

656

About the Author

Until we met Lyn, I thought grammar was only a way for snobbish bourgeois and upper-class twits to make artistic rebels and the salt of the earth working-class people feel inferior. Mrs. McGowen, my third-grade teacher, convinced me that, if I spoke and wrote correctly, I would end up like her, and would be dull, gray, and flabby. Then Lyn appeared in my life: vibrant, colorful, and fit, someone who can both speak correctly and not correct my pronunciation or grammar. (Though there are occasional blank looks when participles dangle too far from whatever they dangle from.) She is a true daughter of GRIMMER, teaching the word by an exemplary life. I have now changed my ways and will follow HER, even if she is crucified by pop culture.

Steven (and Judy-the-human-spellchecker) Rock
Geneva, Switzerland

She made my eyes brighter and my coat shinier.

Marina Nims
San Rafael, California

This is an author that tries to always avoid all alliteration, but seems not to always be able to get his knickers in a twist when put on the spot to write things that others might read and chide him about, since there are so many errors in such short silly sentences that may put you off to the extent that you cannot follow the line of thought. But at least in reading this book you too may learn how to write good: I can write a billion times better now, and that's coming from someone who never overstates his case.

Gregory Provan
Palo Alto, California

BEFORE: "To acknowledge assistance, editorial advice has been sought."

Yech! Ugh! Doug, no! You must learn not to mix passive and active voice.

AFTER: "My thanks to Lyn Dupré (again) for outstanding editorial advice."

Douglas K. Owens
Palo Alto, California

657

Incredibly exciting ... will keep you up all night ... leaves you wanting more ... great book too!

Richard Adamo
Palo Alto, California

One night when Lyn was 5 years old, I noticed that she was starting to eat her dinner with a knife. Holding out a fork, I said "Try this, kid." The next night, when we were having dinner with friends, Lyn watched with a frown as our friends' 3-year-old daughter tried to master her meal with only a knife. Lyn slipped off her chair and carried her fork over to the toddler. "More effective!" she asserted, handing over the implement. Thus began Lyn's career as an editor.

Garrett Oppenheim
Tappan, New York

I am a hard-working linguist at UTAustin, and I certainly don't have time to write copy for other people's books.

Nicola Bessell
Austin, Texas

We are generally pleased with the way that Lyn has turned out. We have found her to be malleable: After 9 years of daily correction and reinforcement, she has learned to provide high-quality meals (she and Max get the leftovers), to keep our beds soft and warm (she and Max are allowed to share the bed with us, provided that they are not overly restless), and to provide appropriate perches throughout the house (for example, she now keeps pillows on her desk, in the windows, and atop her computer). We have also encouraged her to limit her friendships to creatures who do not object to large quantities of white fur adhering to their skin or garments, and certainly to animals who do not sneeze frenetically in our vicinity, as we find the sneezing disturbingly reminiscent of hissing. We granted her permission to undertake this project because it kept her seated at her desk for many hours each day, giving us an abundance of lap time. We have proofread the manuscript carefully. Please excuse us if we did not always remember to wipe our feet carefully before we looked over the pages.

Red and BB
Woodside, California

My daughter is not a bureaucrat. She doesn't put up with crap. She expects the best. She is not a wimp. She is not lots of things, thank goodness!

Dona DeP. Oppenheim
Quadra Island, British Columbia

658

Dear Lyn,

Friendship goes only so far, Lyn. Do you actually expect me to read a grammar book?! The reason that I became a scientist was that I wanted to ensure that I would never have to write or speak correctly. I think I will start another trashy novel instead.

Name withheld by request
Planet Earth

Lyn has always been more interested in teaching us how to think (and thus to write) in good English than in keeping her job as an editor. She possesses an exquisitely perceptive sensor, tuned to the subtlest bugs in writing, which are frequently symptoms of missing logical links or of superficial thinking. By eradicating such bugs from countless papers, Lyn has surely made many important (and not always fully appreciated) contributions to science and technology.

Misha Pavel and Holly Jimison
Portland, Oregon

Dear Lyn,

I just adore your book. It's like a lazy sunny morning, breeze blowing curtains through an open window onto a wide white-sheeted bed. There's something new and fascinating at every inch, and I find myself going slower and slower, examining every freckle, every fold.

Your book I'm talking about. That's maybe why I'm reading it so slowly. I don't want it to end.

Nicholas D. Iversen
New York City

With this book, Ms. Dupré has developed an entirely new literary form: exemplary fiction.

Professor Schmöe Angstrom
University of the Hebrides

Lyn Dupré was born in Manhattan, where her father was an editor for the Wall Street Journal, *the* Herald Tribune, *and* The New York Times, *and where her mother was the buyer for the Teacher's College Bookstore. She studied philosophy and law at Barnard College and at Cambridge University. She has had over 15 years of experience as a freelance copy editor and developmental editor, specializing in computer-science, science, and medical textbooks. She has edited over 400 books for various major publishers, and has worked for numerous academic institutions. She also works directly with graduate students and other authors to help them improve their writing. Lyn edits and writes during breaks from her serious work as a wood carver and photographer. She wrote this book under the close supervision of her cats, BB and Red. Her fondest hope is that the availability of this book will eliminate any future need for her work as a copy editor.*

Max Henrion
Woodside, California

660

661

BUGS in Writing
A Guide to Debugging Your Prose

John Gamache designed the book cover and the icons used in the interior design.

Lyn Dupré and Joseph Norman designed the book.

Lyn Dupré did the page makeup for the book.

The primary fonts that we used are Linotype-Hell Sabon, Garamond Three, and Cochin; and Monotype Gill Sans, Castellar, and Typewriter.

Phoenix Color Corporation printed the cover on 10' Carolina C1S cover stock.

The Courier Companies, Inc., printed the book on Finch Opaque 50# text stock; they also bound the book.

665

666